AF377672

Electrospun Nanomaterials

Electrospun Nanomaterials

Applications in Food, Environmental Remediation, and Bioengineering

Editors

Ricardo Mallavia
Alberto Falco

MDPI • Basel • Beijing • Wuhan • Barcelona • Belgrade • Manchester • Tokyo • Cluj • Tianjin

Editors
Ricardo Mallavia
Instituto de Biología Molecular y Celular,
Universidad Miguel Hernández
Spain

Alberto Falco
Instituto de Biología Molecular y Celular,
Universidad Miguel Hernández
Spain

Editorial Office
MDPI
St. Alban-Anlage 66
4052 Basel, Switzerland

This is a reprint of articles from the Special Issue published online in the open access journal *Nanomaterials* (ISSN 2079-4991) (available at: https://www.mdpi.com/journal/nanomaterials/special_issues/Electrospun_Nanomaterials).

For citation purposes, cite each article independently as indicated on the article page online and as indicated below:

LastName, A.A.; LastName, B.B.; LastName, C.C. Article Title. *Journal Name* **Year**, *Article Number*, Page Range.

ISBN 978-3-03943-226-4 (Hbk)
ISBN 978-3-03943-227-1 (PDF)

Contents

About the Editors

Ricardo Mallavia (full professor) is a chemist and a polymer specialist, having obtained his PhD in 1994 at the University Autónoma of Madrid (UAM, Spain). He is currently a professor at Miguel Hernández University (UMH, Spain) and a member of the Spanish Chemical Society (RSEQ), sections Polymers (POL) and Nanoscience and Molecular Materials (MAM). He has participated in more than 20 research projects in the last 15 years; six projects as principal investigator. He completed two stays as a visiting professor at the University of California in Santa Barbara, in 2002 and 2013. He has co-authored a hundred articles (h = 26). His research activity has mainly focused on polymer science, mostly in the synthesis and characterization of conjugated polyfluorenes with interest for potential biological applications, and recently in the preparation of nanostructures, particularly in nanofibers, based on polymeric biomaterials.

Alberto Falco (senior researcher), after studying Biological Sciences, completed his PhD studies at the Miguel Hernandez University of Elche (Spain) on animal antimicrobial peptides with antiviral activity, in 2008 (summa cum laude). From 2009 to 2011, he worked as a postdoctoral research assistant at the School of Life Sciences of Keele University (United Kingdom) and, from 2011 to 2013, at the Wageningen Institute of Animal Sciences of Wageningen University (Holland). Since 2014, Dr. Alberto Falco has held a position as senior scientist at the Institute of Research, Development and Innovation in Biotechnology of Elche (IDiBE) back to the Miguel Hernández University of Elche. Overall, so far, he has more than 15 years of experience in both public research agencies and industrial R&D organizations, and has authored over 40 publications in peer-reviewed journals, 4 chapters and 1 book. His expertise involves the innate immune responses to animal viruses, and his current main research interests comprise the encapsulation technologies of natural bioactive compounds with applications in antiviral treatment.

Editorial

Electrospun Nanomaterials: Applications in Food, Environmental Remediation, and Bioengineering

Alberto Falco * and Ricardo Mallavia *

Institute of Research, Development and Innovation in Biotechnology of Elche (IDiBE), Miguel Hernández University (UMH), 03202 Elche, Spain
* Correspondence: alber.falco@umh.es (A.F.); r.mallavia@umh.es (R.M.)

Received: 24 August 2020; Accepted: 28 August 2020; Published: 29 August 2020

Among the large number of methods to fabricate nanofibers, electrospinning stands out because of its simplicity and versatility. The formation of nanoscaled fibers via electrospinning is based on the application of high voltage (usually ranging from 1 to 30 kV) to generate an electrostatic field that induces the formation and stretching of a jet from a viscoelastic polymer solution or melt. The nanofibers are finally formed by either evaporation of solvent or freezing of the melt. Regarding the setup, one of the electrodes can be placed either directly in this solution, or onto the metal needle attached to the tip of the syringe feeding the solution at a constant and controllable flux by means of an infusion pump. The other electrode is connected to a metal object that can work as collector (that can be covered by a fabric), usually a static plane surface that is located perpendicular and at a certain distance from the spinneret. As a result of the forces involved, a highly electrified continuous jet is ejected from the pendent drop of solution at the top of the spinneret and deposited on the collector as randomly distributed nanofibers. In addition, by modifying the basic setup of electrospinning and/or the composition of the electrospinnable solution, the morphology (including porosity), diameter and functionality of the final outcome can be controlled. For instance, nanofibers can even be aligned by adapting the collector to a rotary cylinder or disposed in a core/shell structure by using a spinneret with two coaxial capillaries supplying two solutions separately [1–4].

The origin of this method, which allows the efficient obtention of long, uniform nanofibers with either solid or hollow interiors, dates back to the beginning of the 20th century, when some essential technical milestones for its development, such as the generation and manipulation of electricity, were reached. However, a series of other preceding scientific advances paved the way towards this invention, which can be considered as a variant of the electrospraying process (i.e., the collapse of liquid jets into droplets by the effect) [3–5]. Among them, the distortion and attraction of liquid droplets when applying electrostatic forces, reported by William Gilbert in 1600, could be considered as the oldest one. In the middle of the 18th century, George Mathias Bose described the generation of aerosols by the application of high electric potentials to fluids, and Giovanni Battista Beccaria observed that when fluids were charged, they evaporated faster. Such discoveries might be considered as the basis for the development of electrospraying. It was not until the verge of 19th century that John William Strutt (Lord Rayleigh) first observed the electrospinning phenomena, and Charles Vernon Boys first designed and constructed an electrospinning device and drew fibers from a number of melts, mostly molten waxes. It was in 1900 and 1902 when John Francis Cooley and William James Morton, respectively, filed the first electrospinning patents on industrial applications, and a bit later when John Zeleny studied in detail the mechanisms underlying the process (mostly electrospraying). The origin of electrospinning was established with broad consensus in 1934, when Anton Formhals started patenting several inventions on the technology associated to this process. After up to 22 patents in about 10 years, Formhals greatly improved the process and made electrospinning an efficient and viable technique.

Later, the work of Sir Geoffrey Ingram Taylor in the 1960s, whose fundamental studies on the jet forming process laid the theory groundwork for electrospinning, is of note. Since then, the conical shape of the jet occurring as a consequence of the distortion of the spinneret droplet when the electrostatic forces exceed its surface tension has been referenced as the "Taylor cone" in later literature [6]. More recently, Larrondo and Manley in the early 1980s, and the Reneker's group in the early 1990s, notably revitalized this technology by demonstrating the possibility of electrospinning a range of molten polymers [7] and organic polymer solutions [8,9], respectively. Reneker also popularized the term "electrospinning", which derives from the former "electrostatic spinning" used until then. In the last decades, the advances in the fabrication, processing, and characterization of electrospun nanofibers have contributed to the wide expansion of this technique across laboratories and industry. This growth is mainly promoted by the surging interest in nanotechnology and the great expectations placed on the unique properties of nanomaterials, with notable support from the outstanding progress of the materials and polymer sciences in recent times [4].

As for the raw materials used for electrospun nanofibers, polymers comprise an unlimited number of molecules with different properties that can even be endowed with extra specific features by means of feasible functionalization protocols. In addition, electrospun nanofibers can be prepared from not only single/pure polymer sources, but also compatible polymer blends to combine the properties of their moieties [10]. Altogether, this family of compounds guarantee an extraordinary diversity of nanofiber compositions and thus properties, which explains the broad application potential of these nanomaterials. Indeed, depending on their specific composition/properties, electrospun nanofibers can be exploited in multiple applications covering areas as different as nanoelectronics, energy storage, catalyst substrates, sensors, nanofilters, protective and smart clothing, and adsorbent and biomedical materials [11–15].

At this point, and regardless of the application, it is worth mentioning that the assessment of the environmental impact of the nanomaterials used, as well as their fabrication and degradation by-products, is critical to avoid possible harmful effects on ecosystems by allowing, for instance, the design of appropriate disposal protocols for these compounds and to preferentially opt for those that are eco-friendly. In this sense, polymers also offer a large collection of both natural, but also synthetic, electrospinnable compounds that are non-toxic and biodegradable, as well as biocompatible [4,10]. Electrospun nanofibers made of such biomaterials are thus suitable for applications involving direct (and indirect) contact with biological systems, which mostly comprise applications within the biomedical [1,4,11,13,16–19], but also the environmental protection [11,16] and the food packaging fields [11,20,21].

The present book compiles the Special Issue "Electrospun Nanomaterials: Applications in Food, Environmental Remediation, and Bioengineering" from the journal "Nanomaterials", and, therefore, it comprises several review and research articles addressing several applications of electrospun nanofibers in these areas. In regard to the application of these nanomaterials to the food field, the implementation of electrospinning in food packaging is thoroughly revised in Zhao et al. (2020) [22], which also includes a summary of the additional characteristics provided by functional food packaging materials, degradability, superhydrophobicity, edibility, antibacterial activity and high barrier protection, as well as the contribution of electrospun nanofibers to their development. In terms of environmental remediation, this topic is tackled by two research articles that converge on the green/sustainable generation of energy by improving two different applications (i.e., microbial fuel cells [23] and solar thermal techniques [24]) using electrospun nanofibers.

The current research and utilization of nanofibers mainly for biomedical applications is proportionally covered in this compilation. In this sense, the biomedical applications of electrospun nanofibers included here can be classified into two broad types: drug delivery systems and tissue scaffolds. Regarding drug delivery, polymers comprise a large number of biocompatible materials with an extraordinary versatility to be structured as different nanomaterials with drug-loading capacity. Thus, compounds with different solubility properties can be encapsulated into polymeric nanomaterials by either changing the polymer source or the nanomaterial type. Here, this is shown by Mira et al. (2020) [25] for the encapsulation of different classes of antibiotics by using two separate derivatives of poly(methyl vinyl ether-alt-maleic anhydride) (PMVE/MA) that can be used (alone or in combination with other polymers such as fluorescent polyfluorenes [26–28]) for the fabrication of both nanoparticles [29] and electrospun nanofibers [30,31]. Polymeric nanofibers also protect loaded compounds from degradation, as described by Cruz-Salas et al. (2019) [32] for electrospun nanofibers made from agave fructans, which thermoprotect and photoprotect encapsulated β-carotene. Another advantage of polymeric nanofibers is their modifiable drug-release kinetics by means of feasible design changes to adjust their degradability or porosity for providing optimal therapeutic drug concentrations. As reported here [33,34], this property is being intensively investigated at present for the development of improved dressings, bandages or coatings with, for example, antibacterial activity. In this sense, the use of functional polymers such as chitosan (with reported protective immunomodulatory properties) is also attracting great interest, as widely reviewed by Maevskaia et al. (2020) [35].

Finally, the current great effort made by the scientific community in the development of tissue scaffolds based on electrospun nanofibers is also addressed here. The work of Miroshnichenko et al. (2019) [36] provides a representative example of the research lines in this area by reporting the cell interaction improvements when coating polycaprolactone nanofibers with covalently bonded platelet-rich plasma. Likewise, Li et al. (2019) [37] broadly review the progress in the particular area of electrospun polyvinylidene fluoride-based materials used for bone and neural tissue engineering.

In summary, the papers collected in this Special Issue entitled "Electrospun Nanomaterials: Applications in Food, Environmental Remediation, and Bioengineering" illustrate the high diversity and potential for implementation of electrospun nanofibers in these fields, including the covering of a wide number of subtopics. Undoubtably, such pieces of fundamental research will contribute to the promotion of electrospinning as the focal point in the future development of technological applications at the interface of biological systems, which promise long-term benefits for both health and the environment.

Author Contributions: Both guest editors conceived, wrote and reviewed this Editorial Letter. All authors have read and agreed to the published version of the manuscript.

Funding: This research was funded by the Spanish Ministerio de Economía y Competitividad, grant number MAT-2017-86805-R, and Spanish Ministerio de Ciencia e Innovación (MCI)—Agencia Estatal de Investigación (AEI)/Fondo Europeo de Desarrollo Regional (FEDER), grant number RTI2018-101969-J-I00.

Acknowledgments: We are grateful to all the authors who contributed to this Special Issue, as well as to the referees who notably helped to improve the quality of all submitted manuscripts. We also acknowledge the editorial staff of Nanomaterials, and especially Tina Tian, for their great support.

References

1. Frenot, A.; Chronakis, I.S. Polymer nanofibers assembled by electrospinning. *Curr. Opin. Colloid Interface Sci.* **2003**, *8*, 64–75. [CrossRef]
2. Li, D.; Xia, Y. Electrospinning of nanofibers: Reinventing the wheel? *Adv. Mater.* **2004**, *16*, 1151–1170. [CrossRef]
3. Subbiah, T.; Bhat, G.S.; Tock, R.W.; Parameswaran, S.; Ramkumar, S.S. Electrospinning of nanofibers. *J. Appl. Polym. Sci.* **2005**, *96*, 557–569. [CrossRef]
4. Bhardwaj, N.; Kundu, S.C. Electrospinning: A fascinating fiber fabrication technique. *Biotechnol. Adv.* **2010**, *28*, 325–347. [CrossRef]
5. Tucker, N.; Stanger, J.J.; Staiger, M.P.; Razzaq, H.; Hofman, K. The history of the science and technology of electrospinning from 1600 to 1995. *J. Eng. Fibers Fabr.* **2012**, *7*, 63–73. [CrossRef]
6. Taylor, G.I. Electrically driven jets. *Proc. R. Soc. Lond. A Math. Phys. Sci.* **1969**, *313*, 453–475.
7. Larrondo, L.; St. John Manley, R. Electrostatic fiber spinning from polymer melts. I. Experimental observations on fiber formation and properties. *J. Polym. Sci. Polym. Phys. Ed.* **1981**, *19*, 909–920. [CrossRef]
8. Doshi, J.; Reneker, D.H. Electrospinning process and applications of electrospun fibers. *J. Electrost.* **1995**, *35*, 151–160. [CrossRef]

9. Reneker, D.H.; Chun, I. Nanometre diameter fibres of polymer, produced by electrospinning. *Nanotechnology* **1996**, *7*, 216. [CrossRef]

10. Gunn, J.; Zhang, M. Polyblend nanofibers for biomedical applications: Perspectives and challenges. *Trends Biotechnol.* **2010**, *28*, 189–197. [CrossRef]

11. Liu, H.; Gough, C.R.; Deng, Q.; Gu, Z.; Wang, F.; Hu, X. Recent advances in electrospun sustainable composites for biomedical, environmental, energy, and packaging applications. *Int. J. Mol. Sci.* **2020**, *21*, 4019. [CrossRef] [PubMed]

12. Xue, J.; Wu, T.; Dai, Y.; Xia, Y. Electrospinning and electrospun nanofibers: Methods, materials, and applications. *Chem. Rev.* **2019**, *119*, 5298–5415. [CrossRef] [PubMed]

13. Ding, J.; Zhang, J.; Li, J.; Li, D.; Xiao, C.; Xiao, H.; Yang, H.; Zhuang, X.; Chen, X. Electrospun polymer biomaterials. *Prog. Polym. Sci.* **2019**, *90*, 1–34. [CrossRef]

14. Ding, Y.; Li, W.; Zhang, F.; Liu, Z.; Zanjanizadeh Ezazi, N.; Liu, D.; Santos, H.A. Electrospun fibrous architectures for drug delivery, tissue engineering and cancer therapy. *Adv. Funct. Mater.* **2019**, *29*, 1802852. [CrossRef]

15. Barhoum, A.; Pal, K.; Rahier, H.; Uludag, H.; Kim, I.S.; Bechelany, M. Nanofibers as new-generation materials: From spinning and nano-spinning fabrication techniques to emerging applications. *Appl. Mater. Today* **2019**, *17*, 1–35. [CrossRef]

16. Thenmozhi, S.; Dharmaraj, N.; Kadirvelu, K.; Kim, H.Y. Electrospun nanofibers: New generation materials for advanced applications. *Mater. Sci. Eng. B* **2017**, *217*, 36–48. [CrossRef]

17. Parham, S.; Kharazi, A.Z.; Bakhsheshi-Rad, H.R.; Ghayour, H.; Ismail, A.F.; Nur, H.; Berto, F. Electrospun nano-fibers for biomedical and tissue engineering applications: A comprehensive review. *Materials* **2020**, *13*, 2153. [CrossRef]

18. Edmans, J.G.; Clitherow, K.H.; Murdoch, C.; Hatton, P.V.; Spain, S.G.; Colley, H.E. Mucoadhesive electrospun fibre-based technologies for oral medicine. *Pharmaceutics* **2020**, *12*, 504. [CrossRef]

19. Teixeira, M.A.; Amorim, M.T.P.; Felgueiras, H.P. Poly (vinyl alcohol)-based nanofibrous electrospun scaffolds for tissue engineering applications. *Polymers* **2020**, *12*, 7. [CrossRef]

20. Becerril, R.; Nerín, C.; Silva, F. Encapsulation systems for antimicrobial food packaging components: An update. *Molecules* **2020**, *25*, 1134. [CrossRef]

21. Wang, M.; Wang, K.; Yang, Y.; Liu, Y.; Yu, D.-G. Electrospun environment remediation nanofibers using unspinnable liquids as the sheath fluids: A review. *Polymers* **2020**, *12*, 103. [CrossRef] [PubMed]

22. Zhao, L.; Duan, G.; Zhang, G.; Yang, H.; He, S.; Jiang, S. Electrospun functional materials toward food packaging applications: A review. *Nanomaterials* **2020**, *10*, 150. [CrossRef] [PubMed]

23. Massaglia, G.; Frascella, F.; Chiadò, A.; Sacco, A.; Marasso, S.L.; Cocuzza, M.; Pirri, C.F.; Quaglio, M. Electrospun nanofibers: From food to energy by engineered electrodes in microbial fuel cells. *Nanomaterials* **2020**, *10*, 523. [CrossRef] [PubMed]

24. Angel, N.; Vijayaraghavan, S.; Yan, F.; Kong, L. Electrospun cadmium selenide nanoparticles-loaded cellulose acetate fibers for solar thermal application. *Nanomaterials* **2020**, *10*, 1329. [CrossRef] [PubMed]

25. Mira, A.; Sainz-Urruela, C.; Codina, H.; Jenkins, S.I.; Rodriguez-Diaz, J.C.; Mallavia, R.; Falco, A. Physico-chemically distinct nanomaterials synthesized from derivates of a poly (anhydride) diversify the spectrum of loadable antibiotics. *Nanomaterials* **2020**, *10*, 486. [CrossRef]

26. Kahveci, Z.; Vázquez-Guilló, R.; Mira, A.; Martinez, L.; Falcó, A.; Mallavia, R.; Mateo, C.R. Selective recognition and imaging of bacterial model membranes over mammalian ones by using cationic conjugated polyelectrolytes. *Analyst* **2016**, *141*, 6287–6296. [CrossRef] [PubMed]

27. Vázquez-Guilló, R.; Martínez-Tomé, M.J.; Kahveci, Z.; Torres, I.; Falco, A.; Mallavia, R.; Mateo, C.R. Synthesis and characterization of a novel green cationic polyfluorene and its potential use as a fluorescent membrane probe. *Polymers* **2018**, *10*, 938. [CrossRef] [PubMed]

28. Vázquez-Guilló, R.; Falco, A.; Martínez-Tomé, M.J.; Mateo, C.R.; Herrero, M.A.; Vázquez, E.; Mallavia, R. Advantageous microwave-assisted suzuki polycondensation for the synthesis of aniline-fluorene alternate copolymers as molecular model with solvent sensing properties. *Polymers* **2018**, *10*, 215. [CrossRef]

29. Ruiz-Gatón, L.; Espuelas, S.; Larrañeta, E.; Reviakine, I.; Yate, L.A.; Irache, J.M. Pegylated poly (anhydride) nanoparticles for oral delivery of docetaxel. *Eur. J. Pharm. Sci.* **2018**, *118*, 165–175. [CrossRef]

30. Mira, A.; Mateo, C.R.; Mallavia, R.; Falco, A. Poly (methyl vinyl ether-alt-maleic acid) and ethyl monoester as building polymers for drug-loadable electrospun nanofibers. *Sci. Rep.* **2017**, *7*, 1–13. [CrossRef]

31. Martínez-Ortega, L.; Mira, A.; Fernandez-Carvajal, A.; Mateo, C.R.; Mallavia, R.; Falco, A. Development of a new delivery system based on drug-loadable electrospun nanofibers for psoriasis treatment. *Pharmaceutics* **2019**, *11*, 14. [CrossRef]

32. Cruz-Salas, C.N.; Prieto, C.; Calderón-Santoyo, M.; Lagarón, J.M.; Ragazzo-Sánchez, J.A. Micro-and nanostructures of agave fructans to stabilize compounds of high biological value via electrohydrodynamic processing. *Nanomaterials* **2019**, *9*, 1659. [CrossRef] [PubMed]

33. Wang, F.; Sun, Z.; Yin, J.; Xu, L. Preparation, characterization and properties of porous pla/peg/curcumin composite nanofibers for antibacterial application. *Nanomaterials* **2019**, *9*, 508. [CrossRef] [PubMed]

34. Rebia, R.A.; Tanaka, T. Natural antibacterial reagents (centella, propolis, and hinokitiol) loaded into poly [(r)-3-hydroxybutyrate-co-(r)-3-hydroxyhexanoate] composite nanofibers for biomedical applications. *Nanomaterials* **2019**, *9*, 1665. [CrossRef]

35. Maevskaia, E.N.; Shabunin, A.S.; Dresvyanina, E.N.; Dobrovol'skaya, I.P.; Yudin, V.E.; Paneyah, M.B.; Fediuk, A.M.; Sushchinskii, P.L.; Smirnov, G.P.; Zinoviev, E.V. Influence of the introduced chitin nanofibrils on biomedical properties of chitosan-based materials. *Nanomaterials* **2020**, *10*, 945. [CrossRef] [PubMed]

36. Miroshnichenko, S.; Timofeeva, V.; Permyakova, E.; Ershov, S.; Kiryukhantsev-Korneev, P.; Dvořaková, E.; Shtansky, D.V.; Zajíčková, L.; Solovieva, A.; Manakhov, A. Plasma-coated polycaprolactone nanofibers with covalently bonded platelet-rich plasma enhance adhesion and growth of human fibroblasts. *Nanomaterials* **2019**, *9*, 637. [CrossRef] [PubMed]

37. Li, Y.; Liao, C.; Tjong, S.C. Electrospun polyvinylidene fluoride-based fibrous scaffolds with piezoelectric characteristics for bone and neural tissue engineering. *Nanomaterials* **2019**, *9*, 952. [CrossRef]

Review

Electrospun Functional Materials toward Food Packaging Applications: A Review

Luying Zhao [1], Gaigai Duan [1,*], Guoying Zhang [2], Haoqi Yang [3,*], Shuijian He [1] and Shaohua Jiang [1,*]

1. Co-Innovation Center of Efficient Processing and Utilization of Forest Resources, College of Materials Science and Engineering, Nanjing Forestry University, Nanjing 210037, China; zhaoluying1@163.com (L.Z.); shuijianhe@njfu.edu.cn (S.H.)
2. College of Chemistry and Molecular Engineering, Qingdao University of Science and Technology, Qingdao 266000, China; zhanggy@qust.edu.cn
3. College of Material Science and Engineering, Jilin University, Changchun 130022, China
* Correspondence: duangaigai@njfu.edu.cn (G.D.); yhq1214@126.com (H.Y.); shaohua.jiang@njfu.edu.cn (S.J.)

Received: 26 November 2019; Accepted: 10 January 2020; Published: 15 January 2020

Abstract: Electrospinning is an effective and versatile method to prepare continuous polymer nanofibers and nonwovens that exhibit excellent properties such as high molecular orientation, high porosity and large specific surface area. Benefitting from these outstanding and intriguing features, electrospun nanofibers have been employed as a promising candidate for the fabrication of food packaging materials. Actually, the electrospun nanofibers used in food packaging must possess biocompatibility and low toxicity. In addition, in order to maintain the quality of food and extend its shelf life, food packaging materials also need to have certain functionality. Herein, in this timely review, functional materials produced from electrospinning toward food packaging are highlighted. At first, various strategies for the preparation of polymer electrospun fiber are introduced, then the characteristics of different packaging films and their successful applications in food packaging are summarized, including degradable materials, superhydrophobic materials, edible materials, antibacterial materials and high barrier materials. Finally, the future perspective and key challenges of polymer electrospun nanofibers for food packaging are also discussed. Hopefully, this review would provide a fundamental insight into the development of electrospun functional materials with high performance for food packaging.

Keywords: electrospinning; food packaging; functional membrane; nanofibers

1. Introduction

Electrospinning is a versatile technique for continuously producing nanofibers with a fiber diameter range from sub-nanometers to micrometers. The electrospun fibers have been broadly applied in nearly all the fields, such as composites [1–5], tissue engineering [6–9], biomaterials [10,11], energy storage and conversion [12–16], food packaging [17–19], drug deliver and release [20,21], catalysts [22–25], sensors [26–29], flexible electronics [30–32], reactors [33,34], environmental protection [35–37], etc. During the fiber preparation process, the polymer solution or melt is induced by a high-voltage power supply device to accelerate injection onto a collecting plate with opposite polarity to form nanofiber membrane. Basically, there are three key components to fulfill the process: a high voltage supplier, a pipette or needle with small diameter, and a metal collector [38]. In details, during the electrospinning process, the polymer solution is extruded from the capillary tube by the electric field force, and a Taylor cone can be formed at the tip of the capillary. As the strength of electric field increases, positive charges could accumulate on the surface of the Taylor cone, which further overcomes the surface tension and cause fluid ejection. When the spinning process proceeds, the injected fluid could be stretched several times longer than the original length, and the solvent evaporates simultaneously to form a continuous ultrafine polymer fiber.

The electrospinning process is a simple and effective strategy for fabricating nanofibers, which can prepare polymer nanofibers directly, continuously and even in a large scale. It has the advantages of mild experimental conditions, low cost, easy operation and function, wide range of raw materials, etc. The spinning process is controllable, and the parameters can be adjusted according to the different requirements in various research fields. For example, electrospun nanofibers can be prepared with custom shapes and various orientations to quantitatively investigate the relationship of mechanical properties and molecular orientation [39]. Generally, the nanofibers obtained by electrospinning would have the characteristics of fine size, large specific surface area, high porosity, large aspect ratio and superior mechanical properties.

Functional packaging materials are gradually evolving into the public eyes, which have the functions of moisture absorbing [40,41], antioxidant releasing [42,43] and flavor or odor absorbing [44]. However, for the functional packaging materials applied in the food field, some additional features must be considered, such as degradable, superhydrophobic, edible, antibacterial and high barrier. By virtue of their submicron to nano-scale diameter and very large surface area, electrospun fibers may offer numerous advantages compared to conventional film and sheet packages, such as being more responsive to changes (e.g., relative humidity and temperature) in the surrounding atmosphere. Furthermore, because the electrospinning process takes place at ambient conditions, electrospun fibers are more suitable for encapsulating thermally labile active agents as compared to the fibers made by conventional melt spinning process. Given these advantages mentioned above, electrospun fiber not only could incorporate into bioactive substances, but also could satisfy the requirements of designers and consumers for packaging materials. Therefore, the development of functional packaging materials based on electrospinning technology has become a hot spot in the food packaging field. In the following sections, different approaches for the preparation of functional electrospun fiber will be introduced and their applications on functional package materials will be described (Figure 1). Furthermore, a conclusion including future perspective and key challenges for electrospun functional packaging materials are also discussed. In a word, we believe electrospun materials are good candidates for food packaging materials, and this review would significantly promote the research on application of electrospun fibrous materials for food packaging materials.

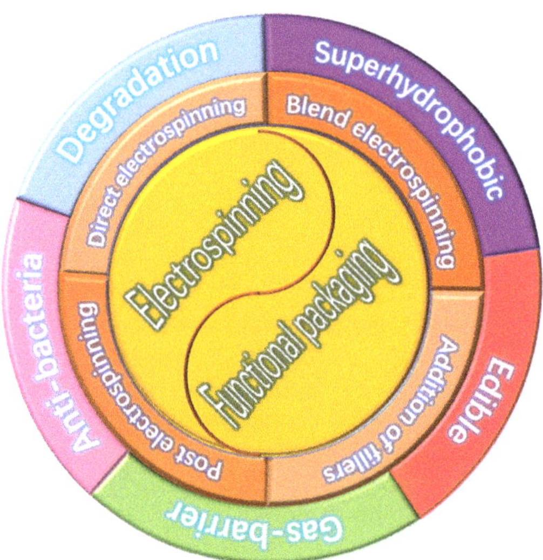

Figure 1. Overview of functional electrospun and food packaging materials diagram.

2. Strategies for the Preparation of Functional Electrospun Materials

In 1934, Formalas invented an experimental device for preparing polymer fibers by electrostatic force and applied for a patent that discloses how a polymer solution forms a jet between two electrodes. The above device could successfully produce a fiber by using high voltage static electricity, which is consequently recognized as the beginning of electrospinning technology [45]. Unfortunately, electrospinning technology did not attract numerous attentions until the middle of the 20th century. With the rapid development of nanomaterials and nanotechnology, electrospinning method has gradually received the attention of scholars from various areas.

So far, the preparation method for nanofibers based on electrospinning technology has been well developed. According to the electrospinning raw materials, it can be divided into melt electrospinning [46–48], solution electrospinning [49,50] and mixed electrospinning [51–53]. According to the design of the spray head, it can be classified as needleless electrospinning [54–56], coaxial or triaxial electrospinning [57–62], multi-jet electrospinning [63–65], etc. The application of electrospinning technology will be described below according to different situations.

2.1. Direct Electrospun Packaging Membrane

Direct electrospinning is defined here as single-component melt electrospinning or single-component solution electrospinning using one jet head. The functional electrospun materials commonly used in packaging field are chitosan (antibacterial), corn protein (edible), polyvinyl alcohol (transparent), etc.

Chitosan (CS) is obtained by deacetylation of chitin, which could form a transparent, elastic and oxygen resistant film. CS film can not only prevent fungi from contaminating and corroding food, but also effectively regulate the composition of oxygen and carbon dioxide around fruits and vegetables, inhibiting the aerobic respiration to a certain extent, so as to improve the shelf life. CS has a huge application potential in the food industry attribute to its advantages of short-time biodegradation, biocompatibility with human tissues, anti-microbial and antifungal activities and non-toxicity. Therefore, chitosan-based nanofiber membrane/film has attracted great attention in food preservation and packaging technology [66,67].

Ohkawa et al. [68] successfully prepared pure CS electrospun nanofibers with trifluoroacetic acid (TFA) as spinning solvent for the first time, because TFA can form salt with amino group in chitosan, effectively reducing the interaction between CS molecules, making electrospinning easier. In addition, the high volatility of TFA is beneficial to the rapid solidification of CS-TFA electrostatic jet. The concentration of CS also affects the morphology of fibers. When the mass fraction of chitosan is 6% or lower, beads and fibers coexist. When the mass fraction of CS is 7%, beads obviously decrease. When the mass fraction of CS is 8%, the spinning effect is better. The diameter of obtained fibers ranges from 390 to 610 nm, with an average diameter of 490 nm, but there still are small beads and interconnected fibers can be seen. To avoid the above phenomena of beads and interconnected fibers and improve the uniformity of electrospun fibers, dichloromethane (DCM) was added into chitosan-TFA solution. Under the optimum conditions, uniform CS nanofibers with an average diameter of 330 nm can be obtained.

In addition to the TFA as solvent, another effective solvent for chitosan is concentrated acetic acid. Geng et al. [69] studied the electrospinning of CS with concentrated acetic acid as solvent. The results show that with the increase of acetic acid concentration, the surface tension of chitosan-acetic acid solution decreases, while the viscosity does not change significantly. At the same time, the charge density of the jet increases. When the mass fraction of acetic acid is 30%, nanofibers begin to appear, accompanied with a large number of beads; when the mass fraction is 90%, uniform fibers with an average diameter of 130 nm can be obtained, and no beads appear. The molecular weight and concentration of CS also affect the formation of fibers. Only when the molecular weight of CS is about 1.06×10^5 g/mol and the mass fraction is 7–7.5%, the beadless nanofibers can be produced. However, high content of CS (more than 90%) cannot be well dissolved in solvent, which is difficult to meet the requirements of spinning viscosity. In addition, the electric field strength also affects the formation of fibers. When the electric field strength is 1 kV/cm, spindle beads and coarse fibers appear. When the electric field strength is 3–4.5 kV/cm, uniform and regular fibers can be formed. However, when the electric field strength is greater than 4.5 kV/cm, a great number of beads could exist in the obtained fine fibers due to the tensile force increases and the charged jet is unstable.

Besides the solvent composition and concentrate, the fiber morphology is also influenced by other parameters, such as applied voltage, needle diameter, receiving distance and feed rate. For instance, Sencadasa et al. [70] investigated the effects of electrospinning parameters on the average diameter and width distribution of CS nanofibers. By controlling the solution parameters and process parameters, the spinning process gets more stable, and the target fiber size could be customized. However, the parameters of the needle tube diameter and feeding speed have no effect on the fiber diameter. When the applied voltage is increased, the diameter of the fibers would decrease. Furthermore, with the reduction of receiving distance, the fibers diameter at the tip of the needle decreases due to the decrease of electrostatic field intensity.

Zein, a kind of storage protein in corn, has a good film-forming property owing to its rich sulfur-containing amino acids that could connect the disulfide bond and water release bond. As a renewable polymer protein, zein has good biocompatibility, which can be used not only as fresh-keeping film, but also as film coating for food preservation [71]. Moreover, it can also be used as edible packaging films in food packaging.

Takanori et al. [72] investigated the parameters such as electric field and polymer solution concentration in the electrospinning process to prepare zein nanofiber membranes. They choose 15 kV and 30 kV as the field strength and 80 wt.% ethanol aqueous solution as solvent. It is found that when the electric field strength is 15 kV, the polymer concentration is mainly 21 wt.%. However, the electric field strength is increased to 30 KV, fibers could be generated at 18 wt.% of polymer concentration. Neo et al. [73] also prepared the electrospun zein membranes with the 80% ethanol as solvent, and further studied the fiber morphology affected by three processing parameters including solution concentration, electric field strength and solution parameters. That is, with the increase of zein concentration, the viscosity of solution would increase, which promotes the formation of bead-free fiber, and the average fiber diameter become larger. They also found that there is an interaction effect between the electric field and feed rate of the solution. At a high voltage, changes of the fiber diameter have a more sensitive response to the feed rate of the solution. Similarly, changes in fiber diameter were also found to be more responsive to the applied voltage at a high feed rate. However, at a low applied voltage, the effect for the feed rate on average fiber diameter will not be significant since the amount of charges are not enough to accelerate the solution.

Polyvinyl alcohol (PVA) film is a kind of antistatic film with good performance, which is widely used in the sales and packaging of textiles and clothing. Compared with polyethylene, polypropylene and other general-purpose films, polyvinyl alcohol film exhibits the advantages of high transparency, good antistatic property, which can significantly reduce the dust absorption effect [74–76]. Furthermore, polyvinyl alcohol is often used in water-soluble packaging.

Tao et al. [77] studied the effects of molecular weight and concentration in a PVA-water system on the morphology and structure of PVA electrospun nanofibers (Figure 2). They found that with the increase of molecular weight, the morphology of PVA fibers can be changed from bead to beaded fibers, then to smooth fibers and finally to flat ribbons at the same concentration. Supaphol et al. [78] further explored the effects of solution concentration and parameters change during electrospinning process (applied voltage and collection distance) on the morphology and fiber diameter of as-spun PVA fiber mats and individual fibers. In the range of solution concentration (6–14%), the average fiber diameter for as-spun PVA fiber mats decreases with the increase of solution concentration and applied voltage, which ranges from 12.5 to 25 kV. At a fixed applied potential of 15 kV, when the receiving distance was changed within a certain range (5–20 cm), the average fiber diameter would become smaller as the receiving distance increased. In addition, there is a decrease of the viscosity that can be observed after sonication, which could result in a decrease of average fiber diameter.

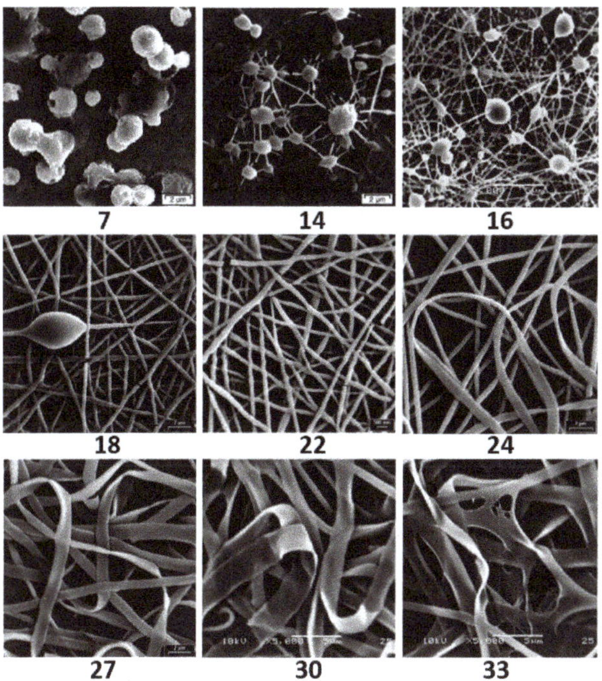

Figure 2. Photographs showing the range of structures that can be produced as the concentration (wt.%) is increased at Mw = 18,000 g/mol [77]. © 2006 Elsevier B.V. All rights reserved.

In another study, Yang et al. [79] employed bubble electrospinning, a special electrospinning method, to prepare PVA nanofibers, and studied the effects of solution concentration and viscosity on the morphology and fiber diameter. They concluded that the higher PVA concentration in solution, the smoother surfaces and larger diameters in the obtained nanofibers. There is an allometric scaling relation between the diameter and PVA aqueous solution concentration: $d \propto C^{5}$. Niu et al. [80] fabricated PVA nanofibers via a needless method, which utilized a cone-shaped metal wire coil as a spinneret instead of needle. This method would increase the fiber production efficiency and achieve a finer average fiber diameter than that of traditional electrospinning. In the same year, this group [81] reported a needleless electrospinning method including a rotating disk and cylinder to prepare PVA nanofibers, and discussed the effect of nozzle shape on spinning process and fiber morphology. On one side, this needleless electrospinning method could improve the productivity, which overcomes the shortcomings of the traditional multi-needle electrospinning method such as large operating space and careful design. On the other side, needleless electrospinning avoids strong charge repulsion between the spinneret and adjacent needles, resulting in an even fiber deposition.

Apart from three polymers mentioned above, some degradable polymers have also been applied in food packaging field, such as polycaprolactone (PCL), poly (propylene carbonate) (PPC), polylactic acid (PLA) and natural cellulose, starch. Moreover, electrospinning plastic materials, such as polyvinyl chloride (PVC), polyethylene (PE), polypropylene (PP), polystyrene (PS), polyethylene terephthalate (PET) and nylon, are commonly used in the field of packaging as well.

2.2. Mixed Electrospun Packaging Membrane

Electrospinning of polymer blends enables the produced nanofibers possess multi-function from each constituent. Up to now, there are many ways for solution mixing, such as electrospinning of multi-components mixed solutions, and design of jet device to obtain multi-components mixed fiber membranes. In the food packaging industry, electrostatic spinning technology is usually employed to prepare composite mats of two or more components.

2.2.1. Blending with Different Polymer Solutions

Before electrospinning, by mixing two or three polymers in spinning solutions, the fiber membranes with better performance could be obtained. Recently, some natural and natural-derived polymers have been used to prepare electrospun fibers through blending with other functional polymers in spinning solutions.

Feng et al. [82] blended gelatin (GT) and PCL (50:50) in trifluoroethanol (TFE) cosolvent, and found a phase-separation behavior of the GT/PCL polymers, which leads to inferior morphology of fibers during the electrospinning. In order to solve this problem, they added a tiny amount of acetic acid to the mixed solution. As a result, the opaque solution became transparent immediately without any obvious precipitation for over one week, and the obtained nanofibers demonstrated a thin, smooth and homogeneous morphology.

Chitosan is soluble in many solvents. It is soluble in organic acids, such as dilute acetic acid, formic acid and lactic acid, and also soluble in the mixture of water and methanol, ethanol or acetone, and its physical and chemical properties can be used for electrospinning, so it is often mixed with other polymers to prepare electrospinning fibers. González et al. [83] studied the thermophysical properties of CS + starch + PET fibers via a blending electrospinning, which could have excellent mechanical properties, thermal stability and biodegradability due to the strong intermolecular force among three components verified by Infrared spectra. Sajeev et al. [84] electrospun PVA and CS in a mixed homogeneous solution to prepare nanofibers and fiber mats. The effects of parameters (such as flow rate, receiving distance and voltage) on the morphology and fibers diameter were also investigated. It was found that the effects of parameters on fibers accorded with a simple extensional creep model. In addition, there is a negative correlation between the content of CS and fibers diameter, which is because the CS is an ionic polyelectrolyte that could enhance the charge density of the surface of jet stream and the received electric field force, causing a decrease of fiber diameter.

2.2.2. Blending by Multiple-Jet Electrospinning

Traditional single-nozzle electrospinning method has restricted its industrial application because of its low output. After years of research and design, the combination of multiple nozzles with different specifications and quantities has been developed.Different nozzles are injected with various solutions, which are mixed and entangled into fibers, then deposited into mats during the electrospinning (as shown in Figure 3, [85]). For example, Ding et al. [86] prepared a series of biodegradable PVA/cellulose acetate (CA) nanofibrous mats via a multi-jet electrospinning method. The weight ratio of the PVA and CA in fiber mats can be controlled by changing the number ratio of PVA/CA jets. Moreover, this multiple-jet electrospinning process ensures a good dispersion of two polymers in the PVA/CA hybrid felts.

The method described below can also be used to prepare hybrid electrospun fibers. Wang et al. [87] utilized a special multi-jet electrospinning method to spin the melted polyethylene and PCL, which actually was an unconfined spinning geometry instead of a needle as performed in Figure 4. In this process, due to the influence of the applied electric field, numerous parallel jets of molten polymer are formed at the plate edge, and finally polymer films are obtained on the collector. Compared with traditional needle electrospinning, unconfined geometries [88–91] rely on electric-field induced spontaneous fluid perturbations to form jet sites, rather than mechanically pumping fluid through a confining nozzle. As the nozzle blockage does not occur in this device, and nanofibers from parallel jets could be prepared synchronously in such an open configuration, so that the production rate has been greatly improved. It is a convenient and high-throughput method for industrial production of microfibers and nanofibers with thermoplastic or other high-viscosity fluids.

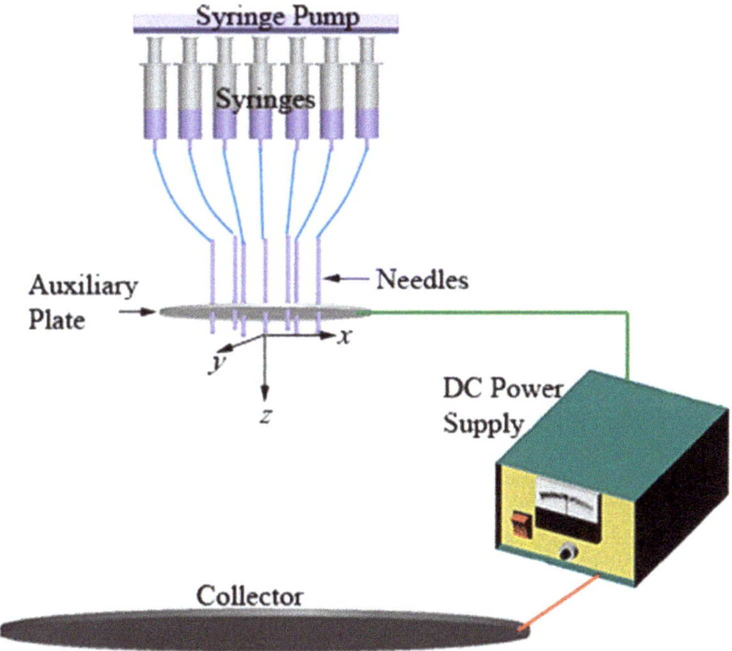

Figure 3. A schematic of the experimental setup used in the multi-jet electrospinning process [85]. © American Chemical Society 2014.

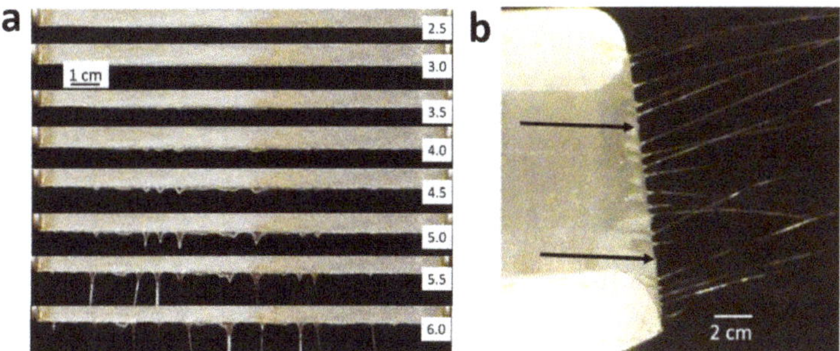

Figure 4. (**a**) Sequential images of the polymer-melt coated plate edge showing the progression of the fluid with time (in minutes, right side) at 180 °C at an applied voltage of −45 kV. The plate is oriented horizontally and being viewed from above. (**b**) Image of electrospinning from a polymer melt coated source plate (PE at 170 °C, −45 kV) in steady state (i.e., after 30 min). The black arrows indicate non-jetting perturbations. This figure is adapted from [87]. © 2014 IOP Publishing Ltd.

2.2.3. Blending by Coaxial Electrospinning

Coaxial electrospinning could produce a new type of nanofibers with a core-shell structure. In the coaxial electrospinning [57], the syringe is designed as two chambers, the outside chamber is generally filled with the polymer solution, and another polymer solution or powders in the inside chamber. During the electrospinning, a core-shell droplet firstly appears at the outlet of the core-shell needle, and then forms the core-shell structure fibers under the force of electric field.

Yao et al. utilized coaxial electrospinning technology to successfully prepare a novel antibacterial composite through encapsulating orange essential oil (OEO) in zein prolamine. The obtained fine fibers ranging in diameter from 0.7 to 2.3 μm depending on the zein prolamine solution concentration and process parameters. Fiber composites exhibit an antibacterial activity against *Escherichia coli* and can be used as food packaging materials for bioactive food preservation, such as extending the shelf life of fruits [92].

It is common that the shell and core are different polymers. Komur et al. [93] fabricated starch and PCL composite nanofibers by coaxial needle electrospinning technique as shown in Figure 5. In this paper, the PCL solution was fed through the inner needle, and the starch solution was located at the outer medium. It was found that processing parameters have a significant influence on the fibers diameter, including polymer concentration, flow rate and voltage. The effect of starch ratio on the physical properties and morphological structures of fiber was also studied. That is, the higher the proportion of starch, the greater of the electrical conductivity, viscosity and density, while the beads would more likely appear in the fibers due to the starch tends to prefer to form beads rather than fibers. When the ratio of the starch increases, the volume of beads would become larger, which will cause the average fibers diameter becomes larger and the fibers strength gets higher. Park et al. [94] prepared levofloxacin-loaded CS and PCL nanofibers by coaxial electrospinning to control the release of antibiotics. They employed CS containing levofloxacin as a core, and PCL as a shell. PCL with different concentrations (8, 12, 16 and 20 wt.%) were set up to explore the effects of nozzle shapes on the sustained release of levofloxacin. As expected, the CS-PCL nanofiber scaffolds with coaxial nozzles had better performance on the sustained release of levofloxacin. Alharbi et al. [95] prepared PLA/PVA and PVA/PLA nanofibers with core/shell-structure to study the mechanical properties of PLA and PVA composite. According to the mechanical test results under static loading, the tensile strength and plasticity of core/shell PLA/PVA nanofibers increased by nearly 233% and 150% respectively compared with the original value. Dynamic loading and creep loading experiments show that there is a strong physical interaction between PLA layer and PVA layer, which could improve the mechanical properties.

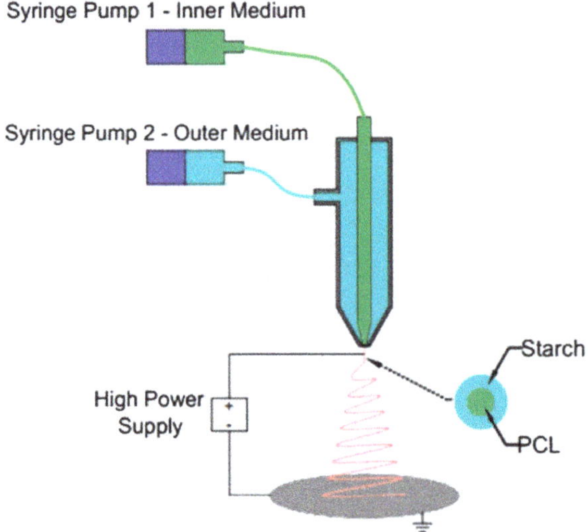

Figure 5. Schematic representation of the coaxial needle electrospinning set-up [94]. © The Korean Society of Pharmaceutical Sciences and Technology 2012.

Sometimes, functional substances such as antibacterial substances are loaded into the core of the coaxial electrospun fiber, which would release later so as to be effective. Korehei et al. [96] directly incorporated the T4 bacteriophage into the fiber core to prepare a core/shell fiber structure via a coaxial electrospinning method. The electrospun fibers were produced using a PEO/chloroform solution as the shell and a T4 bacteriophage/buffer suspension as the core. Coaxial electrospinning could produce continuous core/shell fibers with bead-free morphology, and the T4 bacteriophage was uniformly incorporated in the core of fibers. The core/shell fiber encapsulated bacteriophage exhibits full bacteriophage viability after storing for several weeks at +4 °C, which can be applied in meat packaging. He et al. [97] employed coaxial electrospinning technology to prepare anti-infective drug delivery carriers (Figure 6), i.e., poly (e-caprolactone)/zein uniform beadless core/shell nanofibers loaded with metronidazole (MNA). The prepared fiber membranes were hydrophobic and could inhibit the growth of anaerobic bacteria by releasing MNA.

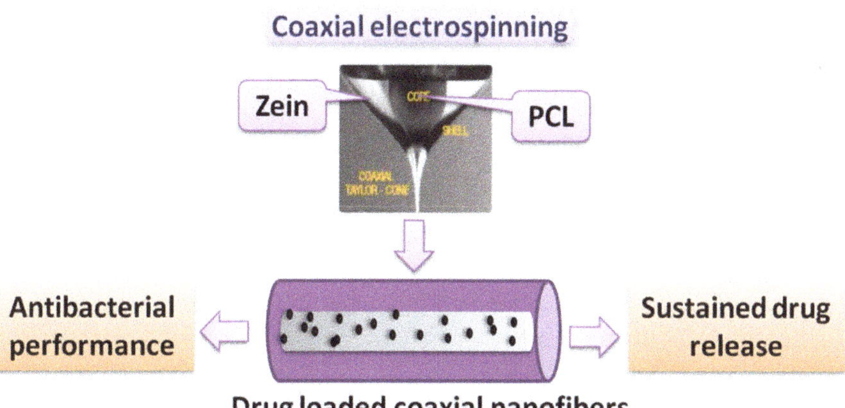

Figure 6. Structure diagram of coaxial fiber [97]. © Elsevier Inc. All rights reserved 2016.

2.3. Addition of Inorganic Fillers

In order to achieve multi-function for nanofiber-based food packaging materials, various fillers have been added into polymer substance, especially inorganic fillers that possess conductivity, magnetism, antibacterial property, etc. [98–101]. Then, a nanofiber-based food packaging material could be prepared through the electrospinning technique.

2.3.1. Conductive Fillers

Electrostatic hazard is a common problem in the packaging industry. Some severe accidents and impacts, such as fire, explosion and dust absorption are related to the electrostatic effect. Generally speaking, the insulating material is easy to produce static electricity. Adding conductive fillers is an efficient way to solve the problem and enhance the antistatic performance. Moreover, conductive packaging materials have a great potential for application in smart packaging due to their combination with some electronic devices.

Carbon-based conductive materials are often used as conductive additives to prepare conductive/anti-static packaging materials with wide application and large usage [102]. Carbon black has been widely adopted, but it has a huge loss of mechanical properties for the composites. By comparison, carbon nanotubes (CNTs) exhibit the advantages of less filling amount and negligible mechanical properties effects, which are emerged as a new generation of carbon-based conductive materials. Yang et al. [103] prepared electrospun biodegradable PLA composites which using CNTs as conductive fillers. They found that the morphology of the fibers is related to the loading of

CNTs. At a low loading level, the CNTs can be well embedded in the PLA matrix to form a fiber axis-oriented nanowire structure. At a high loading level, the CNTs are mainly dispersed in the form of bundles along with the fiber axis, and the resulting fibers are tortuous or misshaped. Meng et al. [104] studied the effects of different multiwalled CNTs (MWCNTs) content (0.1%, 0.5%, 1%, 2% and 5%) on the properties of non-bead PCL-MWCNTs nanofiber membranes, which found that the diameter distribution, average diameter, crystallinity and tensile strength are increased with the addition of MWCNTs. Moreover, the electrospun PCL–MWCNTs nanofiber membranes also exhibited a good degradability. Giner et al. [105] obtained electrospun nanocomposite fibers mats by embedding graphene nanoplatelets (GNPs) in poly (ethylene-co-vinyl alcohol) (EVOH) fibers. The heat-treated fibers mats at 158 °C could produce continuous, contact and transparent films, which have a great potential to be applied in the intelligent packaging field. For instance, smart labels or tags can be anticipated.

2.3.2. Magnetic Fillers

For various electronic components, such as electronic precision instruments, medical devices, computers, automatic office equipment and other products, are sensitive to electromagnetic radiation, therefore, the anti-electromagnetic radiation packaging is urgently needed. Adding magnetic filler is an effective way to endow materials with anti-electromagnetic radiation performance. Recently, Iron oxides or compounds have been often employed as the magnetic fillers for electrospun nanofibers.

Song et al. [106] encapsulated self-assembled iron-platinum (FePt) magnetic nanoparticles in PCL nanofibers by coaxial electrospinning. The discrete FePt nanoparticles array can be arranged in a long-range order along the fiber axis at 3000 nm. Wang et al. [107] prepared Fe_3O_4/PVA composite nanofibers by combining in-situ polymerization and electrospinning technology. The Fe_3O_4 magnetic fluids were synthesized through chemical co-precipitation method in the presence of 6 wt% PVA aqueous solution, which avoids the agglomeration of magnetic nanoparticles due to the stabilizing effect of PVA. Wei et al. [108] prepared a porous magnetic biodegradable Fe_3O_4/CS/PVA nanofiber membranes, and explored the influence of electrospinning parameters (polymer concentration, Fe_3O_4 content and magnitude of applied voltage) on the morphology of fibers. When the polymer concentration was 4.5 wt.%, the applied voltage was 20 kV and the loading of Fe_3O_4 nanoparticles was less than 5 wt.%, the uniform, smooth and continuous Fe_3O_4/CS/PVA nanofiber membrane can be obtained. Stylios [109] reported the effect of processing parameters on the morphology of PVA/$FeCl_3$ magnetic composite fibers, such as applied voltage, receiving distance, flow rate and solution concentration. This research provides guidance for the subsequent manufacture of flexible magnetic composite membranes. Kumar et al. [110] electrospun the PLA, PEG and magnetic nanoparticles (Fe_3O_4@SiO_2 core-shell NPs) mixed solution at room temperature to prepare a series of fibers. With other parameters remaining unchanged, the fiber diameter was reduced from 6 to 3 micrometers after the PLA solution was added to PEG. When the PLA solution mixed with PEG + MNPs (magnetic nanoparticles), the fiber diameter was decreased from 3 to 1 μm.

2.3.3. Photocatalytic Fillers

Photocatalytic fillers have the ability to impart the self-cleaning properties to functional packaging materials. Many scholars demonstrate the self-cleaning ability of materials by testing the rate of dyes degradation with different colors.

TiO_2 is a non-toxic, low-cost and good photocatalytic filler, which is often used to treat pollutants. Bedford et al. [111] prepared a photocatalytic self-cleaning fiber membrane through a coaxial electrospinning device, which employed a CA solution as a core and the TiO_2 dispersion as a shell. The obtained self-cleaning fiber membrane had been tested to completely degrade the blue dye within 7–8 h when exposed to a halogen lamp (Figure 7). Nasikhudin et al. [112] prepared PVA/TiO_2 composite nanofibers by electrospinning method, and studied the photocatalytic activity for the degradation of methylene blue under ultraviolet light. It was found that the PVA/TiO$_2$ composite nanofibers suspended in dye solution could degrade 70% of the dye within 5 h.

ZnO is often doped as the filler into photocatalytic composite fibers because of its simple synthesis, non-toxic and pollution-free. Liu et al. [113] electrospun precursor solution of zinc acetate (ZnAc)/CA in mixed-solvent of N, N-dimethylformamide/acetone, then calcined that to obtain photo-catalytically active ZnO nanofibers. According to experiments, they found that ZnO nanofiber mats could degrade nearly 60% of Rhodamine B within 2 h under visible light irradiation. Khan et al. [114] successfully manufactured ZnO/poly (1,4-cyclohexanedimethylene isosorbide terephthalate) (PICT) nanofibers by electrospinning technique. When the load concentration of ZnO is 9% and the concentration of PICT is 10%, the composite fiber can achieve 99% self-cleaning efficiency within three hours exposed to UV light (Figure 7a). Apart from endowing the composite electrospun fiber membrane with self-cleaning ability, some photocatalytic fillers can also remove some organic compounds in the package. For example, Zhu et al. [115] prepared the electrospun PP films, which can be potentially be used as packaging material for bananas, because TiO$_2$ loaded in the fiber membrane would catalyze the degradation of ethylene and delay the excessive ripeness and deterioration of banana fruits (Figure 7b).

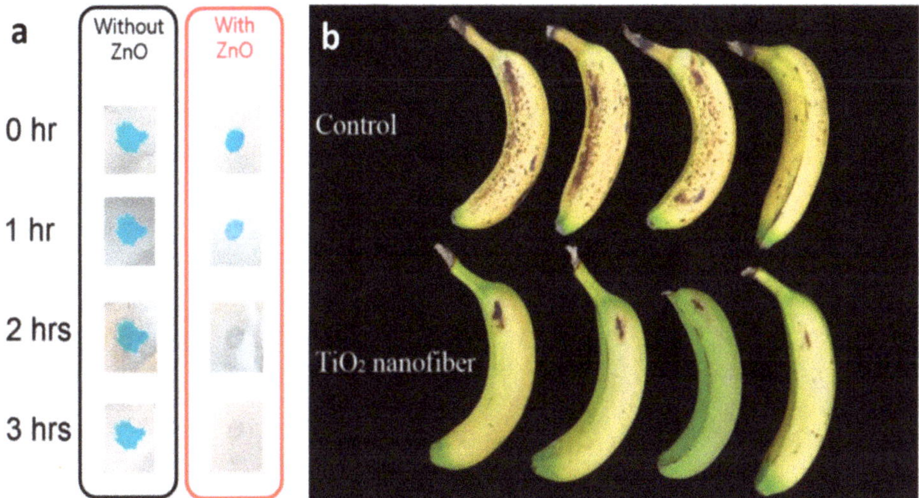

Figure 7. (**a**) Comparison of photocatalytic activity of fibers with and without ZnO [114], © 2018 SAGE Publications. (**b**) photographs of bananas stored for 10 days covered with PP film (control) or PP film and nanofiber containing 5 wt% TiO$_2$ [115]. © Springer Science+Business Media, LLC, part of Springer Nature 2018.

2.3.4. Antibacterial Fillers

Antimicrobial packaging materials has played a majority role in the field of food packaging, which could limit or prevent the growth of food spoilage bacteria or pathogens by prolonging the stagnation period of microorganisms, slowing down the growth rate or reducing the survival number of microorganisms, so as to ensure the quality and extend the shelf life of food. So far, there are three kinds of common antibacterial agents, which are the inorganic antibacterial agent, organic chemical antibacterial agent and natural biological antibacterial agent, respectively. Inorganic antibacterial agents mainly refer to some metal ions (such as Ag$^+$, Cu^{2+}, Zn^{2+}, etc.) or their compounds that would have antibacterial activity, and also the atomic oxygen sterilization, which is produced by a photochemical reaction. Among them, zinc oxide and titanium dioxide are the most investigated in previous reports.

The main characteristics of inorganic antibacterial agents are good heat resistance, wide antibacterial range, long effective antibacterial period and not easy to produce drug resistance [116].

Amna et al. [117] reported for the first time the fabrication of olive oil/polyurethane (PU) composite nanofibrous packaging mats decorated with ZnO nanoparticles by electrospinning. PU is thermoplastic polymer that demonstrates outstanding mechanical properties and water insolubility. Meanwhile, PU possesses good barrier properties, oxygen permeability and more like to the FDA (Food and Drug Administration) point. Olive oil, a natural material, is often loaded with various antibacterial ingredients, flavonoids and antioxidants. Likewise, zinc oxide nanoparticles (ZnO) possess antibacterial activity, which is low toxicity and biodegradability. These kinds of biodegradable packaging materials have displayed a potential antimicrobial activity against *S. aureus* and *S. typhimurium*. Therefore, it can be used for packaging fresh or processed meat and meat-based products (Figure 8). The synthesized nanofibers have the opportunity to replace non-degradable films and overcome the complexity of recycling.

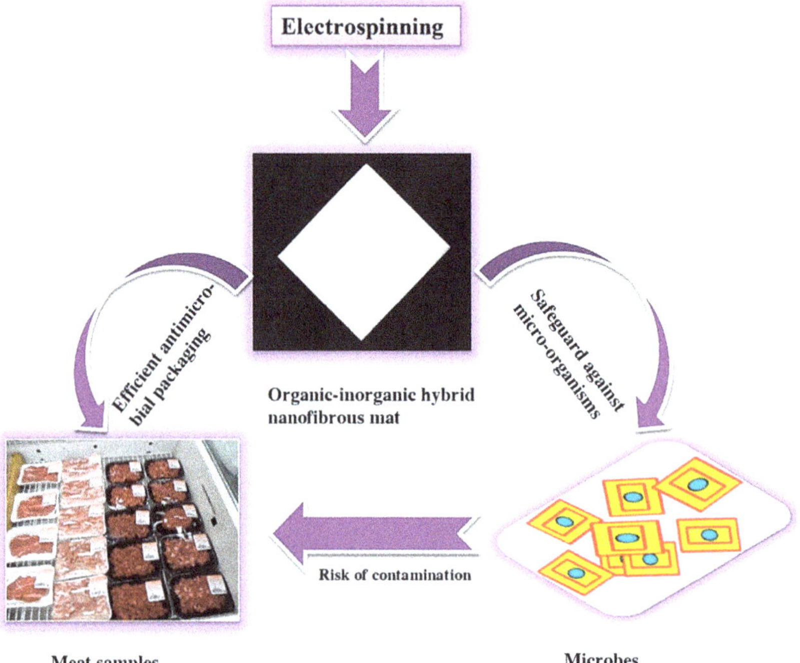

Figure 8. Electrospun hybrid mats; its antimicrobial concept and projected future applications as packaging material for meat and meat-based products [117]. © Association of Food Scientists & Technologists (India) 2014.

2.4. Post-Treatments of Electrospinning Membrane

Post-treatments, such as hot working, surface modification, dip coating, etc., could greatly enhance the performance of electrospun membranes, even expand the application range to meet the high requirements of designers.

2.4.1. Thermal Treatments

By heating the electrospun membrane, the nanofibers in the electrospun membrane can be crosslinked, and the properties such as mechanical properties can be further improved. For example,

Lee et al. [118] fabricated PVA nanofibers crosslinked with blocked isocyanate prepolymer (BIP) by the electrospinning process and subsequent thermal treatment. Their experiment showed that a chemical cross-linking reaction occurs between the hydroxyl group of PVA and the isocyanate group of BIP during the thermal treatment. After the chemical cross-linking finished, PVA/BIP nanofibers not only have a significant improvement in mechanical properties and water resistance, but also have an increased thermal stability. In addition, the fibers membrane has the general advantages of electrospun nanofibers, which is owing to a large surface area and high porosity, resulting in a wider application potential. Lee et al. [119] employed an electrospinning technology to prepare PCL fibers. In order to improve the mechanical properties, the fibers were placed in Pluronic F127 solution and heated at a various temperature ranging from 54 to 60 °C for 30 min. After the heat treatment, bonding occurs between fibers, and the biomechanical properties are improved significantly. Thus, these fibers possess adequate tensile properties, suture retention strength and burst pressure strength.

Theoretically, the solvent needs to volatilize in the electrospinning process, but there is always a little solvent remains in the fiber. As an effective strategy, thermal treatments could promote the volatilization of residual solvent. D'Amato et al. [120] have developed a new method for removing residual solvents to prolong the release time of small molecules in electrospun fibers (Figure 9). Firstly, degradable poly(L-lactic acid) (PLLA) composite nanofibers containing hydrophobic drug 6-aminonicotinamide (6AN) were prepared by using electrospinning technology. Then the composite nanofibers were placed in laboratory environment for 28 days, and heat-treated at fixed temperature and environment in incubator that was maintained at 37 °C, 5% CO_2, and 90–95% relative humidity. Compared with the untreated blank sample releasing 6AN over 9 days, the drug release time of PLLA composite nanofibers with heat-treatment can be extended to over 44 days.

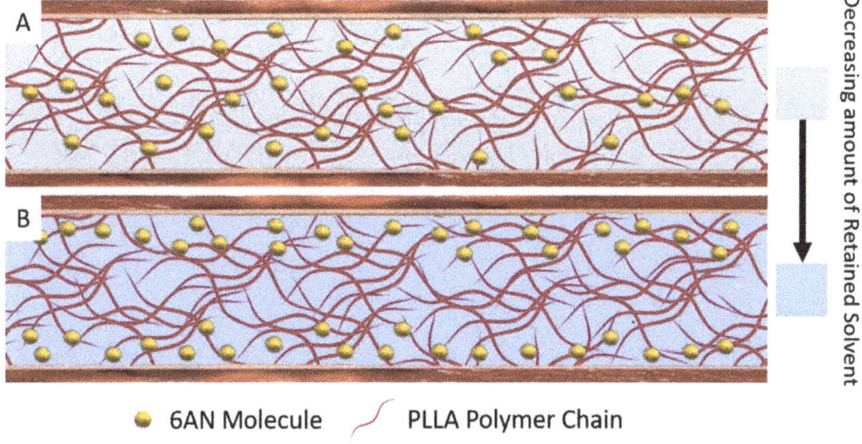

Figure 9. Schematic of the effects of solvent removal on the 6AN distribution inside of an individual electrospun nanofiber. (**A**) A single PLLA fiber immediately after electrospinning with uniformly distributed 6AN. (**B**) A single PLLA fiber after solvent removal shuttled 6AN from inaccessible fiber core towards the fiber surface [120]. © Elsevier Ltd. All rights reserved 2017.

2.4.2. Surface Modifications

The surface properties of nanofibers play an important role in packaging materials. Although electrospinning technology can produce fiber membranes with special internal structure, the surface properties of fiber membranes still cannot meet the application in food packaging fields, such as wettability, adsorption, etc. Thus, the subsequent modification is needed to change or modify surface of electrospun nanofibers in order to meet the packaging field characteristics and special application requirements [121]. There are two usual strategies to modify the surface of electrospun film, one is grafting on the surface and the other is plasma treatment on the surface.

In the field of food packaging, the surface modifications for electrospun fiber membrane are often used to endow the membrane with antibacterial properties. For example, Li et al. [122] prepared the surface of electrospun poly(D,L-lactide) (PDLLA) membrane by plasma pretreatment, UV initiated graft copolymerization of 4-vinylpyridine (4VP) and quaternization of the grafted pyridine groups with hexylbromide. The antibacterial rate of the surface modified electrospun fiber membrane against gram-positive *Staphylococcus aureus* and gram-negative *Staphylococcus* was 99.999%.

Additionally, surface modifications can also improve the physical properties of electrospun fiber membrane, such as mechanical properties. Surucu et al. [123] reported a dielectric barrier discharge (DBD) Ar + O_2 and Ar + N_2 method to modify the surface of PCL/chitosan/PCL nanofibers. The plasma modifications based on Ar + O_2 had improved the mechanical properties and oxygen functionality.

2.4.3. Dip-Coating of Electrospinning Membrane

The dip coating method is to immerse the electrospun fiber membrane in a container for a period of time and take it out, so that the coating is attached to the electrospun fiber. The advantages of dip coating are high production efficiency and simple operation.

Some researchers coated electrospun fibers by the solution or suspension of antibacterial substances to make the composite fibers antibacterial properties. Ignatova et al. [124] combined electrospinning and impregnation techniques to prepare new materials of caffeic acid phenethyl ester (CAPE)/PVP-n-poly (3-hydroxybutyrate) (PHB). The prepared new CAPE-containing material has good antioxidant activity, suggesting that addition or coating of CAPE to the fiber can completely kill Gram-positive *S. aureus* and inhibit the growth of Gram-negative *E. coli*, which is expected to be used in the field of antimicrobial packaging.

Goha et al. [125] prepared beadless, smooth surface PLA/CS nanofibers by electrospinning, and then coated PLA nanofibers with cerium-doped bioactive glass (CeBG), copper-doped bioactive glass (CuBG) and silver-doped bioactive glass (AgBG). Then they tested the bacteriostatic activity against *Escherichia coli* (ATCC 25,922 strains) by the disk diffusion method, and found that the CeBG and CuBG modified PLA/CS nanofibers did not produce bacteriostatic areas against *E. coli* in three samples.

Yakub et al. [126] also combined electrospinning technology with the impregnation method to prepare PCL/CS composite nanofiber by using the natural phenolic acid ferulic acid (FA) as raw materials. The experimental results showed that a combination of FA and CS in the fibers is more effective for killing *Staphylococcus aureus* than FA-containing mats or CS-coated mats, and all the composite fibers containing CS and FA have a good antimicrobial activity.

This method can also modify the electrospun fiber membranes from hydrophobicity to hydrophilicity. Hu et al. [127] fabricated PLA/beta-tricalcium phosphate (b-TCP) composite fiber membranes, after that, the membrane was immersed in 5.0% (w/v) polyethylene oxide solution at room temperature for 5 minutes. After the dried, hot pressed and other steps, the surface modified composite fiber membranes were obtained, which exhibited an enhanced degradation rate, and a change of surface characteristic from hydrophobic to hydrophilic.

3. Functional Materials for Food Packaging Applications

The purpose of food packaging is to ensure the quality and safety of food, provide convenience for users, highlight the appearance and mark of commodity packaging and improve the value of merchandise. Most importantly, food packaging can prevent food deterioration and ensure food quality. With the progress of science and technology and the improvement of people's consumption level, functional food packaging is more and more crucial in people's daily life.

Fresh keeping, environmental protection and convenience are the new development direction of food packaging. Choosing proper packaging materials should not only consider the requirements of consumers and the needs of producers, but also consider the coordination and influence with the environment. In principle, the electrospun nanofibers used in food packaging must have biocompatibility and low toxicity, even non-toxic. Based on these above considerations, functional food packaging materials gradually require some special characteristics like degradable, super-hydrophobic, self-cleaning, edible, antibacterial and high barrier.

3.1. Degradable Electrospun Packaging Membrane

At present, most of the packaging materials are polymers, such as polyethylene, polypropylene, polystyrene, polyvinyl chloride and so on, which are very stable in nature and difficult to degrade. The employment of these materials in packaging industry has caused a serious white pollution due to numerous consumptions of disposable goods, such as disposable tableware and plastic bags. Therefore, in the field of functional packaging, many scholars began to study polymers that could be able to degrade in the natural environment.

The degradable packaging materials generally refer to degradable plastics. According to the traditional classification, degradable plastics can be divided into two categories: photodegradable plastics and biodegradable plastics. So far, biodegradable polymers are very popular in the functional packaging field. Some chemically synthesized polymers, such as PCL, PPC, PLA and PVA are often used in food packaging materials through electrospinning. The presence of these polymers with good biocompatibility can provide a basic comprehension for functional modifications of packaging materials.

Biodegradable polymer materials synthesized by chemical method are similar to natural polymers or polymers with easily degradable functional groups, which can be designed and adjusted to meet the actual needs.

PCL possesses high biocompatibility and biodegradability, which always is applied as degradable packaging. Katsube et al. [128] prepared electrospun PCL nanofibers by using a solution of 12 wt. % PCL in acetone with a capillary flow rate of 24 mL/h, and an electric field of 1.005 kV/cm. Furthermore, the mechanical behavior of electrospun PCL under tensile loading was investigated. Subramanian et al. [129] studied the formation of nanofibers during electrospinning process by using melting PCL as spinning solution, and explored the effects of process parameters such as applied voltage, electrode spacing and molecular weight on fibers diameter. In this work, the diameter distribution was 5–20 microns. By controlling process parameters, the proportion of fine fibers in melt-electrospun PCL mesh was increased. Gaudio et al. [130] used 14% w/v solutions in mixture 1:1 of tetrahydrofuran and N, N-dimethylformamide and chloroform to fabricate micrometric and submicrometric fibrous PCL. The experiments have proved that PCL was nontoxic. Additionally, Bhullar et al. [131] used environmentally friendly and non-toxic melt spinning to obtain PCL micro-fibrous structure, followed by impregnation of bioactive rosemary extract. As expected, the rosemary extract was uniformly dispersed in the PCL microfiber structure. Finally, the bioactive packaging materials with satisfactory antimicrobial, structural and thermal properties can be obtained. The active bio-composite is biodegradable and biocompatible, which is able to replace traditional packaging materials.

CO_2 can be copolymerized with epoxy monomers to obtain a series of biodegradable polycarbonates, among which PPC is an alternating copolymer of carbon dioxide and propylene oxide (PO). Using CO_2 as monomer to synthesize PPC can not only overcome the shortage of petroleum resources, but also help to reduce carbon dioxide pollution. Generally speaking, PPC has good tensile toughness, transparency, biocompatibility and biodegradability. Park et al. [132] used sol–gel electrospinning to prepare pure PPC nanofibers, and found that the mechanical properties of pure PPC were significantly improved after the heat treatment at 60 °C due to the highly bonded structure of nanofibers, which was further interpreted by SEM diagram. Nagiah et al. [133] obtained PPC ultrathin fibers with 10% w/v polymer solution, which have good thermal stability, mechanical properties and high porosity.

PLA is defined as the most promising new packaging material in the new century by the industry, which has a series of advantages such as complete biodegradation, environmental friendliness and recyclability. PLA can be made into film products with high transparency, excellent processability and mechanical properties, which are often used in food packaging. Li et al. [134] studied the morphology, structure and tensile properties of electrospun PLA porous nanofibers with different crystallinity. It was reported that rigid porous PLA nanofibers with high tensile modulus, high strength and small strain can be obtained by the denser structure and enhanced molecular orientation during electrospinning. The study from Casasola et al. [135] demonstrated that the effects of solution properties on the morphology and diameter of PLA nanofibers. It was found that there was the highest fiber productivity in the solvent system of acetone/dimethylformamide, which could prepare the defect-free nanofibers. In this system, the effect of polymer concentration on the formation of PLA nanofibers was also studied. The structure of PLA nanofibers obtained from the critical chain entanglement concentration (CE) was beaded and the defect-free nanofibers were obtained by increasing the concentration about twice higher than that of the entanglement concentration. Critical chain entanglement concentration has positive correlation with its elastic modulus and plastic modulus. In addition, the higher the conductivity of the solvent, the finer the diameter of the fiber.

Tang et al. [136] obtained a kind of electrospun PLA nanofibers with remarkable nano-porous surface and super high specific surface area. The formation mechanism of nanopores on the surface of fibers is solvent phase separation. Huang et al. [137] further studied the effect of different solvents on the surface morphology and internal porous structure of electrospun PLA fibers. Both surface and internal pore can be achieved through different mechanisms. It could be clearly found in Figure 10 that by choosing different solvent systems, electrospun PLA nanofibers with porous surface or inner surface and their combination can be obtained under suitable humidity and environment conditions.

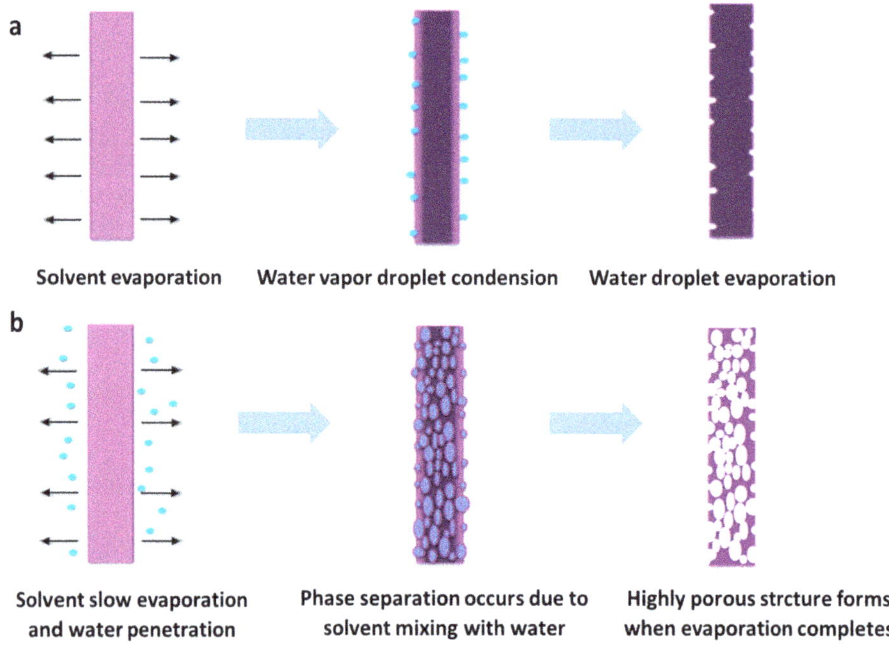

Figure 10. (**a**) Schematic diagram of surface pore formation induced by breath figures mechanism. (**b**) Schematic diagram of porosity induced by a vapor induced phase separation (VIPS) mechanism [137]. © Elsevier Ltd. All rights reserved 2018.

Tomaszewski et al. [138] studied the effects of molecular weight of PLA and spinning solution viscosity on the thickness and quality of fiber felt. In addition, the thermal and tensile properties of fiber felt were also studied. Li et al. [139] came to a similar but more detailed conclusion, which are the uniform fibers that can only be obtained when the concentration of low molecular weight PLA solution is above the entanglement concentration. On the contrary, when the concentration of high molecular weight PLA solution is below entanglement concentration, uniform fibers cannot be obtained. Meanwhile, due to the low viscosity and deformation resistance of precursor solution, the PLA nanofiber has remarkable molecular alignment, which also leads to a rapid cold crystallization and high modulus in the nanofibers.

In the field of food packaging, biodegradable PVA is often combined with other substances to achieve the target functionality owing to some simple electrospinning conditions. Kayaci et al. [140] successfully produced PVA nanowebs incorporating vanillin/cyclodextrin inclusion complex (vanillin/CD–IC) via an electrospinning technique with the goal to obtain functional nanowebs containing flavor/fragrance molecules with enhanced thermal stability and durability. Therefore, PVA/vanillin/CD–IC nanowebs can be quite applicable in active food packaging. Wen et al. [141] fabricated electrospun PVA/cinnamon essential oil/b-cyclodextrin (PVA/CEO/b-CD) antimicrobial nanofibrous film, which can effectively prolong the shelf-life of strawberry, indicating it is potential for the application in active food packaging (Figure 11).

Figure 11. Appearance changes of strawberries stored at 21 °C. (**a**) Control; (**b**) packed with fresh-keeping film and (**c**) packed with PVA/cinnamon essential oil /b-CD nanofilm [141]. © Elsevier Ltd. All rights reserved 2015.

3.2. Superhydrophobic Electrospun Packaging Membrane

The superhydrophobic surface has the advantages of self-cleaning and anti-adhesion, which could delay the deterioration of food and prevent the propagation of microorganisms in food packaging, so the superhydrophobic packaging membrane is very popular in this field.

Ding et al. [142] obtained superhydrophobic fibrous PVA/ZnO composite films by electrospinning and surface treatment with fluoroalkylsilane (FAS). First, ZnO with nanostructured surface was formed by calcining the electrospun composite nanofiber membranes, and then the surface was modified by fluoroalkylsilane coating to obtain the superhydrophobic surface, and the contact angle of the rest with water has changed from 0 to 165 degrees, which means that the surface of composite was also changed from superhydrophilic to superhydrophobic. The comparative experiments showed that the superhydrophobic surface of the composite fiber membrane is the result of combination of high surface roughness and hydrophobic FAS modification.

Actually, the rough surface has played an important role on the superhydrophobic surface, which involves a principle denoted as lotus leaf effect that is an important part for the superhydrophobic surface because there are many micron scale protrusions on the lotus leaf surface. The rough structure increases the air fraction in the space–time contact between the surface and water, and greatly reduces the actual contact area between water and lotus leaf [143]. Therefore, many research groups use the so-called nano plate making technology to prepare the surface with artificial lotus leaf. For instance, Kang et al. [144] utilized the volatility of solvent during electrospinning to prepare PS films with unique protuberances, which the structure is similar to the surface of the lotus leaf, so the electrospun fiber membrane has a superhydrophobic property, the water contact angle is 154.2 ± 0.7° (Figure 12).

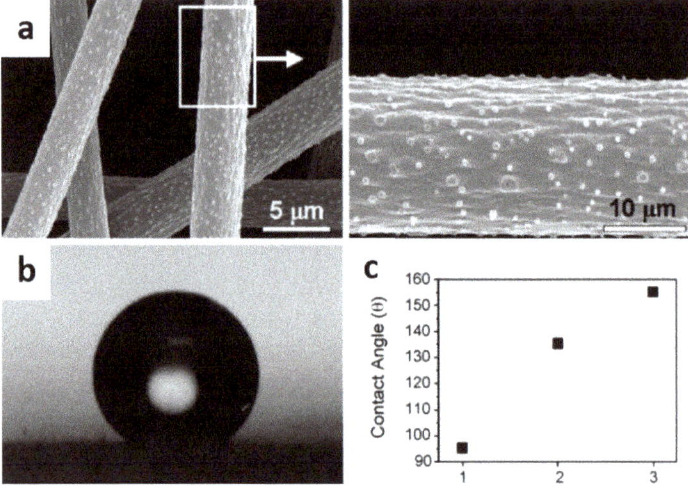

Figure 12. (a) FESEM images of electrospun PS fibers from 35 wt.% solution in DMF, (b) water droplet on electrospun PS fibers from 35 wt.% solution in DMF and (c) variation of water contact angles depending on surface structures (1: PS film; 2: electrospun PS fibers using THF; 3: electrospun PS fibers from DMF) [144]. © Elsevier B.V. All rights reserved 2007.

3.3. Edible Electrospun Packaging Membrane

The application of edible packaging in food packaging has a long history. For decades, the familiar glutinous rice paper used in candy packaging and the corn baking packaging cup used in ice cream packaging were typical edible packaging. With the improvement of people's requirements for food quality and preservation period, as well as the enhancement of people's awareness of environmental protection, edible films consisted of natural biological materials are becoming a research hotspot in the field of food packaging. Generally, edible packaging membranes have different kinds of ingredients according to their compositions. The first one is polysaccharide membranes, such as starch, cellulose derivative, pectin, chitosan, etc. the second one is protein membranes, such as collagen, gel, and the third one is lipid membranes, such as beeswax, paraffin, the fourth kinds is compound membranes, which are obtained through the combination of three above substances.

Tang et al. [145] prepared edible gelatin based composite fibers by electrospinning. The combination of gelatin nanofibers with peppermint essential oil (PO) and chamomile essential oil (CO) could enhance the hydrophobicity of membrane surface, and all the gelatin nanofibers containing PO, CO or PO/CO had better antibacterial properties against *Escherichia coli* and *Staphylococcus aureus*, and had certain antioxidant properties (Figure 13). In particular, the addition of PO leads to a better antibacterial activity of the fiber membrane, while the oxidation resistance of fiber membrane containing CO is better.

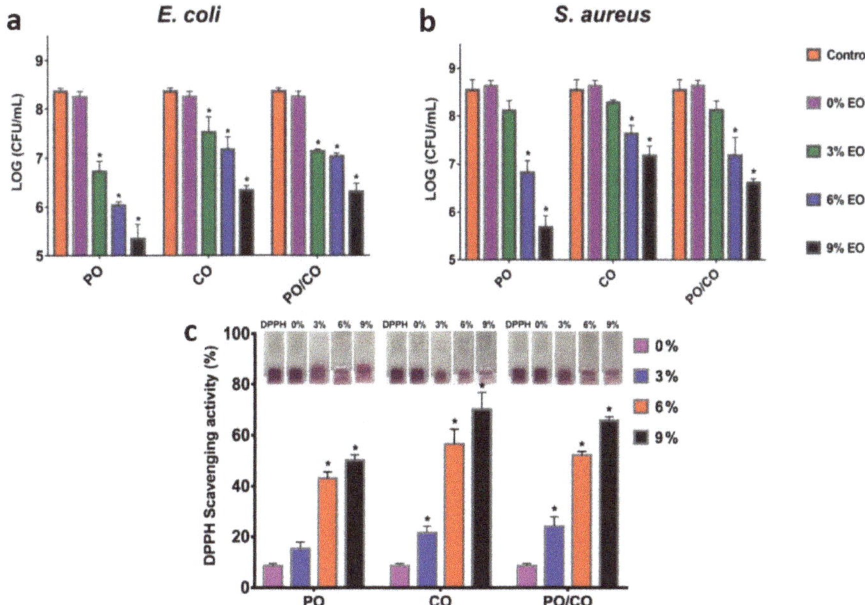

Figure 13. (**a**,**b**) The results of dynamic contact assays against *Escherichia coli* and *Staphylococcus aureus*. (**c**) The antioxidant performance of gelatin/EOs nanofibers was determined using the DPPH radical scavenging method, (∗) *p* (in Tukey's post hoc test) < 0.05 versus the control group [145]. © American Chemical Society 2019.

Based on one-step electrospinning technology, Mascheronia et al. [146] proposed an edible polysaccharide system that can be applied to food packaging. In this system, the cyclodextrin crystal is surrounded by aromatic compounds and fixed on the Prussian nanofiber mesh. The composite, combined with the membrane's pullulan nanofibers, is stable for several months without loss of volatiles when it was stored in ambient relative humidity and at different temperatures. Due to the ability of encapsulation and release of antibacterial aromatic compounds, the system has potential applicability for improving microbial safety, especially for fresh food.

3.4. Antibacterial Electrospun Packaging Membrane

According to the previous description, there are three kinds of antibacterial substances, including natural biological antibacterial agents (essential oil, etc.), organic chemical antibacterial agents (organic acid, etc.) and inorganic antibacterial agents.

Cristina et al. [147] successfully encapsulated a naturally occurring antimicrobial compound, allyl isothiocyanate (AITC) into soy protein isolate (SPI) and PLA fibers by electrospinning technology, and studied its effect on the fiber properties. By elaborately manipulating the formulation of solution, the morphology of composite nanofibers can be adjusted. Most of all, it was found that AITC released from SPI and PLA electrospun fibers could be controlled by changing relative humidity, and the increase of relative humidity of air can triggers the release of AITC in fibers.

Neo et al. [148] evaluated the applicability of gallic acid loaded zein (Ze-GA) electrospun fiber mats towards potential active food packaging. As expected, the Ze-GA fiber mats demonstrated outstanding antibacterial activity and properties consistent with those considered desirable for active packaging material in the food industry.

As a natural mineral, montmorillonite (MMT) can also be applied in food packaging industry. Agarwal et al. [149] coated PP membranes with MMT and nylon 6 nanofibers by electrospinning. The effect of these membranes in the packaging of popular foods such as crisps and bread were investigated (Figure 14). When the membranes were used in the packaging of potato chips, the oxygen barrier property was significantly reduced, which was attributed to the MMT-N6 coating that slightly improves the water vapor barrier. The reason for decrease in microbial spoilage of bread was similar. Therefore, the coated MMT composite membranes can be applied in the packaging industry due to its non-toxic, which can realize the extension of shelf life of packed food.

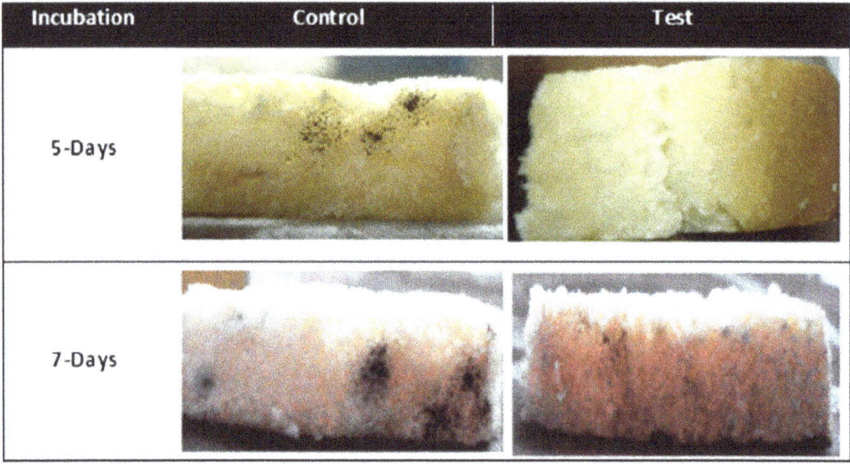

Figure 14. Effect of MMT-N6 nanofibrous membrane coating on PP packets on the natural flora of bread [149]. © Elsevier Ltd. All rights reserved 2014.

As mentioned above, chitosan is a natural antibacterial polymer, so it has attracted numerous attentions for the food packaging industry. Arkoun et al. [150] firstly studied the antibacterial potential of electrospun chitosan-based nanofibers (CNF) by storing with the actual foods, and further investigated its ability to reduce spoilage and food loss. They successfully obtained highly antibacterial CNF-based packaging (CNFP) materials by direct electrospinning. It was found that an advantageous potential for antimicrobial packaging materials is that the quality and freshness of unprocessed or minimally processed and perishable foods could be perfectly preserved along with the extension of meat shelf life to 1 week (Figure 15).

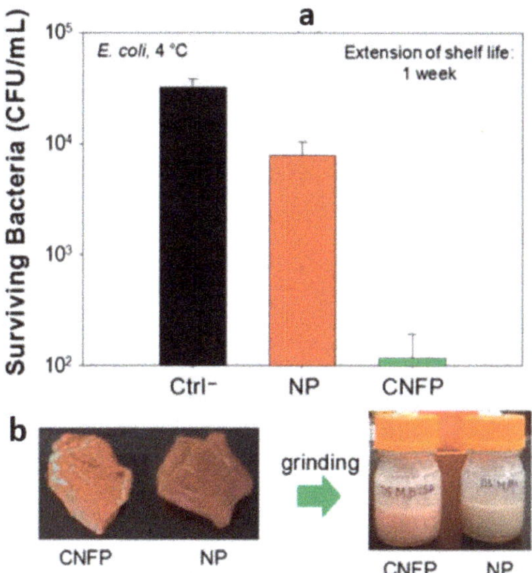

Figure 15. (**a**) In situ antibacterial activity of CNFP against *E. coli* after 7-day storage at 4 °C. (**b**) Appearance of packed red meat with and without CNFP, before and after grinding [150]. © John Wiley & Sons, Ltd 2018.

3.5. Barrier Electrospun Packaging Membrane

The barrier property of packaging materials refers to polymer materials with certain shielding ability for small molecule gas, liquid, water vapor, fragrance and drug taste. It can effectively prevent oxygen and water vapor from seeping into the environment, keep the specific gas composition in the package, and significantly improve the shelf life. At present, in order to meet the market demands, many countries have developed multi-functional high barrier packaging materials and multi-functional packaging materials in recent years.

Fabra et al. [151] developed a fully biodegradable multi-layer system based on a high barrier adhesive sandwich of electrospun zein nanofibers between poly hydroxybutyrate-co-valerate 5% (PHBV5) and poly-(3-hydroxybutyrate) (PHB) homopolymers with low valerate content. By adding zein nanofibers, the water and oxygen barrier in the multi-layer system of PHA was improved, and the flexibility was also enhanced. After that, electrospinning technology was employed to improve the barrier performance of packaging materials [152]. The effects of PCL, PHA and PLA on the oxygen and water resistance of TPCS membrane were compared. The result revealed that the PHB was the most effective in reducing the water and oxygen permeability.

The previous chapter has shown that zein can be prepared as an edible packaging film. In recent years, many scholars have also used electrospun zein film as the interlayer to enhance the barrier properties of food packaging. Fabra et al. [153] employed the electrospun zein as the middle layer and cast PHBV12 as the outer layer to prepare a multilayer structure food packaging films, which can significantly improve the oxygen barrier performance. They also compared the enhancement of the oxygen and water barrier properties of multi-layer packaging films with PHBV3 as the outer layers and different electrospun hydrocolloid films as the interlayers [153]. The result showed that oxygen and water vapor permeability values were significantly improved by employing electrospun zein film as the interlayer.

4. Conclusions and Challenges

In summary, the electrospinning technology is an effective method to prepare nanofibers with various nanostructure and surface characteristics, which can meet the functional requirements for packaging materials in the food fields depends on the different design of device and wide range of raw materials for selection, especially polymers. This article reviews the potential use of electrospinning technology in the food packaging field, not only to prepare pure degradable polymer packaging membranes, but also to obtain mixed polymer mats by designing the spinneret, such as multiple-jet and coaxial electrospinning. In addition, different inorganic fillers and other bioactive particles can be incorporated into the fibers to enhance the functionality and intend to be used in broader applications. The properties of common polymers and fillers mentioned above are summarized in Table 1. Alternatively, the resulting electrospun films are post-treated by the thermal treatments, surface modifications or dip-coating to obtain superior satisfactory performance. In the field of food packaging, based on some special requirements, it is possible to prepare packaging materials that own functions of degradable, superhydrophobic, edible, antibacterial and high barrier by electrospinning technology.

Table 1. Summary of promising properties related to packaging using electrospinning.

Categories	Materials	Properties or Function
Polymer	Chitosan (CS)	biodegradation, biocompatibility, anti-microbial, antifungal activities, and non-toxicity.
	Zein	good film-forming property, biocompatibility, biodegradation, renewable, edible
	Polyvinyl alcohol (PVA)	transparency, gantistatic property, biodegradation, biocompatibility
	Gelatin (GT)	biodegradation, biocompatibility, edible, good toughness
	Polycaprolactone (PCL)	biocompatibility, biodegradability, good mechanical properties, better solvent solubility
	Polyethylene terephthalate (PET)	non-toxic, good mechanical properties, high transparency, good toughness
	Cellulose acetate (CA)	non-toxic, biodegradable, low price, good transparency, high impact resistance
	Polylactic acid (PLA)	biodegradation, biocompatibility, easy to process, good mechanical properties and transparency
	Poly (propylene carbonate) (PPC)	good tensile toughness, transparency, biocompatibility and biodegradability
	Polystyrene (PS)	High transparency, non-toxic, easy to process
Inorganic fillers	Metronidazole (MNA)	hydrophobic, antibacterial
	Carbon nanotubes (CNTs)	conductive, antistatic, smart packaging
	FePt, Fe_3O_4, $FeCl_3$ nanoparticles	radiation protection
	TiO_2	photocatalytic, self-cleaning, photocatalytic degradation of ethylene
	ZnO	photocatalytic, self-cleaning, antibacterial
	Cerium-doped bioactive glass (CeBG), copper-doped bioactive glass (CuBG), silver-doped bioactive glass (AgBG)	antibacterial
	Montmorillonite (MMT)	antibacterial
Active substance	Orange essential oil (OEO)	antibacterial
	Metronidazole (MNA)	antibacterial
	Peppermint essential oil (PO), chamomile essential oil (CO)	antibacterial
	Vanillin/cyclodextrin inclusion complex (vanillin/CD-IC)	containing flavor/fragrance, enhancing thermal stability and durability
	Cinnamon essential oil/b-cyclodextrin (PVA/CEO/b-CD)	prolonging the shelf-life

Although the application of electrospinning technology in functional packaging has a wide range of development prospects, there are still some challenges that need to be faced and require more attention and studies. The biggest challenge is the choice of solvent. Some polymers are insoluble in non-toxic solvents, which would cause environmental damage and pollution. Furthermore, some organic solvents are harmful to human beings. Therefore, it is necessary to find more suitable and non-toxic solvents to prepare electrospun fibers, especially green electrospinning, a novel and aqueous strategy to overcome the disadvantages of organic solvents. Up to now, some commercially available polymers in packaging industry can only be dissolved in toxic solvents. Therefore, water or other non-toxic solvents with environment friendly property would be applied to prepare food packaging films in the future. In fact, during the electrospinning process, the solvents are mostly evaporated. Unfortunately, the above organic solvents are toxic and harmful to the environment, so it is crucial to replace the conventional toxic solvents with non-toxic or low-toxic solvents, such as emulsion electrospinning method [154,155], that will minimize the presence of toxic solvents in the final packaging material. Another challenge is some natural polymers cannot be directly electrospun into nanofibers, so it is crucial to select a suitable polymer, which is also a problem of electrospun material in food packaging. Most of all, the majority of research is still in the laboratories. So, as for the real food packaging industry, it is important to produce electrospun fibers on a large scale, and the subsequent development must be focused on high yields and industrialization. Ordinary multi-nozzle electrospinning or needless electrospinning is suitable for industrial production. There are already some manufacturers that can provide large-scale devices for producing electrospinning fibers or membranes. For example, Elmarco has developed needleless electrospinning device, which they claim is suitable for industrial applications (https://www.elmarco.com). Some factories already have begun large-scale production of nanofiber membranes, such as Jiangxi Xiancai Nanofibers Technology Co., Ltd. in China for the large scale production of electrospun polyimide fibrous membranes, yarns and short nanofibers (http://www.hinanofiber.com/english/). All of these have laid some foundation for the future electrospinning production of food packaging materials, and the focus of electrospinning food packaging materials also need to be transferred from laboratory research to industrial production in the future.

Funding: This research received no external funding.

Acknowledgments: The work was financially supported by National Natural Science Foundation of China (51803093, and 51903123), Natural Science Foundation of Jiangsu Province (BK20180770, and BK20190760), Priority Academic Program Development of Jiangsu Higher Education Institutions (PAPD), Opening Project of Division of Chemistry from Qingdao University of Science and Technology (QUSTHX201921).

Conflicts of Interest: The authors declare no conflict of interest.

References

1. Jayakumar, R.; Prabaharan, M.; Shalumon, K.T.; Chennazhi, K.P.; Nair, S.V. *Biomedical Applications of Polymer/Silver Composite Nanofibers*; Springer: Berlin/Heidelberg, Germany, 2011; Volume 246, pp. 263–282. [CrossRef]
2. Hou, H.; Xu, W.; Ding, Y. The recent progress on high-performance polymer nanofibers by electrospinning. *J. Jiangxi Norm. Univ. (Nat. Sci.)* **2018**, *42*, 551–564. [CrossRef]
3. Jiang, S.; Chen, Y.; Duan, G.; Mei, C.; Greiner, A.; Agarwal, S. Electrospun nanofiber reinforced composites: A review. *Polym. Chem.* **2018**, *9*, 2685–2720. [CrossRef]
4. Yao, K.; Chen, J.; Li, P.; Duan, G.; Hou, H. Robust strong electrospun polyimide composite nanofibers from a ternary polyamic acid blend. *Compos. Commun.* **2019**, *15*, 92–95. [CrossRef]
5. Liao, X.; Dulle, M.; de Souza e Silva, J.M.; Wehrspohn, R.B.; Agarwal, S.; Förster, S.; Hou, H.; Smith, P.; Greiner, A. High strength in combination with high toughness in robust and sustainable polymeric materials. *Science* **2019**, *366*, 1376–1379. [CrossRef] [PubMed]
6. Sundaramurthi, D.; Krishnan, U.M.; Sethuraman, S. Electrospun Nanofibers as Scaffolds for Skin Tissue Engineering. *Polym. Rev.* **2014**, *54*, 348–376. [CrossRef]

7. Yang, F.; Murugan, R.; Wang, S.; Ramakrishna, S. Electrospinning of nano/micro scale poly (l-lactic acid) aligned fibers and their potential in neural tissue engineering. *Biomaterials* **2005**, *26*, 2603–2610. [CrossRef]

8. Wu, T.; Ding, M.; Shi, C.; Qiao, Y.; Wang, P.; Qiao, R.; Wang, X.; Zhong, J. Resorbable polymer electrospun nanofibers: History, shapes and application for tissue engineering. *Chin. Chem. Lett.* **2019**. [CrossRef]

9. Buttafoco, L.; Kolkman, N.G.; Engbers-Buijtenhuijs, P.; Poot, A.A.; Dijkstra, P.J.; Vermes, I.; Feijen, J. Electrospinning of collagen and elastin for tissue engineering applications. *Biomaterials* **2006**, *27*, 724–734. [CrossRef]

10. Gao, S.; Tang, G.; Hua, D.; Xiong, R.; Han, J.; Jiang, S.; Zhang, Q.; Huang, C. Stimuli-responsive bio-based polymeric systems and their applications. *J. Mater. Chem. B* **2019**, *7*, 709–729. [CrossRef]

11. Ajalloueian, F.; Tavanai, H.; Hilborn, J.; Donzel-Gargand, O.; Leifer, K.; Wickham, A.; Arpanaei, A. Emulsion electrospinning as an approach to fabricate PLGA/chitosan nanofibers for biomedical applications. *BioMed Res. Int.* **2014**, *2014*, 475280. [CrossRef]

12. Dong, Z.; Kennedy, S.J.; Wu, Y. Electrospinning materials for energy-related applications and devices. *J. Power Sources* **2011**, *196*, 4886–4904. [CrossRef]

13. Han, J.; Wang, S.; Zhu, S.; Huang, C.; Yue, Y.; Mei, C.; Xu, X.; Xia, C. Electrospun Core–Shell Nanofibrous Membranes with Nanocellulose-Stabilized Carbon Nanotubes for Use as High-Performance Flexible Supercapacitor Electrodes with Enhanced Water Resistance, Thermal Stability, and Mechanical Toughness. *ACS Appl. Mater. Interfaces* **2019**. [CrossRef] [PubMed]

14. Sun, G.; Sun, L.; Xie, H.; Liu, J. Electrospinning of Nanofibers for Energy Applications. *Nanomaterials* **2016**, *6*, 129. [CrossRef] [PubMed]

15. Li, S.; Cui, Z.; Li, D.; Yue, G.; Liu, J.; Ding, H.; Gao, S.; Zhao, Y.; Wang, N.; Zhao, Y. Hierarchically structured electrospinning nanofibers for catalysis and energy storage. *Compos. Commun.* **2019**, *13*, 1–11. [CrossRef]

16. Hwang, T.H.; Lee, Y.M.; Kong, B.S.; Seo, J.S.; Choi, J.W. Electrospun core-shell fibers for robust silicon nanoparticle-based lithium ion battery anodes. *Nano Lett.* **2012**, *12*, 802–807. [CrossRef] [PubMed]

17. Drosou, C.G.; Krokida, M.K.; Biliaderis, C.G. Encapsulation of bioactive compounds through electrospinning/electrospraying and spray drying: A comparative assessment of food-related applications. *Dry. Technol.* **2016**, *35*, 139–162. [CrossRef]

18. Anu Bhushani, J.; Anandharamakrishnan, C. Electrospinning and electrospraying techniques: Potential food based applications. *Trends Food Sci. Technol.* **2014**, *38*, 21–33. [CrossRef]

19. Rezaei, A.; Nasirpour, A.; Fathi, M. Application of Cellulosic Nanofibers in Food Science Using Electrospinning and Its Potential Risk. *Compr. Rev. Food Sci. Food Saf.* **2015**, *14*, 269–284. [CrossRef]

20. Venugopal, J.; Prabhakaran, M.; Low, S.; Aw, T.; Gupta, D.; Venugopa, R.; Deepika, G. Continuous Nanostructures for the Controlled Release of Drugs. *Curr. Pharm. Des.* **2009**, *15*, 1799–1808. [CrossRef]

21. Duan, G.; Bagheri, A.R.; Jiang, S.; Golenser, J.; Agarwal, S.; Greiner, A. Exploration of Macroporous Polymeric Sponges As Drug Carriers. *Biomacromolecules* **2017**, *18*, 3215–3221. [CrossRef]

22. Ouyang, W.; Liu, S.; yao, K.; Zhao, L.; Cao, L.; Jiang, S.; Hou, H. Ultrafine hollow TiO2 nanofibers from core-shell composite fibers and their photocatalytic properties. *Compos. Commun.* **2018**, *9*, 76–80. [CrossRef]

23. Choi, S.J.; Chattopadhyay, S.; Kim, J.J.; Kim, S.J.; Tuller, H.L.; Rutledge, G.C.; Kim, I.D. Coaxial electrospinning of WO3 nanotubes functionalized with bio-inspired Pd catalysts and their superior hydrogen sensing performance. *Nanoscale* **2016**, *8*, 9159–9166. [CrossRef] [PubMed]

24. Li, Z.; Liu, S.; Song, S.; Xu, W.; Sun, Y.; Dai, Y. Porous ceramic nanofibers as new catalysts toward heterogeneous reactions. *Compos. Commun.* **2019**, *15*, 168–178. [CrossRef]

25. Han, C.; Wang, Y.; Lei, Y.; Wang, B.; Wu, N.; Shi, Q.; Li, Q. In situ synthesis of graphitic-C3N4 nanosheet hybridized N-doped TiO2 nanofibers for efficient photocatalytic H2 production and degradation. *Nano Res.* **2015**, *8*, 1199–1209. [CrossRef]

26. Qiao, Y.; Shi, C.; Wang, X.; Wang, P.; Zhang, Y.; Wang, D.; Qiao, R.; Wang, X.; Zhong, J. Electrospun Nanobelt-Shaped Polymer Membranes for Fast and High-Sensitivity Detection of Metal Ions. *ACS Appl. Mater. Interfaces* **2019**, *11*, 5401–5413. [CrossRef]

27. Chen, Y.; Lu, K.; Song, Y.; Han, J.; Yue, Y.; Biswas, S.K.; Wu, Q.; Xiao, H. A Skin-Inspired Stretchable, Self-Healing and Electro-Conductive Hydrogel with a Synergistic Triple Network for Wearable Strain Sensors Applied in Human-Motion Detection. *Nanomaterials* **2019**, *9*, 1737. [CrossRef]

28. Zheng, C.; Yue, Y.; Gan, L.; Xu, X.; Mei, C.; Han, J. Highly Stretchable and Self-Healing Strain Sensors Based on Nanocellulose-Supported Graphene Dispersed in Electro-Conductive Hydrogels. *Nanomaterials* **2019**, *9*, 937. [CrossRef]

29. Zhang, N.; Qiao, R.; Su, J.; Yan, J.; Xie, Z.; Qiao, Y.; Wang, X.; Zhong, J. Recent Advances of Electrospun Nanofibrous Membranes in the Development of Chemosensors for Heavy Metal Detection. *Small* **2017**, *13*, 1604293. [CrossRef]

30. Sun, B.; Long, Y.-Z.; Chen, Z.-J.; Liu, S.-L.; Zhang, H.-D.; Zhang, J.-C.; Han, W.-P. Recent advances in flexible and stretchable electronic devices via electrospinning. *J. Mater. Chem. C* **2014**, *2*, 1209–1219. [CrossRef]

31. Han, J.; Lu, K.; Yue, Y.; Mei, C.; Huang, C.; Wu, Q.; Xu, X. Nanocellulose-templated assembly of polyaniline in natural rubber-based hybrid elastomers toward flexible electronic conductors. *Ind. Crops Prod.* **2019**, *128*, 94–107. [CrossRef]

32. Yan, J.; Jeong, Y.G. High Performance Flexible Piezoelectric Nanogenerators based on BaTiO3 Nanofibers in Different Alignment Modes. *ACS Appl. Mater. Interfaces* **2016**, *8*, 15700–15709. [CrossRef] [PubMed]

33. Zhang, L.; Li, Y.; Zhang, Q.; Wang, H. Formation of the modified ultrafine anatase TiO2 nanoparticles using the nanofiber as a microsized reactor. *CrystEngComm* **2013**, *15*, 1607. [CrossRef]

34. Jiang, S.; Gruen, V.; Rosenfeldt, S.; Schenk, A.S.; Agarwal, S.; Xu, Z.-K.; Greiner, A. Virtually Wall-Less Tubular Sponges as Compartmentalized Reaction Containers. *Research* **2019**, *2019*, 4152536. [CrossRef] [PubMed]

35. Lv, D.; Zhu, M.; Jiang, Z.; Jiang, S.; Zhang, Q.; Xiong, R.; Huang, C. Green Electrospun Nanofibers and Their Application in Air Filtration. *Macromol. Mater. Eng.* **2018**, *303*, 1800336. [CrossRef]

36. Zhu, M.; Han, J.; Wang, F.; Shao, W.; Xiong, R.; Zhang, Q.; Pan, H.; Yang, Y.; Samal, S.K.; Zhang, F.; et al. Electrospun Nanofibers Membranes for Effective Air Filtration. *Macromol. Mater. Eng.* **2017**, *302*, 1600353. [CrossRef]

37. Lv, D.; Wang, R.; Tang, G.; Mou, Z.; Lei, J.; Han, J.; De Smedt, S.; Xiong, R.; Huang, C. Ecofriendly Electrospun Membranes Loaded with Visible-Light-Responding Nanoparticles for Multifunctional Usages: Highly Efficient Air Filtration, Dye Scavenging, and Bactericidal Activity. *ACS Appl. Mater. Interfaces* **2019**, *11*, 12880–12889. [CrossRef]

38. Huang, Z.-M.; Zhang, Y.Z.; Kotaki, M.; Ramakrishna, S. A review on polymer nanofibers by electrospinning and their applications in nanocomposites. *Compos. Sci. Technol.* **2003**, *63*, 2223–2253. [CrossRef]

39. Yang, H.; Jiang, S.; Fang, H.; Hu, X.; Duan, G.; Hou, H. Molecular orientation in aligned electrospun polyimide nanofibers by polarized FT-IR spectroscopy. *Spectrochim. Acta Part. A Mol. Biomol. Spectrosc.* **2018**, *200*, 339–344. [CrossRef]

40. Chen, C.-W.; Xie, J.; Yang, F.-X.; Zhang, H.-L.; Xu, Z.-W.; Liu, J.-L.; Chen, Y.-J. Development of moisture-absorbing and antioxidant active packaging film based on poly (vinyl alcohol) incorporated with green tea extract and its effect on the quality of dried eel. *J. Food Process. Preserv.* **2018**, *42*, e13374. [CrossRef]

41. Kim, J.-K.; Cho, B.-G.; Han, Y.-K.; Kim, Y.B. Modification of a crosslinked poly (acrylic acid) based new dehumidifying agent and its moisture absorbing characteristics. *Macromol. Res.* **2009**, *17*, 544–548. [CrossRef]

42. Ashwar, B.A.; Shah, A.; Gani, A.; Shah, U.; Gani, A.; Wani, I.A.; Wani, S.M.; Masoodi, F.A. Rice starch active packaging films loaded with antioxidants-development and characterization. *Starch Stärke* **2015**, *67*, 294–302. [CrossRef]

43. López de Dicastillo, C.; Bustos, F.; Guarda, A.; Galotto, M.J. Cross-linked methyl cellulose films with murta fruit extract for antioxidant and antimicrobial active food packaging. *Food Hydrocoll.* **2016**, *60*, 335–344. [CrossRef]

44. Soto-Cantu, C.D.; Graciano-Verdugo, A.Z.; Peralta, E.; Islas-Rubio, A.R.; Gonzalez-Cordova, A.; Gonzalez-Leon, A.; Soto-Valdez, H. Release of butylated hydroxytoluene from an active film packaging to Asadero cheese and its effect on oxidation and odor stability. *J. Dairy Sci.* **2008**, *91*, 11–19. [CrossRef] [PubMed]

45. Li, D.; Xia, Y.N. Electrospinning of Nanofibers: Reinventing the Wheel? *Adv. Mater.* **2004**, *16*, 1151–1170. [CrossRef]

46. Liu, Y.I.; Tan, J.; Yu, S.A.; Yousefzadeh, M.; Ramakrishna, S. High-fficiency preparation of polypropylene nanofiber by melt differential centrifugal electrospinning. *J. Appl. Polym. Sci.* **2019**, *137*, 48299.

47. Abdal-hay, A.; Abbasi, N.; Gwiazda, M.; Hamlet, S.; Ivanovski, S. Novel polycaprolactone/hydroxyapatite nanocomposite fibrous scaffolds by direct melt-electrospinning writing. *Eur. Polym. J.* **2018**, *105*, 257–264. [CrossRef]

48. Zeng, J.; Wang, H.; Lin, Y.; Zhang, J.; Liang, F.; Fang, F.; Yang, F.; Wang, P.; Zhu, Z.; Chen, X.; et al. Fabrication of microfluidic channels based on melt-electrospinning direct writing. *Microfluid. Nanofluid.* **2018**, *22*. [CrossRef]

49. Acik, G.; Altinkok, C. Polypropylene microfibers via solution electrospinning under ambient conditions. *J. Appl. Polym. Sci.* **2019**, *136*, 48199. [CrossRef]

50. Surip, S.N.; Abdul Aziz, F.M.; Bonnia, N.N.; Sekak, K.A. Effect of Pineapple Leaf Fibers (PALF) concentration on nanofibers formation by electrospinning. *IOP Conf. Ser. Mater. Sci. Eng.* **2018**, *290*, 012003. [CrossRef]

51. Wang, G.; Sun, X.; Bai, J.; Han, L. Preparation of Fe–C nanofiber composites by metal organic complex and potential application in supercapacitors. *J. Mater. Sci. Mater. Electron.* **2019**, *30*, 4665–4675. [CrossRef]

52. Han, J.; Branford-White, C.J.; Zhu, L.-M. Preparation of poly (ε-caprolactone)/poly (trimethylene carbonate) blend nanofibers by electrospinning. *Carbohydr. Polym.* **2010**, *79*, 214–218. [CrossRef]

53. Liu, S.; Song, Y.; Ma, C.; Shi, J.-l.; Guo, Q.-g.; Liu, L. The electrochemical performance of porous carbon nanofibers produced by electrospinning. *Carbon* **2012**, *50*, 3963. [CrossRef]

54. Jegina, S.; Kukle, S.; Gravitis, J. Evaluation of aloe vera extract loaded polyvinyl alcohol nanofiber webs obtained via needleless electrospinning. *IOP Conf. Ser. Mater. Sci. Eng.* **2018**, *459*, 012016. [CrossRef]

55. Wortmann, M.; Frese, N.; Sabantina, L.; Petkau, R.; Kinzel, F.; Gölzhäuser, A.; Moritzer, E.; Hüsgen, B.; Ehrmann, A. New Polymers for Needleless Electrospinning from Low-Toxic Solvents. *Nanomaterials* **2019**, *9*, 52. [CrossRef] [PubMed]

56. Sriyanti, I.; Jauhari, J. Synthesis of polyvinyl acetate (PVAc) fibers using needleless electrospinning technique with straight wire electrode. *J. Phys. Conf. Ser.* **2019**, *1166*, 012012. [CrossRef]

57. Yarin, A.L. Coaxial electrospinning and emulsion electrospinning of core-shell fibers. *Polym. Adv. Technol.* **2011**, *22*, 310–317. [CrossRef]

58. Lallave, M.; Bedia, J.; Ruiz-Rosas, R.; Rodríguez-Mirasol, J.; Loscertales, I.G. Filled and Hollow Carbon Nanofibers by Coaxial Electrospinning of Alcell Lignin without Binder Polymers. *Adv. Mater.* **2010**, *19*, 4292–4296. [CrossRef]

59. Han, D.; Steckl, A.J. Superhydrophobic and oleophobic fibers by coaxial electrospinning. *Langmuir ACS J. Surf. Colloids* **2009**, *25*, 9454–9462. [CrossRef]

60. Han, D.; Sherman, S.; Filocamo, S.; Steckl, A.J. Long-term antimicrobial effect of nisin released from electrospun triaxial fiber membranes. *Acta Biomater.* **2017**, *53*, 242–249. [CrossRef]

61. Jiang, S.; Duan, G.; Zussman, E.; Greiner, A.; Agarwal, S. Highly flexible and tough concentric triaxial polystyrene fibers. *ACS Appl. Mater. Interfaces* **2014**, *6*, 5918–5923. [CrossRef]

62. Yu, D.G.; Li, X.Y.; Wang, X.; Yang, J.H.; Bligh, S.W.; Williams, G.R. Nanofibers Fabricated Using Triaxial Electrospinning as Zero Order Drug Delivery Systems. *ACS Appl. Mater. Interfaces* **2015**, *7*, 18891–18897. [CrossRef]

63. Yoon, J.W.; Park, Y.; Kim, J.; Park, C.H. Multi-jet electrospinning of polystyrene/polyamide 6 blend: Thermal and mechanical properties. *Fash. Text.* **2017**, *4*. [CrossRef]

64. Zhang, Y.; Cheng, Z.; Han, Z.; Zhao, S.; Zhao, X.; Kang, L. Stable multi-jet electrospinning with high throughput using the bead structure nozzle. *RSC Adv.* **2018**, *8*, 6069–6074. [CrossRef]

65. Varesano, A.; Rombaldoni, F.; Mazzuchetti, G.; Tonin, C.; Comotto, R. Multi-jet nozzle electrospinning on textile substrates: Observations on process and nanofibre mat deposition. *Polym. Int.* **2010**, *59*, 1606–1615. [CrossRef]

66. Aider, M. Chitosan application for active bio-based films production and potential in the food industry: Review. *LWT Food Sci. Technol.* **2010**, *43*, 837–842. [CrossRef]

67. Ignatova, M.; Manolova, N.; Rashkov, I. Electrospun Antibacterial Chitosan-B ased Fibers. *Macromol. Biosci.* **2013**, *13*, 860–872. [CrossRef]

68. Ohkawa, K.; Cha, D.; Kim, H.; Nishida, A.; Yamamoto, H. Electrospinning of Chitosan. *Macromol. Rapid Commun.* **2004**, *25*, 1600–1605. [CrossRef]

69. Geng, X.; Kwon, O.H.; Jang, J. Electrospinning of chitosan dissolved in concentrated acetic acid solution. *Biomaterials* **2005**, *26*, 5427–5432. [CrossRef]

70. Sencadas, V.; Correia, D.M.; Areias, A.; Botelho, G.; Fonseca, A.M.; Neves, I.C.; Gomez Ribelles, J.L.; Lanceros Mendez, S. Determination of the parameters affecting electrospun chitosan fiber size distribution and morphology. *Carbohydr. Polym.* **2012**, *87*, 1295–1301. [CrossRef]

71. Torres-Giner, S.; Gimenez, E.; Lagaron, J.M. Characterization of the morphology and thermal properties of Zein Prolamine nanostructures obtained by electrospinning. *Food Hydrocoll.* **2008**, *22*, 601–614. [CrossRef]

72. Miyoshi, T.; Toyohara, K.; Minematsu, H. Preparation of ultrafine fibrous zein membranes via electrospinning. *Polym. Int.* **2005**, *54*, 1187–1190. [CrossRef]

73. Neo, Y.P.; Ray, S.; Easteal, A.J.; Nikolaidis, M.G.; Quek, S.Y. Influence of solution and processing parameters towards the fabrication of electrospun zein fibers with sub-micron diameter. *J. Food Eng.* **2012**, *109*, 645–651. [CrossRef]

74. Ding, Q.; Xu, X.; Yue, Y.; Mei, C.; Huang, C.; Jiang, S.; Wu, Q.; Han, J. Nanocellulose-Mediated Electroconductive Self-Healing Hydrogels with High Strength, Plasticity, Viscoelasticity, Stretchability, and Biocompatibility toward Multifunctional Applications. *ACS Appl. Mater. Interfaces* **2018**, *10*, 27987–28002. [CrossRef] [PubMed]

75. Han, J.; Ding, Q.; Mei, C.; Wu, Q.; Yue, Y.; Xu, X. An intrinsically self-healing and biocompatible electroconductive hydrogel based on nanostructured nanocellulose-polyaniline complexes embedded in a viscoelastic polymer network towards flexible conductors and electrodes. *Electrochim. Acta* **2019**, *318*, 660–672. [CrossRef]

76. Han, J.; Yue, Y.; Wu, Q.; Huang, C.; Pan, H.; Zhan, X.; Mei, C.; Xu, X. Effects of nanocellulose on the structure and properties of poly (vinyl alcohol)-borax hybrid foams. *Cellulose* **2017**, *24*, 4433–4448. [CrossRef]

77. Tao, J.; Shivkumar, S. Molecular weight dependent structural regimes during the electrospinning of PVA. *Mater. Lett.* **2007**, *61*, 2325–2328. [CrossRef]

78. Supaphol, P.; Chuangchote, S. On the electrospinning of poly (vinyl alcohol) nanofiber mats: A revisit. *J. Appl. Polym. Sci.* **2008**, *108*, 969–978. [CrossRef]

79. Yang, R.R.; He, J.H.; Xu, L.; Yu, J.Y. Effect of solution concentration on diameter and morphology of PVA nanofibres in bubble electrospinning process. *Mater. Sci. Technol.* **2013**, *26*, 1313–1316. [CrossRef]

80. Wang, X.; Niu, H.; Lin, T.; Wang, X. Needleless electrospinning of nanofibers with a conical wire coil. *Polym. Eng. Sci.* **2009**, *49*, 1582–1586. [CrossRef]

81. Niu, H.; Lin, T.; Wang, X. Needleless electrospinning. I. A comparison of cylinder and disk nozzles. *J. Appl. Polym. Sci.* **2009**, *114*, 3524–3530. [CrossRef]

82. Feng, B.; Tu, H.; Yuan, H.; Peng, H.; Zhang, Y. Acetic-acid-mediated miscibility toward electrospinning homogeneous composite nanofibers of GT/PCL. *Biomacromolecules* **2012**, *13*, 3917–3925. [CrossRef]

83. Espindola-Gonzalez, A.; Martinez-Hernandez, A.L.; Fernandez-Escobar, F.; Castano, V.M.; Brostow, W.; Datashvili, T.; Velasco-Santos, C. Natural-synthetic hybrid polymers developed via electrospinning: The effect of PET in chitosan/starch system. *Int. J. Mol. Sci.* **2011**, *12*, 1908–1920. [CrossRef]

84. Sajeev, U.S.; Anoop Anand, K.; Menon, D.; Nair, S. Control of nanostructures in PVA, PVA/chitosan blends and PCL through electrospinning. *Bull. Mater. Sci.* **2008**, *31*, 343–351. [CrossRef]

85. Zheng, Y.; Zhuang, C.; Gong, R.H.; Zeng, Y. Electric Field Design for Multijet Electropsinning with Uniform Electric Field. *Industrial & Engineering Chemistry Research* **2014**, *53*, 14876–14884. [CrossRef]

86. Ding, B.; Kimura, E.; Sato, T.; Fujita, S.; Shiratori, S. Fabrication of blend biodegradable nanofibrous nonwoven mats via multi-jet electrospinning. *Polymer* **2004**, *45*, 1895–1902. [CrossRef]

87. Wang, Q.; Curtis, C.K.; Thoppey, N.M.; Bochinski, J.R.; Gorga, R.E.; Clarke, L.I. Unconfined, melt edge electrospinning from multiple, spontaneous, self-organized polymer jets. *Mater. Res. Express* **2014**, *1*, 045304. [CrossRef]

88. Thoppey, N.M.; Bochinski, J.R.; Clarke, L.I.; Gorga, R.E. Edge electrospinning for high throughput production of quality nanofibers. *Nanotechnology* **2011**, *22*, 345301. [CrossRef]

89. Thoppey, N.M.; Gorga, R.E.; Bochinski, J.R.; Clarke, L.I. Effect of Solution Parameters on Spontaneous Jet Formation and Throughput in Edge Electrospinning from a Fluid-Filled Bowl. *Macromolecules* **2012**, *45*, 6527–6537. [CrossRef]

90. Roman, M.P.; Thoppey, N.M.; Gorga, R.E.; Bochinski, J.R.; Clarke, L.I. Maximizing Spontaneous Jet Density and Nanofiber Quality in Unconfined Electrospinning: The Role of Interjet Interactions. *Macromolecules* **2013**, *46*, 7352–7362. [CrossRef]

91. Thoppey, N.M.; Bochinski, J.R.; Clarke, L.I.; Gorga, R.E. Unconfined fluid electrospun into high quality nanofibers from a plate edge. *Polymer* **2010**, *51*, 4928–4936. [CrossRef]
92. Yao, Z.C.; Chen, S.C.; Ahmad, Z.; Huang, J.; Chang, M.W.; Li, J.S. Essential Oil Bioactive Fibrous Membranes Prepared via Coaxial Electrospinning. *J. Food Sci.* **2017**, *82*, 1412–1422. [CrossRef] [PubMed]
93. Komur, B.; Bayrak, F.; Ekren, N.; Eroglu, M.S.; Oktar, F.N.; Sinirlioglu, Z.A.; Yucel, S.; Guler, O.; Gunduz, O. Starch/PCL composite nanofibers by co-axial electrospinning technique for biomedical applications. *Biomed. Eng. Online* **2017**, *16*, 40. [CrossRef] [PubMed]
94. Park, H.; Yoo, H.; Hwang, T.; Park, T.-J.; Paik, D.-H.; Choi, S.-W.; Kim, J.H. Fabrication of levofloxacin-loaded nanofibrous scaffolds using coaxial electrospinning. *J. Pharm. Investig.* **2012**, *42*, 89–93. [CrossRef]
95. Alharbi, H.F.; Luqman, M.; Fouad, H.; Khalil, K.A.; Alharthi, N.H. Viscoelastic behavior of core-shell structured nanofibers of PLA and PVA produced by coaxial electrospinning. *Polym. Test.* **2018**, *67*, 136–143. [CrossRef]
96. Korehei, R.; Kadla, J. Incorporation of T4 bacteriophage in electrospun fibres. *J. Appl. Microbiol.* **2013**, *114*, 1425–1434. [CrossRef]
97. He, M.; Jiang, H.; Wang, R.; Xie, Y.; Zhao, C. Fabrication of metronidazole loaded poly (epsilon-caprolactone)/zein core/shell nanofiber membranes via coaxial electrospinning for guided tissue regeneration. *J. Colloid Interface Sci.* **2017**, *490*, 270–278. [CrossRef]
98. George, J.; Ishida, H. A review on the very high nanofiller-content nanocomposites: Their preparation methods and properties with high aspect ratio fillers. *Prog. Polym. Sci.* **2018**, *86*, 1–39. [CrossRef]
99. Yue, Y.; Wang, X.; Wu, Q.; Han, J.; Jiang, J. Assembly of Polyacrylamide-Sodium Alginate-Based Organic-Inorganic Hydrogel with Mechanical and Adsorption Properties. *Polymers* **2019**, *11*, 1239. [CrossRef]
100. Wolf, C.; Angellier-Coussy, H.; Gontard, N.; Doghieri, F.; Guillard, V. How the shape of fillers affects the barrier properties of polymer/non-porous particles nanocomposites: A review. *J. Membr. Sci.* **2018**, *556*, 393–418. [CrossRef]
101. Yue, Y.; Wang, X.; Han, J.; Yu, L.; Chen, J.; Wu, Q.; Jiang, J. Effects of nanocellulose on sodium alginate/polyacrylamide hydrogel: Mechanical properties and adsorption-desorption capacities. *Carbohydr. Polym.* **2019**, *206*, 289–301. [CrossRef]
102. Han, J.; Wang, H.; Yue, Y.; Mei, C.; Chen, J.; Huang, C.; Wu, Q.; Xu, X. A self-healable and highly flexible supercapacitor integrated by dynamically cross-linked electro-conductive hydrogels based on nanocellulose-templated carbon nanotubes embedded in a viscoelastic polymer network. *Carbon* **2019**, *149*, 1–18. [CrossRef]
103. Yang, T.; Wu, D.; Lu, L.; Zhou, W.; Zhang, M. Electrospinning of polylactide and its composites with carbon nanotubes. *Polym. Compos.* **2011**, *32*, 1280–1288. [CrossRef]
104. Meng, Z.X.; Zheng, W.; Li, L.; Zheng, Y.F. Fabrication and characterization of three-dimensional nanofiber membrane of PCL–MWCNTs by electrospinning. *Mater. Sci. Eng. C* **2010**, *30*, 1014–1021. [CrossRef]
105. Torres-Giner, S.; Echegoyen, Y.; Teruel-Juanes, R.; Badia, J.D.; Ribes-Greus, A.; Lagaron, J.M. Electrospun Poly (ethylene-co-vinyl alcohol)/Graphene Nanoplatelets Composites of Interest in Intelligent Food Packaging Applications. *Nanomaterials* **2018**, *8*, 745. [CrossRef]
106. Song, T.; Zhang, Y.; Zhou, T.; Lim, C.T.; Ramakrishna, S.; Liu, B. Encapsulation of self-assembled FePt magnetic nanoparticles in PCL nanofibers by coaxial electrospinning. *Chem. Phys. Lett.* **2005**, *415*, 317–322. [CrossRef]
107. Wang, S.; Wang, C.; Zhang, B.; Sun, Z.; Li, Z.; Jiang, X.; Bai, X. Preparation of Fe3O4/PVA nanofibers via combining in-situ composite with electrospinning. *Mater. Lett.* **2010**, *64*, 9–11. [CrossRef]
108. Wei, Y.; Zhang, X.; Song, Y.; Han, B.; Hu, X.; Wang, X.; Lin, Y.; Deng, X. Magnetic biodegradable Fe3O4/CS/PVA nanofibrous membranes for bone regeneration. *Biomed. Mater.* **2011**, *6*, 055008. [CrossRef]
109. Chowdhury, M.; Stylios, G. Process optimization and alignment of PVA/FeCl3 nano composite fibres by electrospinning. *J. Mater. Sci.* **2011**, *46*, 3378–3386. [CrossRef]
110. Kumar, M.; Unruh, D.; Sindelar, R.; Renz, F. Preparation of Magnetic Polylactic Acid Fiber Mats by Electrospinning. *Nano Hybrids Compos.* **2017**, *14*, 39–47. [CrossRef]
111. Bedford, N.; Steckl, A. Photocatalytic Self Cleaning Textile Fibers by Coaxial Electrospinning. *ACS Appl. Mater. Interfaces* **2010**, *2*. [CrossRef]

112. Nasikhudin; Ismaya, E.P.; Diantoro, M.; Kusumaatmaja, A.; Triyana, K. Preparation of PVA/TiO2 Composites Nanofibers by using Electrospinning Method for Photocatalytic Degradation. *IOP Conf. Ser. Mater. Sci. Eng.* **2017**, *202*, 012011. [CrossRef]

113. Liu, H.; Yang, J.; Liang, J.; Huang, Y.; Tang, C. ZnO Nanofiber and Nanoparticle Synthesized Through Electrospinning and Their Photocatalytic Activity Under Visible Light. *J. Am. Ceram. Soc.* **2008**, *91*, 1287–1291. [CrossRef]

114. Khan, M.Q.; Lee, H.; Koo, J.M.; Khatri, Z.; Sui, J.; Im, S.S.; Zhu, C.; Kim, I.S. Self-cleaning effect of electrospun poly (1,4-cyclohexanedimethylene isosorbide terephthalate) nanofibers embedded with zinc oxide nanoparticles. *Text. Res. J.* **2017**, *88*, 2493–2498. [CrossRef]

115. Zhu, Z.; Zhang, Y.; Shang, Y.; Wen, Y. Electrospun Nanofibers Containing TiO2 for the Photocatalytic Degradation of Ethylene and Delaying Postharvest Ripening of Bananas. *Food Bioprocess Technol.* **2018**, *12*, 281–287. [CrossRef]

116. Sawai, J.; Igarashi, H.; Hashimoto, A.; Kokugan, T.; Shimizu, M. Evaluation of Growth Inhibitory Effect of Ceramics Powder Slurry on Bacteria by Conductance Method. *J. Chem. Eng. Jpn.* **1995**, *28*, 288–293. [CrossRef]

117. Amna, T.; Yang, J.; Ryu, K.S.; Hwang, I.H. Electrospun antimicrobial hybrid mats: Innovative packaging material for meat and meat-products. *J. Food Sci. Technol.* **2015**, *52*, 4600–4606. [CrossRef]

118. Lee, J.-H.; Lee, U.-S.; Jeong, K.-U.; Seo, Y.-A.; Park, S.-J.; Kim, H.-Y. Preparation and characterization of poly(vinyl alcohol) nanofiber mats crosslinked with blocked isocyanate prepolymer. *Polym. Int.* **2010**, *59*, 1683–1689. [CrossRef]

119. Lee, S.J.; Oh, S.H.; Liu, J.; Soker, S.; Atala, A.; Yoo, J.J. The use of thermal treatments to enhance the mechanical properties of electrospun poly (epsilon-caprolactone) scaffolds. *Biomaterials* **2008**, *29*, 1422–1430. [CrossRef]

120. D'Amato, A.R.; Schaub, N.J.; Cardenas, J.M.; Fiumara, A.S.; Troiano, P.M.; Fischetti, A.; Gilbert, R.J. Removal of Retained Electrospinning Solvent Prolongs Drug Release from Electrospun PLLA Fibers. *Polymer (Guildf.)* **2017**, *123*, 121–127. [CrossRef]

121. Wang, H.-S.; Fu, G.-D.; Li, X.-S. Functional Polymeric Nanofibers from Electrospinning. *Recent Pat. Nanotechnol.* **2009**, *3*, 21–31. [CrossRef]

122. Yao, C.; Li, X.-s.; Neoh, K.G.; Shi, Z.-l.; Kang, E.T. Antibacterial poly (D,L-lactide) (PDLLA) fibrous membranes modified with quaternary ammonium moieties. *Chin. J. Polym. Sci.* **2010**, *28*, 581–588. [CrossRef]

123. Surucu, S.; Turkoglu Sasmazel, H. DBD atmospheric plasma-modified, electrospun, layer-by-layer polymeric scaffolds for L929 fibroblast cell cultivation. *J. Biomater. Sci. Polym. Ed.* **2016**, *27*, 111–132. [CrossRef] [PubMed]

124. Ignatova, M.; Manolova, N.; Rashkov, I.; Markova, N. Antibacterial and antioxidant electrospun materials from poly (3-hydroxybutyrate) and polyvinylpyrrolidone containing caffeic acid phenethyl ester—"in" and "on" strategies for enhanced solubility. *Int. J. Pharm.* **2018**, *545*, 342–356. [CrossRef] [PubMed]

125. Goh, Y.-f.; Akram, M.; Alshemary, A.; Hussain, R. Antibacterial polylactic acid/chitosan nanofibers decorated with bioactive glass. *Appl. Surf. Sci.* **2016**, *387*, 1–7. [CrossRef]

126. Yakub, G.; Ignatova, M.; Manolova, N.; Rashkov, I.; Toshkova, R.; Georgieva, A.; Markova, N. Chitosan/ferulic acid-coated poly (epsilon-caprolactone) electrospun materials with antioxidant, antibacterial and antitumor properties. *Int. J. Biol. Macromol.* **2018**, *107*, 689–702. [CrossRef]

127. Hu, H.-T.; Lee, S.-Y.; Chen, C.-C.; Yang, Y.-C.; Yang, J.-C. Processing and properties of hydrophilic electrospun polylactic acid/beta-tricalcium phosphate membrane for dental applications. *Polym. Eng. Sci.* **2013**, *53*, 833–842. [CrossRef]

128. Duling, R.R.; Dupaix, R.B.; Katsube, N.; Lannutti, J. Mechanical characterization of electrospun polycaprolactone (PCL): A potential scaffold for tissue engineering. *J. Biomech. Eng.* **2008**, *130*, 011006. [CrossRef]

129. Subramanian, C.; Ugbolue, S.C.; Warner, S.B.; Patra, P. *The Melt Electrospinning of Polycaprolactone (PCL) Ultrafine Fibers*; MRS Online Proceedings Library Archive: Warrendale, PA, USA, 2008; Volume 1134, p. 1134-BB1108.

130. Del Gaudio, C.; Bianco, A.; Folin, M.; Baiguera, S.; Grigioni, M. Structural characterization and cell response evaluation of electrospun PCL membranes: Micrometric versus submicrometric fibers. *J. Biomed. Mater. Res. Part A* **2009**, *89*, 1028–1039. [CrossRef]

131. Bhullar, S.K.; Kaya, B.; Jun, M.B.-G. Development of Bioactive Packaging Structure Using Melt Electrospinning. *J. Polym. Environ.* **2015**, *23*, 416–423. [CrossRef]

132. Park, J.-y.; Lee, E.-S.; Amna, T.; Jang, Y.; Park, D.H.; Kim, B.-S. Effects of heat-treatment on surface morphologies, mechanical properties of nanofibrous poly (propylene carbonate) biocomposites and its cell culture. *Colloids Surf. A Physicochem. Eng. Asp.* **2016**, *492*, 138–143. [CrossRef]

133. Nagiah, N.; Sivagnanam, U.T.; Mohan, R.; Srinivasan, N.T.; Sehgal, P.K. Development and Characterization of Electrospun Poly (propylene carbonate) Ultrathin Fibers as Tissue Engineering Scaffolds. *Adv. Eng. Mater.* **2012**, *14*, B138–B148. [CrossRef]

134. Li, Y.; Lim, C.T.; Kotaki, M. Study on structural and mechanical properties of porous PLA nanofibers electrospun by channel-based electrospinning system. *Polymer* **2015**, *56*, 572–580. [CrossRef]

135. Casasola, R.; Thomas, N.L.; Trybala, A.; Georgiadou, S. Electrospun poly lactic acid (PLA) fibres: Effect of different solvent systems on fibre morphology and diameter. *Polymer* **2014**, *55*, 4728–4737. [CrossRef]

136. Tang, X.P.; Xu, L.; Liu, H.Y.; Si, N. Fabrication of PLA Nanoporous Fibers by DMF/CF Mixed Solvent via Electrospinning. *Adv. Mater. Res.* **2014**, *941–944*, 400–403. [CrossRef]

137. Huang, C.; Thomas, N.L. Fabricating porous poly (lactic acid) fibres via electrospinning. *Eur. Polym. J.* **2018**, *99*, 464–476. [CrossRef]

138. Tomaszewski, W.; Duda, A.; Szadkowski, M.; Libiszowski, J.; Ciechanska, D. Poly (L-lactide) Nano- and Microfibers by Electrospinning: Influence of Poly (L-lactide) Molecular Weight. *Macromol. Symp.* **2008**, *272*, 70–74. [CrossRef]

139. Li, S.; Lv, R.; Liu, H.; Na, B.; Zhou, H.; Ge, L. Uniform high-molecular-weight polylactide nanofibers electrospun from a solution below its entanglement concentration. *J. Appl. Polym. Sci.* **2017**, *134*. [CrossRef]

140. Kayaci, F.; Uyar, T. Encapsulation of vanillin/cyclodextrin inclusion complex in electrospun polyvinyl alcohol (PVA) nanowebs: Prolonged shelf-life and high temperature stability of vanillin. *Food Chem.* **2012**, *133*, 641–649. [CrossRef]

141. Wen, P.; Zhu, D.-H.; Wu, H.; Zong, M.-H.; Jing, Y.-R.; Han, S.-Y. Encapsulation of cinnamon essential oil in electrospun nanofibrous film for active food packaging. *Food Control* **2016**, *59*, 366–376. [CrossRef]

142. Ding, B.; Ogawa, T.; Kim, J.; Fujimoto, K.; Shiratori, S. Fabrication of a super-hydrophobic nanofibrous zinc oxide film surface by electrospinning. *Thin Solid Films* **2008**, *516*, 2495–2501. [CrossRef]

143. Barthlott, W.; Neinhuis, C. Purity of the sacred lotus, or escape from contamination in biological surfaces. *Planta* **1997**, *202*, 1–8. [CrossRef]

144. Kang, M.; Jung, R.; Kim, H.-S.; Jin, H.-J. Preparation of superhydrophobic polystyrene membranes by electrospinning. *Colloids Surf. A Physicochem. Eng. Asp.* **2008**, *313–314*, 411–414. [CrossRef]

145. Tang, Y.; Zhou, Y.; Xingzi, L.; Huang, D.; Luo, T.; Ji, J.; Mafang, Z.; Miao, X.; Wang, H.; Wang, W. Electrospun Gelatin Nanofibers Encapsulated with Peppermint and Chamomile Essential Oils as Potential Edible Packaging. *J. Agric. Food Chem.* **2019**, *67*. [CrossRef]

146. Mascheroni, E.; Fuenmayor, C.A.; Cosio, M.S.; Di Silvestro, G.; Piergiovanni, L.; Mannino, S.; Schiraldi, A. Encapsulation of volatiles in nanofibrous polysaccharide membranes for humidity-triggered release. *Carbohydr. Polym* **2013**, *98*, 17–25. [CrossRef]

147. Vega-Lugo, A.-C.; Lim, L.-T. Controlled release of allyl isothiocyanate using soy protein and poly (lactic acid) electrospun fibers. *Food Res. Int.* **2009**, *42*, 933–940. [CrossRef]

148. Neo, Y.P.; Swift, S.; Ray, S.; Gizdavic-Nikolaidis, M.; Jin, J.; Perera, C.O. Evaluation of gallic acid loaded zein sub-micron electrospun fibre mats as novel active packaging materials. *Food Chem.* **2013**, *141*, 3192–3200. [CrossRef]

149. Agarwal, A.; Raheja, A.; Natarajan, T.S.; Chandra, T.S. Effect of electrospun montmorillonite-nylon 6 nanofibrous membrane coated packaging on potato chips and bread. *Innov. Food Sci. Emerg. Technol.* **2014**, *26*, 424–430. [CrossRef]

150. Arkoun, M.; Daigle, F.; Holley, R.A.; Heuzey, M.C.; Ajji, A. Chitosan-based nanofibers as bioactive meat packaging materials. *Packag. Technol. Sci.* **2018**, *31*, 185–195. [CrossRef]

151. Fabra, M.J.; Lopez-Rubio, A.; Lagaron, J.M. Nanostructured interlayers of zein to improve the barrier properties of high barrier polyhydroxyalkanoates and other polyesters. *J. Food Eng.* **2014**, *127*, 1–9. [CrossRef]

152. Fabra, M.J.; Lopez-Rubio, A.; Cabedo, L.; Lagaron, J.M. Tailoring barrier properties of thermoplastic corn starch-based films (TPCS) by means of a multilayer design. *J. Colloid Interface Sci.* **2016**, *483*, 84–92. [CrossRef]

153. Fabra, M.J.; Lopez-Rubio, A.; Lagaron, J.M. High barrier polyhydroxyalcanoate food packaging film by means of nanostructured electrospun interlayers of zein. *Food Hydrocoll.* **2013**, *32*, 106–114. [CrossRef]
154. Fabra, M.J.; López-Rubio, A.; Lagaron, J.M. On the use of different hydrocolloids as electrospun adhesive interlayers to enhance the barrier properties of polyhydroxyalkanoates of interest in fully renewable food packaging concepts. *Food Hydrocoll.* **2014**, *39*, 77–84. [CrossRef]
155. Klimov, E.; Raman, V.; Venkatesh, R.; Heckmann, W.; Stark, R. Designing Nanofibers via Electrospinning from Aqueous Colloidal Dispersions: Effect of Cross-Linking and Template Polymer. *Macromolecules* **2010**, *43*, 6152–6155. [CrossRef]

Article

Electrospun Nanofibers: from Food to Energy by Engineered Electrodes in Microbial Fuel Cells

Giulia Massaglia [1,2,*], **Francesca Frascella** [1], **Alessandro Chiadò** [1], **Adriano Sacco** [2], **Simone Luigi Marasso** [1,3], **Matteo Cocuzza** [1,3], **Candido F. Pirri** [1,2] and **Marzia Quaglio** [1,*]

[1] Department of Applied Science and Technology (DISAT), Politecnico di Torino, Corso Duca degli Abruzzi 24, 10129 Torino, Italy; francesca.frascella@polito.it (F.F.); alessandro.chiado@polito.it (A.C.); simone.marasso@polito.it (S.L.M.); matteo.cocuzza@infm.polito.it (M.C.); fabrizio.pirri@polito.it (C.F.P.)

[2] Center for Sustainable Future Technologies (CSFT)@Polito, Istituto Italiano di Tecnologia, Environment Park, Building B2 Via Livorno 60, 10144 Torino, Italy; adriano.sacco@iit.it

[3] IMEM-CNR, Parco Area delle Scienze 37, 43124 Parma, Italy

* Correspondence: giulia.massaglia@polito.it (G.M.); marzia.quaglio@polito.it (M.Q.)

Received: 31 January 2020; Accepted: 10 March 2020; Published: 14 March 2020

Abstract: Microbial fuel cells (MFCs) are bio-electrochemical devices able to directly transduce chemical energy, entrapped in an organic mass named fuel, into electrical energy through the metabolic activity of specific bacteria. During the last years, the employment of bio-electrochemical devices to study the wastewater derived from the food industry has attracted great interest from the scientific community. In the present work, we demonstrate the capability of exoelectrogenic bacteria used in MFCs to catalyze the oxidation reaction of honey, employed as a fuel. With the main aim to increase the proliferation of microorganisms onto the anode, engineered electrodes are proposed. Polymeric nanofibers, based on polyethylene oxide (PEO-NFs), were directly electrospun onto carbon-based material (carbon paper, CP) to obtain an optimized composite anode. The crucial role played by the CP/PEO-NFs anodes was confirmed by the increased proliferation of microorganisms compared to that reached on bare CP anodes, used as a reference material. A parameter named recovered energy (Erec) was introduced to determine the capability of bacteria to oxidize honey and was compared with the Erec obtained when sodium acetate was used as a fuel. CP/PEO-NFs anodes allowed achieving an Erec three times higher than the one reached with a bare carbon-based anode.

Keywords: electrospun nanofibers; polyethylene oxide nanofibers PEO-NFs; microbial fuel cells; honey; food industry; recovered energy (Erec)

---

1. Introduction

Microbial fuel cells (MFCs) are bio-electrochemical devices that directly convert the chemical energy embedded in an organic compound (i.e., the fuel) into electrical energy by the metabolic action of a particular class of microorganisms, named exo-electrogens.

Basically, the process is based on the ability of exo-electrogens to oxidize organic matter, acting as carbon energy sources [1], and to directly transfer the produced electrons outside their cells exogenously [2]. In an MFC, electrons are firstly released to the anode by the microorganisms arranged in a biofilm in intimate contact with the anode surface and successively they flow through an external circuit to reach the cathode side, where the terminal electron acceptor (TEA), usually oxygen, is finally reduced. Potential applications of MFCs have been foreseen in several areas [3,4], ranging from energy harvesting [5,6] to wastewater treatment [7] and sensing [8–12]. Since MFCs can operate their energy recovery and conversion processes starting from a wide range of molecules, even quite complex, during the last decades, they have attracted an ever-increasing interest for application in the food industry [13]. In this perspective, MFCs can contribute to the overall energetic efficiency of the process by combining the treatment of wastewater streams to energy production. Interesting examples exist, concerning applications in the treatment of brewery wastewater [14,15], olive mill wastewater [16], winery wastes [17,18], and dairy wastewater [19].

The use of MFC for wastewater treatment plays a crucial role as an alternative to traditional treatment processes, such as anaerobic digestion with methane fermentation, which include indirect energy recovery from wastes [20]. Knowledge in this area has grown significantly, and information about the amount of energy than can be recovered from many substrates, such as urban wastewaters [21], short-chain volatile fatty acid [22], as well as fermentable and non-fermentable reference substrates, such as sodium acetate and glucose, respectively, is now available [23].

Nonetheless, further improvements are necessary to increase power production [24], which is currently hindering the marketability of MFCs. In this view, the anodic electrode and its interface with the bacteria biofilm play a crucial role, since they are responsible for energy conversion and the electron transfer process.

In the present work, a new nanofiber-based interface between a carbon paper (CP) anode and a bacterial biofilm is investigated with the aim to optimize the adhesion of the biofilm to the anode. Furthermore, optimized adhesion of the biofilm to the anode plays a crucial role in improving the biofilm–anode electrochemical interaction, thus effectively ameliorating the overall MFC performance in terms of energy production. The idea is based on the hypothesis that the larger is the number of bacteria on the anode, the higher is the resulting electron transfer rate, which is true only if the interface is designed in such a way to keep electrical resistance low. Therefore, in the present work, new nanostructured polymeric mats with high specific area were designed to modify the surface of carbon-based materials in order to improve their ability to create effective interfaces with bacterial biofilms.

During the last decades, some works in the literature [25–27] investigated the modification of carbon-based electrodes by the application of a polymeric matrix on their surfaces to increase the contact between bacteria and anodes in MFCs. Several polymeric matrices have been developed, using both natural polymers, such as agar, alginate, or agarose, and synthetic polymers [28–30]. Despite being interesting because of their sustainable origin and biocompatibility, natural polymers suffer from poor mechanical strength and durability, while synthetic polymers show an opposite behavior, since they combine higher mechanical strength and durability with a higher risk of toxicity for bacteria proliferation. Among the synthetic polymers proposed up to now in the literature, a promising one is poly(vinyl) alcohol (PVA) [25–27]. It has been investigated in different arrangements, as foams, nanofibers, and film, with the aim to reach a high density of immobilized microorganisms while ensuring good mass transfer properties. Currently, the best performing PVA anode-to-biofilm interface system is the one proposed by Bai et al. [27]. They developed a highly porous foam that showed good results in terms of immobilization of microorganisms, but the fabrication process required the use of boric acid, which caused side problems, such as agglomeration of PVA beads and residuals of toxic boric acid and enhanced swelling of the polymer foam. In order to overcome the limits of such technological approach, in the present work, polyethylene oxide nanofibers mats (PEO-NFs) are proposed as a biomass carrier. Electrospinning is selected as the process to obtain the highly porous matrix, and PEO works as the reference polymer. PEO shows a wide range of intriguing properties: (1) it is not cytotoxic, therefore it allows bacteria proliferation, (2) it is a sustainable and environmentally friendly material that can be processed using water as the only solvent, and (3) it is an important solid polymer electrolyte in electrochemical applications [28,29].

PEO-NFs were directly deposited onto CP-based electrodes, according to the electrospinning-on-electrode process, a binder-free method for nanofiber assembly onto an electrode that we introduced in our previous work [30].

The resulting anodes, named CP/PEO-NFs, were investigated in single-chamber MFCs (SCMFCs) and compared with reference devices using pure CP as anodes [30].

We demonstrate a huge improvement of microorganisms' proliferation toward the desired optical density (OD), thus confirming the possibility to use CP/PEO-NFs as a biomass carrier for bacteria entrapment. CP decorated by PEO-NFs interfaces were then tested in SCMFCs. A mixed bacterial consortium extracted from a marine sediment sample was used as the inoculum source [31]. A current density as high as 12 mA m^{-2} was reached when CP/PEO-NFs was used as the anode electrode. This value is higher than the one reached with CP anodes, indicating how the presence of PEO-NFs interfaces does not affect the electron transfer process, simultaneously inducing the proliferation of microorganisms.

Finally, we tested the potential of the newly designed anodic electrode to be used in MFCs for energy recovery in the agro-food industry. To demonstrate the robustness of our nanofiber-based interface, we selected honey as a new and more complex electron donor than those tested up to now in MFCs. The use of honey by humans traces back to ancient times, and today honey is a crucial ingredient in several products ranging from foods to beverages, as well as in medical products and cosmetics [32,33]. Honey can be classified as a natural sweetener with a complex composition [34]. In particular, honey is a saturated-sugar solution, where the carbohydrates amount to 95% of its chemical components. Moreover, other important compounds are naturally contained into honey, such as proteins, amino acids, enzymes, organic acids, minerals, and vitamins [34–36].

Given both its complex composition, that makes it a complete food from a nutritional point of view, and its natural antibiotic behavior, it is quite interesting to analyze the behavior of MFCs fed with water streams containing a limited amount of honey. Indeed, since MFCs' power production is based on the metabolic activity of exo-electrogenic microorganisms, all substances able to alter and/or modify the biofilm metabolism could alter the power output. Therefore, we tested the newly designed CP/PEO-NFs anodes in SCMFCs in which a honey-in-water solution at 0.02 wt % was used as the electrolyte. Given the aforementioned antibiotic behavior of honey, the first goal of this study was to demonstrate that MFCs can convert the organic matter of honey into electricity without damaging the biofilms, and the second goal was to demonstrate that new nanofiber-based anodes are more efficient than the commonly used CP references.

Results are analyzed in terms of recovered energy (E_{rec}) per unit volume [18,21,37]. E_{rec} was obtained by the integration of the measured power output over the batch treatment time. In particular, E_{rec} obtained for honey was compared with E_{rec} obtained when sodium acetate was used as a fuel in SCMFCs, demonstrating the microorganisms' capability to oxidize honey. A further comparison is then proposed to appreciate the different behavior of CP/PEO-NFs anodes in comparison to CP reference anodes.

Finally, we demonstrate that the SCMFCs' response, in terms of power output, changes with honey concentration. Different concentrations of honey (from 0.83 to 2 gL^{-1}) were tested and correlated with the corresponding energy recovery parameter.

2. Materials and Methods

2.1. Materials and Nanofibers Synthesis

A new nanofiber-based interface between a carbon paper-based anode and a bacteria biofilm was investigated. In particular, new nanostructured polymeric mats with high specific area were designed to modify the surface of carbon-based materials. Nanofibers mats based on PEO, were directly electrospun on a carbon-based material (named CP, Fuel Cell Earth, Woburn, Massachusetts, USA), used as reference material since it is the most employed anode in MFCs. As already demonstrated in our previous work [30], the morphology of CP, characterized by several conductive micro-protrusions, plays a crucial role in tuning a selective patterned deposition of PEO nanofibers (CP/PEO-NFs), thus leading to ensure a binder-free deposition of nanofibers. In particular, CP/PEO-NFs were obtained by electrospinning, starting from a polymeric solution containing PEO (purchased from Sigma Aldrich, with an average molecular weight Mw = 600 kDa) dissolved in deionized water. Electrospinning was performed by the NANON 01A machine from MECC Co. Ltd. The polymeric solution was loaded into a syringe, and nanofiber were obtained by applying a high positive voltage equal to 20 kV and a flow rate of 0.5 mL h^{-1} at a working distance of 12 cm. The duration of the electrospinning process was close to 10 min to ensure at the same time an ordered distribution of nanofibers, which were directly collected onto CP without a binder, and a high surface area-to-volume ratio. As demonstrated in our previous work [30], the ordered distribution is mainly due to the conductive protrusions of CP that induce electric field enhancement. This ordered distribution can be optimized when the thickness of the nanofibers is close to few micrometers. Indeed, a higher thickness of nanofiber mats could induce an insulator effect onto a carbon-based surface, minimizing the electric field variations that rule the nanofibers' distribution onto a CP surface. Moreover, in the present work, the surface area was defined by implementing Brunauer–Emmett–Tell (BET) measurements. In order to establish the lack of cytotoxicity of PEO for microorganisms, optical density (OD) measurements were carried out by means of a LAMBDA 850+ UV/Vis spectrophotometer.

In particular, OD measurement estimated the growth and metabolic activity of bacteria. Both CP/PEO-NFs and bare CP pieces were put into a tube containing the inoculum source and the electrolyte solution. The electrolyte solution was based on sodium acetate ($C_2H_3NaO_2$), ammonium chloride (0.31 gL^{-1} of NH4Cl) used as a nitrogen source to aid bacteria growth, and phosphate-buffered solution (PBS) that maintained a neutral pH. Every day, total cell density (dead and alive cells) was established by measuring OD at 600 nm using the spectrophotometer.

2.2. MFC Architecture and Configuration

The MFCs used in the present work were squared single-chamber microbial fuel cells (SCMFCs) with an open-air cathode, fabricated by micro milling (Al.Tip srl). [38]. In particular, our devices were composed of 3 compartments: the anodic part, the intermediate compartment, and the cathodic compartment. The internal volume of SCMFCs was 12.5 mL, and both anode and cathode had a geometric area equal to 5.76 cm^2. Furthermore, in the present work, two different anodes were investigated and compared: (1) CP/PEO-NFs obtained by direct deposition of PEO nanofiber mats on a carbon-based electrode, employed as a carbon backbone to ensure the electron transfer generated and released by the microorganisms; (2) a carbon-based material (CP) used as a control. The cathode was based on CP, properly modified in order to present gas diffusion layers (DLs) based on polytetrafluoroethylene (PTFE) on its outer side and a catalyst layer based on Platinum (Pt/C 0.5 mg/cm^2, from Sigma Aldrich, St.Louis, Missouri, USA) and Nafion (5 wt % Nafion, from Sigma Aldrich) on its inner side. Titanium wires (Goodfellow Cambridge Limited) were fixed onto the anode and cathode through a conductive paste made of carbon cement (Leit-C Cement). Two different electrolyte solutions were used. The first was an electrolyte solution based on sodium acetate ($C_2H_3NaO_2$), used as a carbon energy source at a concentration of 2 gL^{-1} together with other compounds able to ensure the optimal operation of the SCMFCs. All these compounds were based on ammonium chloride (0.31 gL^{-1} of NH4Cl), used as a nitrogen source to support bacterial growth, and PBS, able to maintain a neutral pH (containing 0.13 gL^{-1} of potassium chloride, 4.28 gL^{-1} of sodium phosphate dibasic, and 2.45 gL^{-1} of sodium phosphate monobasic monohydrate). The second electrolyte contained honey in the same amount defined for sodium acetate, close to 2 gL^{-1}, dissolved in deionized water and PBS. In the second electrolyte, PBS was added to ensure a neutral pH of the solution; due to the complexity of honey, the second electrolyte did not require any further addition of nutritional compounds for the microorganisms. Both solutions were autoclaved prior to use. All experiments were conducted in duplicate for each tested electrolyte: 2 SCMFCs with CP/PEO-NFs anode and 2 SCMFCs with CP anodes were studied. All SCMFCs were inoculated with a mixed culture of bacteria from a seawater sediment. All SCMFCs worked under batch mode, meaning that all devices are filled with *"new electrolyte"* when the drop of power output reached its minimum value. Anodes and cathodes of the SCMFCs were connected to a multichannel data acquisition unit (Agilent 34972A), and two different values of external resistance were applied. At the beginning of the experiments, to ensure the formation of biofilm on both anodes, i.e., CP/PEO-NFs and CP, an external load equal to 47 Ω was applied, and the electrolyte containing sodium acetate was used. Successively, to evaluate the overall SCMFCs' performance and to demonstrate that SCMFCs can convert honey organic matter into electricity without damaging the biofilms, an external load of 1000 Ω was applied. Furthermore, to establish and confirm the possibility to employ honey as a fuel in SCMFCs and demonstrate that its natural antibiotic behavior does not affect the overall SCMFCs' performance, we introduced a physical parameter, i.e., energy recovery (E_{rec}), calculated for both honey and sodium acetate [18,21,37]. The E_{rec} was defined according to Equation (1):

$$E_{rec} = \frac{\int Pdt}{V_{int}} \tag{1}$$

where E_{rec} ($J\ m^{-3}$) is energy recovery, $V_{int}(m^3)$ is the internal volume of SCMFCs, and $\int Pdt$ (J) is the integral of the generated power over time. We were able to demonstrate the possibility of using honey as a fuel in SCMFCs. In order to simultaneously verify the SCMFCs' response in terms of power output as honey concentration changed, different honey concentrations (from $0.83gL^{-1}$ to $2\ gL^{-1}$) were tested and correlated with the values of power output.

Internal resistance of the SCMFCs was evaluated through Nyquist plots, using electrochemical impedance spectroscopy (EIS). EIS was performed in open-circuit voltage (OCV) over the range of frequency between 150 and 200 mHz, with a sinusoidal signal with an amplitude of 25 mV.

3. Results and Discussion

3.1. Electrospun Nanofibers Onto Carbon-Based Materials and Their Role as a Biomass Carrier

As already demonstrated in our previous work [30] and in order to optimize the deposition of nanofiber mats onto carbon-based materials, we selected CP as a carbon-based material. The morphology of bare CP is reported in Figure 1a. In particular, it is possible to appreciate that the morphology of CP shows many naturally occurring conductive protrusions, based on carbon fibers, whose diameter was over 10 µm. All these protrusions, as also demonstrated in our previous work [30], play a crucial role in modulating the electric field during the electrospinning process, inducing an intensity enhancement and granting an ordered distribution of PEO-NFs. As represented in the Figure 1b, CP/PEO-NFs nanofibers preferentially arranged themselves on top of the conductive protrusions of CP, thus optimizing the interface between the nanofibers mats and the carbon-based electrode without the presence of a binder, commonly used to fix the nanostructures onto the electrode.

Figure 1a,b shows that PEO-NFs were mostly deposited onto the CP surface in correspondence of the interconnections between the conductive protrusions, covering completely the CP surface. Moreover, the final CP/PEO-NFs anode resulted to be an engineered electrode with a high surface area-to-volume ratio. As indicated by the BET results, PEO-NFs showed a surface area close to $40\ m^2\ g^{-1}$.

As presented in Figure 1c, the OD measurements confirmed the capability of CP/PEO-NFs to create an effective interface with bacterial biofilms, thus ensuring bacterial proliferation onto the CP-based anode. This result also confirmed the possibility to apply CP/PEO-NFs as a biomass carrier. The high bacterial proliferation is due to the capability exhibited by CP/PEO-NFs to entrap the microorganisms into its nanostructures.

3.2. SCMFCs Performance

Thanks to CP/PEO-NFs biocompatibility, in terms of microorganisms' proliferation and their morphological properties, CP/PEO-NFs mats were produced and directly deposited onto the carbon-based material. During the first phase of the experiment, known as the acclimation phase, we applied an external resistance close to 47 Ω with the main aim to ensure biofilm formation onto the electrodes. The duration of this acclimation phase was 1 month. As shown in Figure 2, the overall performance reached by the CP/PEO-NFs composite anode was three times higher than the one obtained with the bare carbon-based material. Since all the other features, i.e., the cathode, SCMFCs' configuration, and the electrolyte, were the same, it was possible to establish that the enhancement of the current density was related to the proliferation extent of the microorganisms, being higher onto PEO–NFs composite anodes. Moreover, by analyzing all peaks reported in Figure 2, it was possible to demonstrate a good electrical output produced by SCMFCs and, consequently, a stable and sustained bacteria proliferation on the anodes.

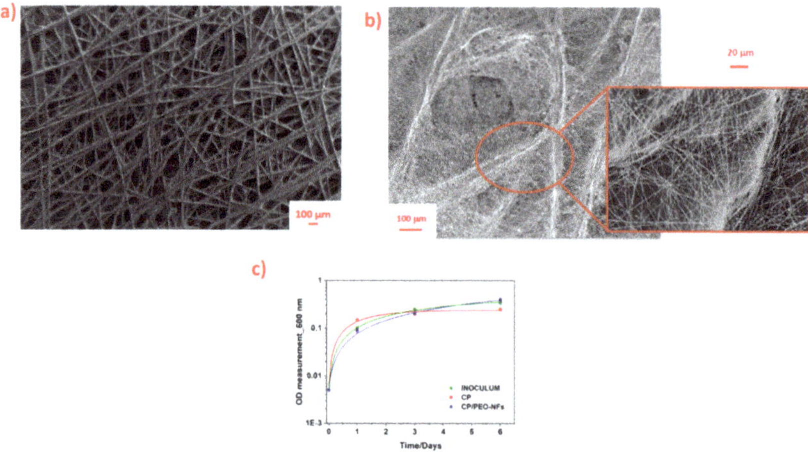

Figure 1. (**a**) FESEM images representing the bare carbon-based material (CP); (**b**) FESEM images representing the ordered distribution of CP/polyethylene oxide nanofibers (PEO-NFs) onto the carbon-based material. In the red box, the preferential distribution of CP/PEO-NFs onto the conductive protrusions of CP is highlighted; (**c**) logarithmic representation of OD measurements performed for all samples: bare inoculum (green), carbon-based material (CP, red), and CP/PEO-NFs (blue).

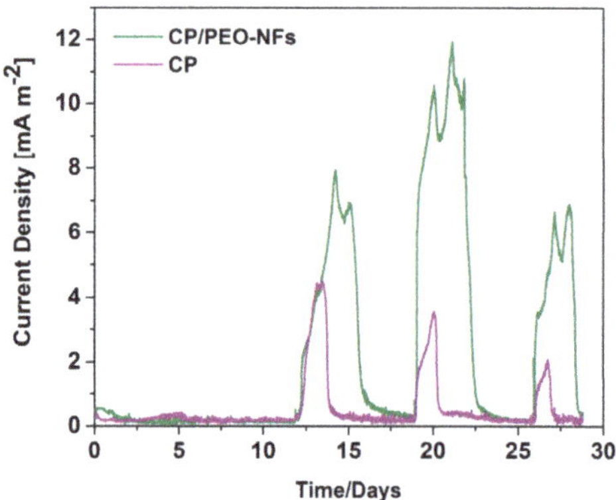

Figure 2. Current density produced in the acclimation phase, during which the biofilm formed onto the anodes: CP/PEO-NFs (green line) and carbon-based material used as a reference electrode (purple line). In this phase, the electrolyte contained sodium acetate; an external load of 47 Ω was applied.

The overall performance obtained with CP/PEO-NFS was compared with that of a reference material consisting of bare carbon-based anodes (CP). Moreover, given the complex composition of honey and its natural antibiotic behavior, a second experiment was implemented to demonstrate that SCMFCs can convert honey organic matter into electricity without damage to a biofilm and to confirm that the new nanofiber-based anodes (CP/PEO-NFs) perform better than the CP reference anodes.

In this experiment, an external load of 1kΩ was applied, and the overall performance, when both electrolytes were used, was analyzed. In this case, the concentration of sodium acetate and honey was 2 gL^{-1}. The current density trends, obtained from all tests in SCMFCs, were defined by normalizing with respect to the anode geometric area (5.76 cm^2) and are reported in Figure 3a,b. It is possible to appreciate that the maximum current densities reached when sodium acetate and honey were used as electrolytes were comparable, demonstrating the capability of bacteria to catalyze the oxidation reaction of honey. These results also suggest the possibility to apply SCMFCs to convert honey organic matter into electricity without damaging the biofilms. Moreover, since all these devices exploited formally identical cathodes and architectures, the current density trends were univocally correlated with the anodic reaction. Furthermore, as presented in Figure 3a,b, the maximum current density of total SCMFCs, reached when CP/PEO-NFs were used as the anode, was equal to (23.2 $\pm$ 0.1) mA m^{-2} and was comparable to the one obtained when CP was used as the reference anode. The presence of PEO-NFs, which basically showed insulating properties from an electrical point of view, not only did not affect the electron transfer rate but also simultaneously sustained the proliferation of microorganisms onto the anode. Moreover, all results are analyzed in terms of recovered energy (E$_{rec}$, mJ m^{-3}). E$_{rec}$ is calculated by integrating the electric power density over the batch treatment time and normalizing with respect to the electrolyte volume. Figure 3c shows that E$_{rec}$ (equal to 100 mJ m^{-3}) obtained when using honey and CP material was close to E$_{rec}$ reached when sodium acetate was employed as the electrolyte. Moreover, for both fuels, E$_{rec}$ achieved when using CP/PEO-NFs was three times higher than the one reached when using the carbon-based material, confirming the crucial role of nanofibers, which ensure a better interface between the anode and the bacteria and enhance the overall performance of SCMFC. This latter result demonstrates that PEO-NFs can act as a biomass carrier. The interface between biofilm and CP/PEO-NFs anode was enhanced, as also confirmed by the OD measurements, underlying a more extensive bacterial proliferation onto CP/PEO-NFs anodes. Both features, the optimized interface between anode and biofilm and the role of PEO as a polymeric electrolyte in electrochemical devices, allowed reaching a higher E$_{rec}$ in comparison with the one reached when using a bare carbon-based material (CP).

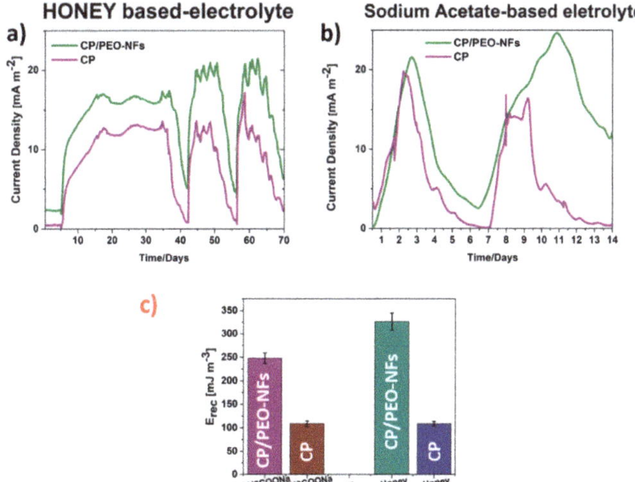

Figure 3. (**a**) Current density obtained when honey was used as a fuel and comparison of the current density values obtained for CP/PEO-NFs and CP; (**b**) current density reached using sodium acetate as a fuel and comparison of the values obtained for CP/PEO-NFs and CP; (**c**) recovered energy (E$_{rec}$) determined for CP/PEO-NFs and CP, using two different fuels.

Figure 4a,b show that the performance of all devices increased with increasing honey concentrations from 0.83 gL^{-1} to 2 gL^{-1}. The maximum value of recovered energy E$_{rec}$ was reached at the highest honey concentration employed (2 gL^{-1}) for both CP/PEO-NFs and CP anodes. At all honey concentrations, CP/PEO-NFs showed a better performance in terms of E$_{rec}$ than the CP anodes.

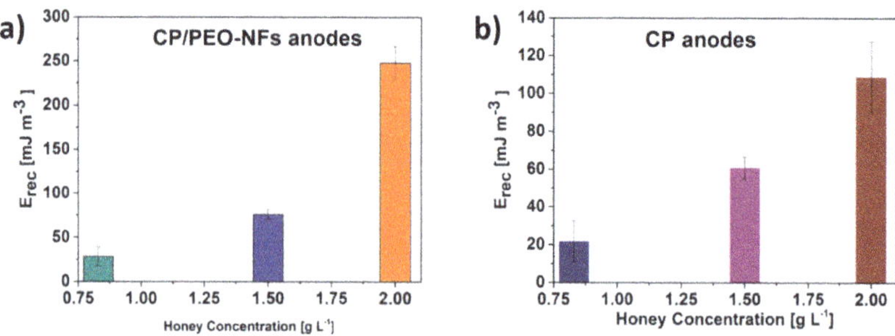

Figure 4. (**a**) Recovered energy defined for CP/PEO-NFs correlated with honey concentration; (**b**) Recovered energy defined for CP correlated with honey concentration.

3.3. Electrochemical Impedance Spectroscopy Results

EIS was performed to investigate the impedance behavior and, in particular, the internal resistance [39] when honey was employed as a fuel. By analyzing the Nyquist plot represented in Figure 5, it can be observed that MFCs exploiting CP/PEO-NFs as the anode were characterized by similar total impedance values with respect to carbon-based devices (982.2 Ω and 916.5 Ω, respectively). Moreover, the results allowed analyzing the resistance related to the charge transfer (R$_{ct}$) of both CP/PEO-NFs and CP anodes. The obtained results confirmed that SCMFCs exploiting CP/PEO-NFs as the anode presented impedance values close to those reached with a CP-based anode. The value of R$_{ct}$ for CP/PEO-NFs, close to 761 Ω, was quite similar to that obtained for CP, equal to about 808 Ω. These results demonstrated that the presence of PEO-NFs, which basically has insulating properties from an electrical point of view, not only did not affect the electron transfer rate but also simultaneously sustained the proliferation of microorganisms onto the anode

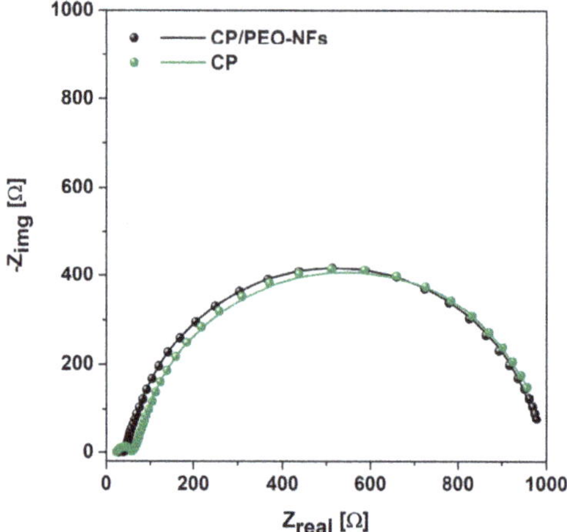

Figure 5. Impedance spectra of CP/PEO NFs (black dots and line, representing the experimental data and their fitting) and CP anode (green dots and line).

This result suggests the possibility to apply CP/PEO-NFs to achieve a large nanofiber-based interface between a CP anode and a bacterial biofilm, thus optimizing the adhesion of the biofilm to the anode.

4. Conclusions

In the present work, a CP/PEO-NFs composite anode, based on PEO-NFs directly electrospun onto a carbon-based material, was designed and optimized. CP/PEO-NFs showed a final ordered arrangement of PEO-NFs, able to grant a large surface area-to-volume ratio. We demonstrated that the resulting structure of PEO/NFs greatly promoted microorganisms' proliferation, thus suggesting the possibility to employ PEO-NFs as a biomass carrier for bacteria entrapment. Therefore, we investigated the behavior of a CP/PEO-NFs anode in SCMFCs, using a mixed bacterial consortium extracted from a marine sediment sample as the biofilm source. In particular, a high current density, close to 20 mA m^{-2}, was reached when CP/PEO-NFs were used as the anode. This value, higher than the one reached when bare CP was used as the anode, allowed us to demonstrate that the presence of PEO-NFs, with their insulating properties from an electrical point of view, did not affect the electron transfer rate and simultaneously sustained the proliferation of the microorganisms. Finally, the results concerning the evaluation of energy recovery confirmed the possibility to use a more complex substrate, such as honey, as a fuel in MFCs. Moreover, we evidenced that the new designed CP/PEO-NFs anodes are able to ensure a three-time higher recovered energy (i.e., 300 mJ m^{-3}) than that obtained when using a bare carbon-based anode.

The results show that nanostructured interfaces made of PEO-based nanofibers are advantageous for the fabrication of robust and efficient electrodes to be used in MFCs as energy conversion tools for the valorization of waste in the food industry.

Author Contributions: M.Q. and G.M. conceived the work. G.M. worked on electrospinning and MFCs. A.S. carried out the electrochemical characterization. M.Q., M.C., S.L.M. worked on the design and preparation of MFCs. F.F. and A.C. designed the experiments with honey. C.F.P. and M.Q. organized the research activity. All authors contributed to the final manuscript. All authors have read and agreed to the published version of the manuscript.

Funding: The present work was performed in the framework of the POLITO BIOMed LAB, an interdepartmental laboratory financed by Politecnico di Torino. The activity was partially financed by SMART3D ("Smart 3D Polymer Devices Production Chain") project, financed by MIUR and Piedmont Region agreement in the framework of "Smart Industry" and by Piedmont Region in the framework of the "FOOD-DRUG-FREE" Platform project. In the frame of the FOOD-DRUG-FREE project, the authors especially acknowledge support from Sebaste Srl for providing the honey samples used in the work.

Conflicts of Interest: The authors declare no conflict of interest

References

1. Slate, A.J.; Whiteland, K.A.; Brownson, D.A.C.; Banks, C.E. Microbial fuel cells: An overview of current technology. *Renew. Sustain. Energy Rev.* **2019**, *101*, 60–81. [CrossRef]
2. Logan, B.E. *Microbial Fuel Cells*; John Wiley & Sons Inc.: New York, NY, USA, 2008.
3. Santoro, C.; Arbizzani, C.; Erable, B.; Ieropoulos, I. Microbial fuel cells: From fundamentals to applications. *A review. J. Power Sources* **2017**, *356*, 225–244. [CrossRef] [PubMed]
4. Harnisch, F.; Aulenta, F.; Schroeder, U. Microbial Fuel Cells and Bioelectrochemical Systems: Industrial and Environmental Biotechnologies Based on Extracellular Electron Transfer. In *Comprehensive Biotechnology*, 2nd ed.; Elsevier: Amsterdam, The Netherlands, 2011; pp. 644–659.
5. Melhuish, C.; Ieropoulos, I.; Greenman, J.; Horsfield, I. Energetically autonomous robots: Food for thought. *Auton. Robots* **2006**, *21*, 187–198. [CrossRef]
6. Sanket, G. From waste to watts in micro-devices: Review on development of Membraned and Membraneless Microfluidic Microbial Fuel Cell. *Appl. Mater. Today* **2018**, *11*, 270–279.
7. Li, W.-W.; Yu, H.-Q.; He, Z. Towards sustainable wastewater treatment by using microbial fuel cells-centered technologies. *Energy Environ. Sci.* **2014**, *7*, 911–924. [CrossRef]

8. Chang, I.S.; Jang, J.K.; Gil, G.C.; Kim, M.; Kim, H.J.; Cho, B.W.; Kim, B.H. Continuous determination of biochemical oxygen demand using microbial fuel cell type biosensor. *Biosens. Bioelectron.* **2004**, *19*, 607–613. [CrossRef]

9. Shantaram, A.; Beyenal, H.; Raajan, R.; Veluchamy, A.; Lewandowski, Z. Wireless sensors powered by microbial fuel cells. *Environ. Sci. Technol.* **2005**, *39*, 5037–5042. [CrossRef]

10. Donovan, C.; Dewan, A.; Heo, D.; Beyenal, H. Batteryless, wireless sensor powered by a sediment microbial fuel cell. *Environ. Sci. Technol.* **2008**, *42*, 8591–8596. [CrossRef]

11. Tender, L.M.; Gray, S.A.; Groveman, E.; Lowy, D.A.; Kauffman, P.; Melhado, J.; Tyce, R.C.; Flynn, D.; Petrecca, R.; Dobarro, J. The first demonstration of a microbial fuel cell as a viable power supply: powering a meteorological buoy. *J. Power Sources* **2008**, *179*, 571–575. [CrossRef]

12. Venkata Mohan, S.; Raghavulu, V.S.; Sarma, P.N. Biochemical evaluation of bioelectricity production process from anaerobic wastewater treatment in a single chambered microbial fuel cell (MFC) employing glass wool membrane. *Biosens. Bioelectron.* **2008**, *23*, 1326–1332. [CrossRef]

13. Cecconet, D.; Molognoni, D.; Callegari, A.; Capodaglio, A.G. Agro-food industry wastewater treatment with microbial fuel cells: Energetic recovery issues. *Int. J. Hydrogen Energy* **2018**, *43*, 500–511. [CrossRef]

14. Feng, Y.; Wang, X.; Logan, B.E.; Lee, H. Brewery wastewater treatment using air-cathode microbial fuel cells. *Appl. Microbiol. Biotechnol.* **2008**, *78*, 873–880. [CrossRef] [PubMed]

15. Wang, X.; Feng, Y.J.; Lee, H. Electricity production from beer brewery wastewater using single chamber microbial fuel cell. *Water Sci. Technol.* **2008**, *57*, 1117–1121. [CrossRef] [PubMed]

16. Sciarra, T.P.; Tenca, A.; D'Epifanio, A.; Mecheri, B.; Merlino, G.; Barbato, M.; Borin, S.; Licoccia, S.; Garaviglia, V.; Adani, F. Using olive mill wastewater to improve performance in producing electricity from domestic wastewater by using single-chamber microbial fuel cell. *Bioresour. Technol.* **2013**, *147*, 246–253. [CrossRef]

17. Sciarra, T.P.; Merlino, G.; Scaglia, B.; D'Epifanio, A.; Mecheri, B.; Borin, S.; Licoccia, S.; Adani, F. Electricity generation using white and red wine lees in air cathode microbial fuel cells. *J. Power Sources* **2015**, *274*, 393–399. [CrossRef]

18. Penteado, E.D.; Fernandez-Marchante, C.M.; Zaiat, M.; Canizares, P.; Gonzales, E.R.; Rodrigo, M.A.R. Energy recovery from winery wastewater using a dual chamber microbial fuel cell. *J Chem. Technol. Biotechnol.* **2016**, *91*, 1802–1808. [CrossRef]

19. Mardanpour, M.M.; Esfahany, M.N.; Behzad, T.; Sedaqatvand, R. Single chamber microbial fuel cell with spiral anode for dairy wastewater treatment. *Biosens. Bioelectron.* **2012**, *38*, 264–269. [CrossRef]

20. Oh, S.T.; Kim, J.R.; Premier, G.C.; Lee, T.H.; Kim, C.; Sloan, W.T. Sustainable wastewater treatment: How might microbial fuel cells contribute. *Biotechnology Advances.* **2010**, *28*, 871–881. [CrossRef]

21. Yang, G.; Wang, J.; Zhang, H.; Jia, H.; Zhang, Y.; Cui, Z.; Gao, F. Maximizing energy recovery from homeostasis in microbial fuel cell by synergistic conversion of short-chain volatile fatty acid. *Bioresour. Technol. Rep.* **2019**, *7*, 100200. [CrossRef]

22. Lee, H.S.; Parameswaran, P.; Kato-Marcus, A.; Torres, C.I.; Rittmann, B.E. Evaluation of energy-conversion efficiencies in microbial fuel cells (MFCs) utilizing fermentable and non-fermentable substrates. *Water Res.* **2008**, *24*, 1501–1510. [CrossRef]

23. Ge, Z.; Ping, Q.; Xiao, L.; He, Z. Reducing effluent discharge and recovering bioenergy in an osmotic microbial fuel cell treating domestic wastewater. *Desalination.* **2013**, *312*, 52–59. [CrossRef]

24. Fornero, J.J.; Rosenbaum, M.; Angenent, L.T. Electric Power Generation from Municipal, Food, and Animal Wastewaters Using Microbial Fuel Cells. *Electroanalysis.* **2010**, *22*, 832–843. [CrossRef]

25. Zhang, L.S.; Wu, W.; Wang, J. Immobilization of activated sludge using improved polyvinyl alcohol (PVA) gel. *J. Environ. Sci.* **2007**, *19*, 1293–1297. [CrossRef]

26. Jiang, D.; Li, B. Novel electrode materials to enhance the bacterial adhesion and increase the power generation in microbial fuel cells (MFCs). *Water Sci. Technol.* **2009**, *59*, 557–563. [CrossRef] [PubMed]

27. Bai, X.; Ye, Z.F.; Li, Y.F.; Zhou, L.C.; Yang, L.Q. Preparation of crosslinked macroporous PVA foam carrier for immobilization of microorganisms. *Process Biochem.* **2010**, *45*, 60–66. [CrossRef]

28. Bruce, P.G. Structure and electrochemistry of polymer electrolytes. *Electrochim. Acta* **1995**, *40*, 2077–2086. [CrossRef]

29. Xie, J.; Duan, R.G.; Han, Y.; Kerr, J.B. Morphological, rheological and electrochemical studies of Poly(ethylene oxide) electrolytes containing fumed silica nanoparticles. *Solid State Ionics* **2004**, *175*, 755–758. [CrossRef]

30. Quaglio, M.; Chiodoni, A.; Massaglia, G.; Delmondo, L.; Sacco, A.; Garino, N.; Castellino, M.; Bianco, S.; Margaria, V.; Salvador, G.P.; et al. Electrospinning-on-Electrode Assembly for Air-Cathodes in Microbial Fuel Cells. *Adv. Mater. Interfaces* **2018**, *5*, 1801107. [CrossRef]
31. Massaglia, G.; Margaria, V.; Sacco, A.; Tommasi, T.; Pentassugli, S.; Ahmed, D.; Mo, R.; Pirri, C.F.; Quaglio, M. In situ continuous current production from marine floating microbial fuel cells. *Appl. Energy* **2018**, *230*, 78–85. [CrossRef]
32. Alvarez-Suarez, J.M.; Tulipani, S.; Romandini, S.; Bertoli, E.; Battino, M. Contribution of honey in nutrition and human health: A review. *Mediterr. J. Nutr. Metab.* **2010**, *3*, 15–23. [CrossRef]
33. Burlando, B.; Cornara, L. Honey in dermatology and skin care: A review. *J. Cosmet. Dermatol.* **2013**, *12*, 306–313. [CrossRef] [PubMed]
34. De-Melo, A.A.M.; Almeida-Muradina, L.B.; Sancho, M.T.; Pascual-Mate, A. Composition and properties of Apis mellifera honey: A review. *J. Apicult. Res.* **2018**, *57*, 5–37. [CrossRef]
35. Doner, L.W. Honey. In *Encyclopedia of Food Sciences and Nutrition*, 2nd ed.; Caballero, B., Finglas, P.M., Trugo, L.C., Eds.; Academic Press: London, UK, 2003; pp. 3125–3130.
36. Sabatini, A.G. Il miele: Origine, composizione e proprieta. In *Conscere il Miele*; Sabatini, A.G., Botolotti, L., Marcazzan, G.L., Eds.; Avenue Media: Bologna/Milano, Italy, 2007; pp. 3–37.
37. Capodaglio, A.G.; Molognoni, D.; Dallago, E.; Liberale, A.; Cella, R.; Longoni, P.; Pantaleoni, L. Microbial Fuel Cells for Direct Electrical Energy Recovery from Urban Wastewaters. *Sci. World J.* **2013**, *2013*, 634738. [CrossRef] [PubMed]
38. Massaglia, G.; Gerosa, M.; Agostino, V.; Cingolani, A.; Sacco, A.; Saracco, G.; Margaria, V.; Quaglio, M. Fluid Dynamic Modeling for Microbial Fuel Cell Based Biosensor Optimization. *Fuel Cells* **2017**, *17*, 627–634. [CrossRef]
39. Hidalgo, D.; Sacco, A.; Hernàndez, S.; Tommasi, T. Electrochemical and impedance characterization of Microbial Fuel Cells based on 2D and 3D anodic electrodes working with seawater microorganisms under continuous operation. *Bioresour. Technol.* **2015**, *195*, 139–146. [CrossRef] [PubMed]

 *nanomaterials*

Communication

Electrospun Cadmium Selenide Nanoparticles-Loaded Cellulose Acetate Fibers for Solar Thermal Application

Nicole Angel [1,†], S. N. Vijayaraghavan [2,†], Feng Yan [2,*] and Lingyan Kong [3,*]

[1] Department of Mechanical Engineering, The University of Alabama, Tuscaloosa, AL 35487, USA; nmangel@crimson.ua.edu
[2] Department of Metallurgical and Materials Engineering, The University of Alabama, Tuscaloosa, AL 35487, USA; vsankaranarayanannair@crimson.ua.edu
[3] Department of Human Nutrition and Hospitality Management, The University of Alabama, Tuscaloosa, AL 35487, USA
*   Correspondence: fyan@eng.ua.edu (F.Y.); lkong@ches.ua.edu (L.K.)
†   These authors contributed equally to this work.

Received: 28 June 2020; Accepted: 4 July 2020; Published: 8 July 2020

Abstract: Solar thermal techniques provide a promising method for the direct conversion of solar energy to thermal energy for applications, such as water desalination. To effectively realize the optimal potential of solar thermal conversion, it is desirable to construct an assembly with localized heating. Specifically, photoactive semiconducting nanoparticles, when utilized as independent light absorbers, have successfully demonstrated the ability to increase solar vapor efficiency. Additionally, bio-based fibers have shown low thermal conductive photocorrosion. In this work, cellulose acetate (CA) fibers were loaded with cadmium selenide (CdSe) nanoparticles to be employed for solar thermal conversion and then subsequently evaluated for both their resulting morphology and conversion potential and efficiency. Electrospinning was employed to fabricate the CdSe-loaded CA fibers by adjusting the CA/CdSe ratio for increased solar conversion efficiency. The microstructural and chemical composition of the CdSe-loaded CA fibers were characterized. Additionally, the optical sunlight absorption performance was evaluated, and it was demonstrated that the CdSe nanoparticles-loaded CA fibers have the potential to significantly improve solar energy absorption. The photothermal conversion under 1 sun (100 mW/cm^2) demonstrated that the CdSe nanoparticles could increase the temperature up to 43 °C. The CdSe-loaded CA fibers were shown as a feasible and promising hybrid material for achieving efficient solar thermal conversion.

Keywords: photoactive nanoparticles; cadmium selenide; cellulose acetate; electrospun fibers; solar thermal

1. Introduction

Currently, clean freshwater scarcity is a significant issue that is closely linked with social and economic development [1–3]. Already, billions of people worldwide lack access to safe drinking water, and millions die annually due to diseases relating to water-borne illnesses [2,4]. Some proposed solutions for addressing water scarcity work at the expense of aggravating present energy problems, while others suggest the implementation of large-scale infrastructure [1,4]. Therefore, developing a method to address water scarcity in a clean, affordable, and sustainable manner is of increasing importance. Solar thermal techniques provide an affordable way to convert solar energy to thermal energy; for instance, solar vapor generation for water desalination has been demonstrated as a potential technique to provide a sustainable solution to water scarcity [1,3,5,6].

One of the main drawbacks of this technique is low solar thermal conversion efficiency [1]. Both the low solar absorption of the water as well as heat loss due to the use of conventional water heating systems have significantly contributed to this low efficiency [1]. Therefore, shifting from a bulk water heating system to a nanoscaled solar absorber system with localized heating capability, specifically through the implementation of photosensitive nanoparticles (NPs), has demonstrated a noticeable increase in conversion efficiency [1,5,6]. To form a localized heating structure, it is desirable to couple high thermal conductive, photoactive NPs with a low thermal conductive polymer matrix (such as a bio-based polymer). The thermal energy generated from the photoexcited carriers in the NPs become localized heat that has difficulty diffusing into the surrounding polymer substrate; thus, the localized heating of water is enabled [7]. This advantage of localized heating can overcome the remarkable heat dissipation that occurs during bulk water heating via conventional semiconductor abosrbers. Beneficially, this process does not require the total surrounding liquid volume to reach its boiling point for successful vapor generation [6].

In this study, we investigate the implementation of electrospun cellulose acetate (CA) fibers carrying photosensitive cadmium selenide (CdSe) NPs as a nanoscaled solar conversion device for solar thermal conversion. CdSe, a Group II-VI compound semiconductor, exhibits extraordinary electronic and optoelectronic properties [8,9]. CdSe displays optimal sunlight absorption with a bandgap of 1.74 eV [10] and a high thermal conductivity (-0.53 W/cmK) [11]. Cellulose acetate is a low-cost cellulose derivative produced via the acetylation of cellulose [12,13] with low thermal conductivity (-0.10 W/cmK) [14]. It was chosen as the fiber material in this study due to its advantageous mechanical properties, excellent fiber-forming ability, biodegradability, stability in water, and cost-effectiveness [13,15]. Electrospinning has become a popular method for the synthesis of micro- and nanofibrous materials due to the use of relatively simple manufacturing equipment, low spinning cost, variety of spinnable materials, and highly controllable processes [16,17]. Moreover, micro- to nanoscale fibers display a high surface-area-to-volume ratio, good flexibility, high porosity, and superior stiffness and tensile strength than when compared to other forms of the spun material [15]. Note that, for the seawater desalination application, the heavy toxic metal element, i.e., Cd, may diffuse into the water via photocorrosion and lead to water contamination. Here, we use the CdSe as a model semiconductor with a suitable sunlight absorption band edge to explore the potential to load into CA fibers through electrospinning technique. Other low-toxic and nontoxic semiconductor particles with desired light absorption capability, such as Sb_2S_3, Fe_2O_3, and Fe_3O_4, will be explored in future studies. In particular, the diffusion behavior of the metal elements during the water desalination will also be investigated later.

2. Materials and Methods

2.1. Materials

Cellulose acetate (MW ~100,000 Da; acetyl content ~39.7 wt%) and acetone were purchased from VWR International (Radnor, PA, USA).

Based on the slow reaction between Cd^{2+} and Se^{2-} ions in an aqueous basic bath with pH > 10, the CdSe nanoparticles were synthesized using a chemical bath deposition process. Cadmium sulfate ($CdSO_4$) and sodium selenosulfite (Na_2SSeO_3) were used as the sources for Cd^{2+} and Se^{2-}, respectively. Na_2SSeO_3 was prepared by dissolving elemental Se in the form of fine powder in an aqueous solution of sodium sulfite heated to 60 °C. The solution was stirred well until the Se was completely dissolved. The pH of the solution was adjusted by adding excess NaOH. CdSe was formed in 2 h at a temperature of 70 °C. The obtained CdSe powder was washed using deionized water, centrifuged repeatedly, and subsequentially dried in a vacuum oven.

Because CdSe is considered toxic to human cells, it was handled with the utmost care during this study [18]. CdSe toxicity has been mainly attributed to the release of Cd^{2+} ions from the CdSe to surrounding cells [18]. The method of exposure by which this diffusion occurs highly alters how significantly the affected cells react to CdSe [18]. Consequently, equipment in contact with CdSe was cleaned thoroughly, waste was properly disposed of, and the CdSe NPs were properly labeled and stored before and after use.

2.2. Methods

2.2.1. Electrospinning

In this study, the CA concentration in the spinning dope was held constant at 12% (w/v). To prepare the spinning solutions, CA and CdSe were mixed in pure acetone via magnetic stirring at room temperature (20 °C) and then ultrasonicated (VWR International, Radnor, PA, USA). The spinning dope was then loaded into a 10-mL syringe (Becton, Dickinson and Company, Franklin Lakes, NJ, USA) with a 22-gauge blunt needle (Hamilton Company, Reno, NV, USA) as the spinneret.

The electrospinning setup was comprised of a high voltage generator (ES30P, Gamma High Voltage Research, Inc., Ormond Beach, FL, USA), a syringe pump (NE-300, New Era Pump Systems, Inc., Farmingdale, NY, USA), and a grounded aluminum foil as the collector. In the present study, the optimal spinning parameters to spin CA fibers were determined from our previous study [15]. Specifically, the spinning distance was held at 8 cm, the feed rate at 3 mL/h, and the voltage at 12 kV. Electrospinning was conducted in a fume hood at 20 °C without airflow. Airflow was removed due to the high evaporation rate of acetone. Eliminating airflow helped to slow the buildup of viscous fluid at the spinneret tip that could directly cause the destabilization of the spinning jet. Periodically, fiber formation can be halted because the spinneret can become clogged and the jet will no longer be considered continuous. Consequently, despite the lack of airflow in this study, the spinneret required constant cleaning during the spinning process to successfully fabricate fibers [19]. After the spinning process, the formed fibers, which had deposited on the aluminum foil directly beneath the spinneret tip, were collected and stored away from light and moisture for further analysis.

2.2.2. Scanning Electron Microscopy (SEM)

The morphology of the electrospun fibers was studied using an Apreo field emission SEM (FE-SEM, Thermo Fisher Scientific, Waltham, MA, USA), equipped with energy-dispersive X-ray spectroscopy (EDS), at an accelerating voltage of 20 kV. Images were subsequently analyzed using ImageJ software (National Institute of Health, Bethesda, MD, USA). Three images were used for each fiber sample. Ten random segments on each image were taken and used to measure the fiber diameters. Through visual inspection, fiber morphology was evaluated as either good, fair, or poor [20]. Good fibers were defined as continuous, uniform, smooth, and defect-free. Fair fibers exhibited a fibrous shape with moderate defects. Poor fibers had significant defects, such as beads or sprayed particles. Notably, the CdSe NPs altered how smooth the fibers appeared, but were not considered to be fiber defects.

2.2.3. Wide-Angle X-ray Diffraction (XRD)

The wide-angle X-ray diffraction (XRD) patterns of the fibers were obtained using a Philips X'Pert Materials Research Diffractometer (Malvern Panalytical, Westborough, MA, USA) operated at 45 kV under 40 mA Cu Kα radiation (λ = 0.15405 nm).

2.2.4. Ultraviolet-Visible (UV-vis) Spectroscopy

The light absorbance of the fiber samples were measured using EnliTech QE measurement system (Kaohsiung City, Taiwan). The incident wavelength was swept from 300 to 1100 nm, and the absorbance of the fibers were recorded. The optical direct bandgap (E_g) of the fibers was determined by using Tauc plot relation to the plot $(\alpha h\nu)^{1/r}$ versus the energy of the photons, where α is the absorption coefficient of the materials, and r represents the nature of the transition of the charge carriers, and r = 1/2 for direct bandgap materials.

2.2.5. Solar Energy Conversion

Using a compact infrared (IR) camera (FLIR C2, FLIR Systems, Inc., Wilsonville, OR, USA), the fibers were tested for solar vapor generation under 1 sun (100 mW/cm^2) illumination. The thermal images were recorded after the nanofibers were exposed to the simulated sunlight for 30 min to stabilize in the air between heating and cooling.

3. Results and Discussion

3.1. Electrospinnability and Fiber Morphology

The CA fibers were fabricated by electrospinning, a simple, scalable, and versatile technique for nanofiber production [21]. Given the potential applications of this study, the ability for our resulting solar vapor generator to be mass scaled is of utmost importance. The basic lab-scale electrospinning setup only requires a high voltage source, a syringe with a blunt needle tip (i.e. the spinneret), and a grounded collector [15,21]. High voltage is applied to the spinneret where the spinning dope is pumped via a syringe pump at a constant rate. Once the applied voltage reaches the critical point, a Taylor cone will form at the spinneret tip. A continuous jet flows from the Taylor cone to the grounded collector. During this process, the electric field causes the jet to stretch and elongate as the solvent evaporates. Fibers form on the grounded collector, e.g. the aluminum foil, as used in this study [15].

The CA fibers were loaded with varying CdSe concentrations (CdSe:CA ratio of 1:4 and 1:1, w/w) to analyze the successful uptake of CdSe into the CA fibers and to observe the resulting photosensitive properties of the fibers as provided by the CdSe NPs. The overall electrospinnability of the CdSe-CA compositions was analyzed through both visual inspection during spinning and the obtained fiber images (Figure 1). All three CdSe-CA dispersions were able to successfully produce fibers in a relatively short amount of time. The time needed for the collector to become completely covered in fibers was about the same for all three runs, suggesting that CdSe did not have a drastic effect on the electrospinnability of CA in the single solvent acetone. The overall quality of the electrospun fibers was also independent of the addition of CdSe. As seen in Figure 1, the CdSe-CA fibers remain continuous and smooth, discounting the CdSe NPs, and display approximately the same amount of beading as the pure CA fibers. These fibers were thus rated as good fibers, albeit their size was on the microscale rather than the nanoscale.

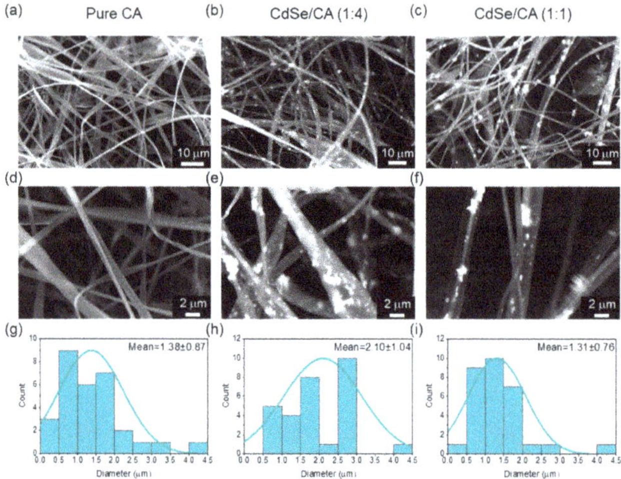

Figure 1. (**a–f**) Field emission scanning electron micrographs (FE-SEM) images and (**g–i**) fiber diameter distributions of CA, CdSe-CA (1:4), and CdSe-CA (1:1) fibers, respectively.

3.2. CdSe Incorporation

As shown in Figure 2, CdSe was successfully incorporated into the CA fibers. The white particles embedded in the fibers were confirmed to be comprised of elements Cd and Se, respectively (Figure 2c–f). As expected, CdSe-CA (1:1) fibers demonstrated a higher quantity of successful CdSe incorporation as compared with CdSe-CA (1:4) fibers. However, in both the (1:1) and (1:4) fibers, the CdSe particles were segregated in clusters rather than being evenly dispersed. This could be a result of the difficulty in achieving a homogeneous distribution and stable suspension of CdSe in the spinning dope. Additionally, during electrospinning, the CdSe particles may be charged and easily segregated together.

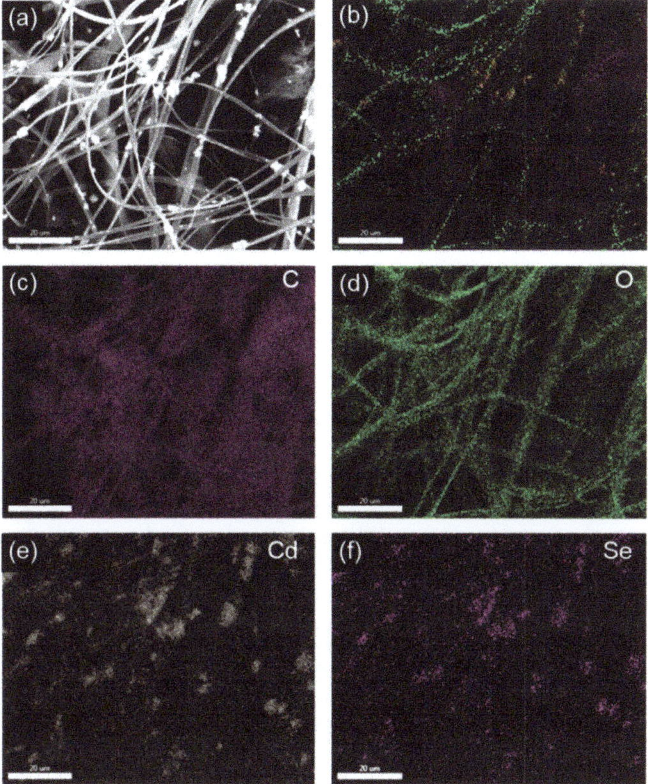

Figure 2. (**a**) FESEM, (**b**) overall elemental distribution, and (**c–f**) C, O, Cd, and Se elemental distribution of CdSe-CA (1:1) fibers obtained using energy dispersive X-ray spectrometry (EDS) mapping.

The structure of the CdSe loaded CA fibers was characterized using X-ray diffraction (XRD), as shown in Figure 3. The XRD peaks of CdSe at $2\theta = 25.3°$, $42°$, and $50°$ were observed in the CdSe-loaded fibers, whereas these peaks did not exist in the pattern of the pure CA fibers. The (002), (110), and (201) peaks of CdSe, corresponding to $2\theta \sim 25°$, $41°$, and $50°$, respectively, can be indexed to the wurtzite phase. Similar to most other biopolymers, CA is semi-crystalline yet largely amorphous. The diffraction peaks around $10°$ and $20°$, corresponding to the (101) and (020) planes, respectively, confirm the semi-crystalline nature of CA [22]. When the CdSe/CA ratio increased, the intensity of the CdSe peaks increased.

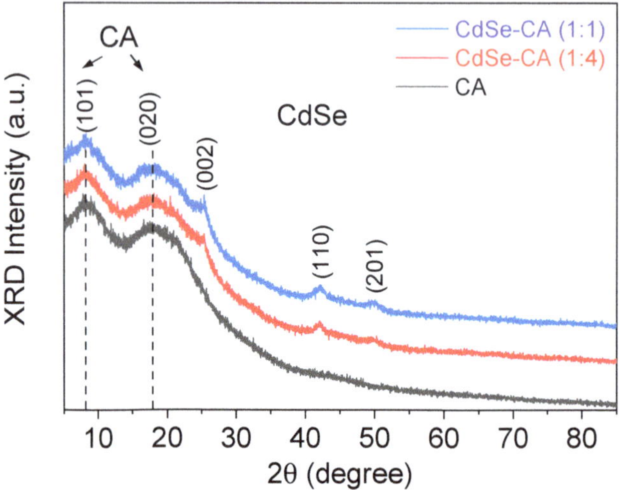

Figure 3. X-ray diffraction patterns of CA, CdSe-CA (1:4), and CdSe-CA (1:1) fibers.

3.3. Optical Properties

As shown in Figure 4a, the absorbance of the CA fibers increased as the CdSe content increased, with the light absorption edge around 700 nm for the CdSe-loaded CA fibers. These fibers have superior sunlight absorption than pure CA fibers which indicates increased solar thermal conversion [23]. From the extrapolation of the Tauc plot, as shown in Figure 4b, CA was determined to have a wide bandgap of ~3.4 eV which corresponds to the ultraviolet light range in the solar spectrum. When CdSe was introduced into CA fibers with varying concentrations, the CdSe-CA (1:4) fibers were found to have a bandgap of ~1.7 eV. When more CdSe was incorporated into CA in the CdSe-CA (1:1) fibers, the bandgap remained constant. Note that pure CdSe is estimated to have a bandgap of 1.74 eV according to previous works [10]. The fact that the bandwidth of the CdSe-loaded fibers is close to that of pure CdSe demonstrates that loading CdSe into CA fibers can significantly increase the sunlight absorption of CA fibers. Therefore, CA fibers show great potential in their ability to act as a supporting matrix for CdSe NPs-based solar thermal conversion for improved localized heating.

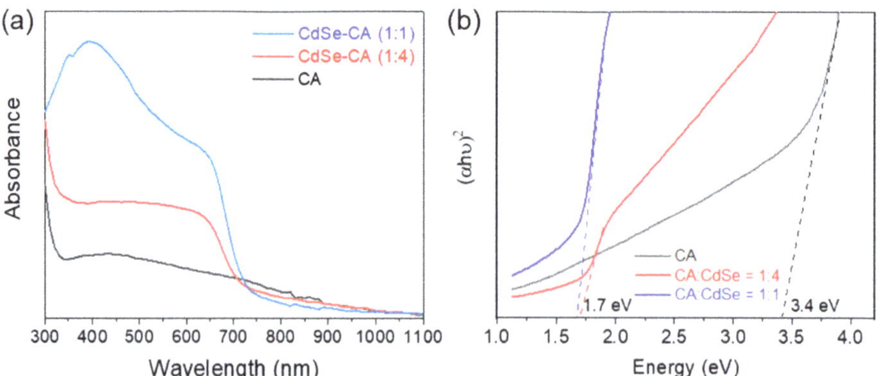

Figure 4. (**a**) UV-vis absorbance spectra (normalized at 1100 nm) and (**b**) Tauc plot of CA, CdSe-CA (1:4), and CdSe-CA (1:1) fibers.

3.4. Photothermal Conversion

To investigate the solar thermal properties of the CA and CA-CdSe fibers, both optical and IR thermal images were captured (Figure 5). As seen in the optical image, the CdSe-CA (1:4) fibers were dark due to the light absorption by CdSe, while the pure CA fibers were white, which is in agreement with the UV-Vis absorbance test above. The maximum temperature in the CdSe-CA (1:4) fibers can be over 40 °C under the 1 sun (100 mW/cm^2) illumination, while the pure CA fibers maintained a low temperature without photothermal conversion. This solar thermal conversion performance is similar to the carbonaceous membrane assisted by localized heating design [24]. It is suggested that the CA-CdSe fibers could effectively absorb sunlight and convert it into thermal energy.

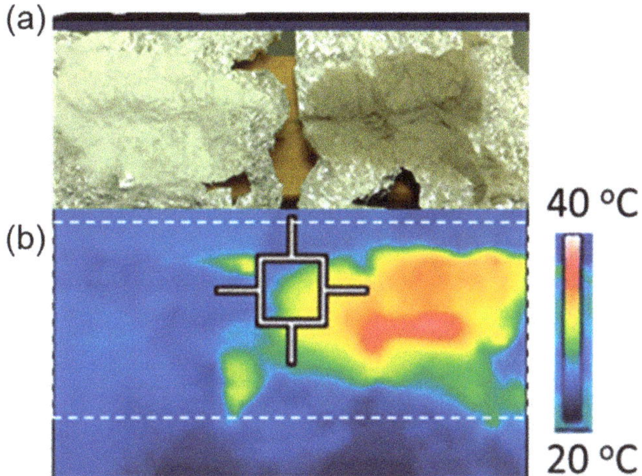

Figure 5. (**a**) Optical picture and (**b**) infrared thermal image of the CA and CA-CdSe fiber mats.

4. Conclusions

In conclusion, the loading of highly thermal conductive CdSe NPs into CA fibers of low thermal conductivity was successfully demonstrated in this study. Overall, fiber morphology, including quality and diameter, was relatively independent of the addition of CdSe. The solar energy absorption of CdSe NPs were virtually unaffected while in the fibers. Specifically, the bandwidth of CdSe NPs remained the same while in the CA fibers, and the final assemblies were highly absorbent in the visible light spectrum. It was demonstrated that electrospinning can effectively introduce photosensitive NPs into the bio-based CA fibers. This low-cost and scalable technique can be used to facilitate water desalination through solar thermal evaporation with localized heating that has the potential to help address the current concerns regarding water scarcity encountered across the globe.

Author Contributions: Conceptualization, F.Y. and L.K.; methodology, N.A., and S.N.V.; writing—original draft preparation, N.A., and S.N.V.; writing—review and editing, F.Y. and L.K.; project administration, F.Y. and L.K.; funding acquisition, F.Y. and L.K. All authors have read and agreed to the published version of the manuscript.

Funding: This research was funded by the USDA National Institute of Food and Agriculture (NIFA), AFRI project award # 2020-67022-31376, and the NSF Award #1944374.

Acknowledgments: The Alabama Water Institute and the College of Human Environmental Sciences at the University of Alabama are acknowledged for their support.

Conflicts of Interest: The authors declare no conflict of interest.

References

1. Gao, M.; Zhu, L.; Peh, C.K.; Ho, G.W. Solar absorber material and system designs for photothermal water vaporization towards clean water and energy production. *Energy Environ. Sci.* **2019**, *12*, 841–864. [CrossRef]
2. Shannon, M.A.; Bohn, P.W.; Elimelech, M.; Georgiadis, J.G.; Marinas, B.J.; Mayes, A.M. Science and technology for water purification in the coming decades. In *Nanoscience and Technology: A Collection of Reviews from Nature Journals*; World Scientific: Singapore, 2010; pp. 337–346.
3. Liu, C.; Huang, J.; Hsiung, C.E.; Tian, Y.; Wang, J.; Han, Y.; Fratalocchi, A. High-performance large-scale solar steam generation with nanolayers of reusable biomimetic nanoparticles. *Adv. Sustain. Syst.* **2017**, *1*, 1600013. [CrossRef]
4. Rijsberman, F.R. Water scarcity: Fact or fiction? *Agric. Water Manag.* **2006**, *80*, 5–22. [CrossRef]
5. Zhao, F.; Zhou, X.; Shi, Y.; Qian, X.; Alexander, M.; Zhao, X.; Mendez, S.; Yang, R.; Qu, L.; Yu, G. Highly efficient solar vapour generation via hierarchically nanostructured gels. *Nat. Nanotechnol.* **2018**, *13*, 489–495. [CrossRef]
6. Neumann, O.; Urban, A.S.; Day, J.; Lal, S.; Nordlander, P.; Halas, N.J. Solar vapor generation enabled by nanoparticles. *Acs Nano* **2013**, *7*, 42–49. [CrossRef] [PubMed]
7. Govorov, A.O.; Richardson, H.H. Generating heat with metal nanoparticles. *Nano Today* **2007**, *2*, 30–38. [CrossRef]
8. Jin, W.; Hu, L. Review on quasi one-dimensional cdse nanomaterials: Synthesis and application in photodetectors. *Nanomaterials* **2019**, *9*, 1359. [CrossRef] [PubMed]
9. Li, P.; Zhu, B.; Li, P.; Zhang, Z.; Li, L.; Gu, Y. A facile method to synthesize cdse-reduced graphene oxide composite with good dispersion and high nonlinear optical properties. *Nanomaterials* **2019**, *9*, 957. [CrossRef]
10. Mahato, S.; Kar, A.K. The effect of annealing on structural, optical and photosensitive properties of electrodeposited cadmium selenide thin films. *J. Sci. Adv. Mater. Devices* **2017**, *2*, 165–171. [CrossRef]
11. Ma, Y.; Liu, M.; Jaber, A.; Wang, R.Y. Solution-phase synthesis and thermal conductivity of nanostructured cdse, in 2 se 3, and composites thereof. *J. Mater. Chem. A* **2015**, *3*, 13483–13491. [CrossRef]
12. Fischer, S.; Thümmler, K.; Volkert, B.; Hettrich, K.; Schmidt, I.; Fischer, K. *Macromolecular Symposia: In Properties and Applications of Cellulose Acetate*; Wiley Online Library: Hoboken, NJ, USA, 2008; pp. 89–96.
13. Nosar, M.N.; Salehi, M.; Ghorbani, S.; Beiranvand, S.P.; Goodarzi, A.; Azami, M. Characterization of wet-electrospun cellulose acetate based 3-dimensional scaffolds for skin tissue engineering applications: Influence of cellulose acetate concentration. *Cellulose* **2016**, *23*, 3239–3248. [CrossRef]
14. Gutiérrez, M.C.; De Paoli, M.-A.; Felisberti, M.I. Cellulose acetate and short curauá fibers biocomposites prepared by large scale processing: Reinforcing and thermal insulating properties. *Ind. Crop. Prod.* **2014**, *52*, 363–372. [CrossRef]
15. Angel, N.; Guo, L.; Yan, F.; Wang, H.; Kong, L. Effect of processing parameters on the electrospinning of cellulose acetate studied by response surface methodology. *J. Agric. Food Res.* **2020**, *2*, 100015. [CrossRef]
16. Yu, J.; Zhao, Z.; Sun, J.; Geng, C.; Bu, Q.; Wu, D.; Xia, Y. Electrospinning highly concentrated sodium alginate nanofibres without surfactants by adding fluorescent carbon dots. *Nanomaterials* **2020**, *10*, 565. [CrossRef]
17. Odhiambo, V.O.; Ongarbayeva, A.; Kéri, O.; Simon, L.; Szilágyi, I.M. Synthesis of tio2/wo3 composite nanofibers by a water-based electrospinning process and their application in photocatalysis. *Nanomaterials* **2020**, *10*, 882. [CrossRef]
18. Wang, L.; Nagesha, D.K.; Selvarasah, S.; Dokmeci, M.R.; Carrier, R.L. Toxicity of cdse nanoparticles in caco-2 cell cultures. *J. Nanobiotechnol.* **2008**, *6*, 1–15. [CrossRef] [PubMed]
19. Rodríguez, K.; Gatenholm, P.; Renneckar, S. Electrospinning cellulosic nanofibers for biomedical applications: Structure and in vitro biocompatibility. *Cellulose* **2012**, *19*, 1583–1598. [CrossRef]
20. Wang, H.; Kong, L.; Ziegler, G.R. Fabrication of starch-nanocellulose composite fibers by electrospinning. *Food Hydrocoll.* **2019**, *90*, 90–98. [CrossRef]
21. Teo, W.E.; Ramakrishna, S. A review on electrospinning design and nanofibre assemblies. *Nanotechnology* **2006**, *17*, R89. [CrossRef] [PubMed]
22. Sanjay, P.; Deepa, K.; Madhavan, J.; Senthil, S. Synthesis and spectroscopic characterization of cdse nanoparticles for photovoltaic applications. *MsE* **2018**, *360*, 012010. [CrossRef]

23. López-Herraiz, M.; Fernández, A.B.; Martinez, N.; Gallas, M. Effect of the optical properties of the coating of a concentrated solar power central receiver on its thermal efficiency. *Sol. Energy Mater. Sol. Cells* **2017**, *159*, 66–72. [CrossRef]

24. Wu, D.; Liang, J.; Zhang, D.; Zhang, C.; Zhu, H. Solar evaporation and electricity generation of porous carbonaceous membrane prepared by electrospinning and carbonization. *Sol. Energy Mater. Sol. Cells* **2020**, *215*, 110591. [CrossRef]

Communication

Physico-Chemically Distinct Nanomaterials Synthesized from Derivates of a Poly(Anhydride) Diversify the Spectrum of Loadable Antibiotics

Amalia Mira [1], Carlos Sainz-Urruela [1], Helena Codina [1], Stuart I. Jenkins [2], Juan Carlos Rodriguez-Diaz [3], Ricardo Mallavia [1,*] and Alberto Falco [1,*]

[1] Institute of Research, Development and Innovation in Biotechnology of Elche (IDiBE), Miguel Hernández University (UMH), 03202 Elche, Alicante, Spain; a.mira@umh.es (A.M.); carlos.sainz@goumh.umh.es (C.S.-U.); helena.codina@goumh.umh.es (H.C.)

[2] Neural Tissue Engineering group: Keele (NTEK), School of Medicine, Keele University, Keele ST5 5BG, Staffordshire, UK; s.i.jenkins@keele.ac.uk

[3] Microbiology Section, University General Hospital of Alicante, Alicante Institute for Health and Biomedical Research (ISABIAL-FISABIO Foundation), Alicante 03010, Spain; rodriguez_juadia@gva.es

* Correspondence: r.mallavia@umh.es (R.M.); alber.falco@umh.es (A.F.)

Received: 17 February 2020; Accepted: 6 March 2020; Published: 8 March 2020

Abstract: Recent advances in the field of nanotechnology such as nanoencapsulation offer new biomedical applications, potentially increasing the scope and efficacy of therapeutic drug delivery. In addition, the discovery and development of novel biocompatible polymers increases the versatility of these encapsulating nanostructures, enabling chemical properties of the cargo and vehicle to be adapted to specific physiological requirements. Here, we evaluate the capacity of various polymeric nanostructures to encapsulate various antibiotics of different classes, with differing chemical structure. Polymers were sourced from two separate derivatives of poly(methyl vinyl ether-*alt*-maleic anhydride) (PMVE/MA): an acid (PMVE/MA-Ac) and a monoethyl ester (PMVE/MA-Es). Nanoencapsulation of antibiotics was attempted through electrospinning, and nanoparticle synthesis through solvent displacement, for both polymers. Solvent incompatibilities prevented the nanoencapsulation of amikacin, neomycin and ciprofloxacin in PMVE/MA-Es nanofibers. However, all compounds were successfully loaded into PMVE/MA-Es nanoparticles. Encapsulation efficiencies in nanofibers reached approximately 100% in all compatible systems; however, efficiencies varied substantially in nanoparticles systems, depending on the tested compound (14%–69%). Finally, it was confirmed that both these encapsulation processes did not alter the antimicrobial activity of any tested antibiotic against *Staphylococcus aureus* and *Escherichia coli*, supporting the viability of these approaches for nanoscale delivery of antibiotics.

Keywords: biomaterials; polymers; PMVE/MA; electrospinning; nanofibers; nanoparticles; nanoencapsulation; antibiotics

---

1. Introduction

So far, antibiotics are the most reliable weapon against infectious diseases of bacterial origin. However, the dramatic diminishment in the rate of discovery of new antibiotics limits our response to pathogens resistant to conventional antibiotics and to new emerging diseases. In response, the scientific community is developing alternative strategies to increase the effectiveness, and/or to overcome the limitations, of existing antibiotics [1,2]. In this context, recent advances in the field of nanotechnology offer new tools such as nanoencapsulation: the loading of pharmaceutical agents within nanomaterials [3,4].

From a commercial and clinical point of view, nanoencapsulation can protect pharmaceutical products, extending both shelf-life and biological half-life (e.g., in circulation). In addition, with the corresponding modifications, these nanostructured systems can facilitate targeted drug delivery and/or specific controlled-release kinetics, thereby increasing the effectiveness of the treatment and, thus, reducing necessary dosage and side effects [3,4].

In line with this, the continuing development of novel biocompatible polymers contributes to the potential versatility of these nanostructures, by enabling the delivery of compounds with solubility limitations. In addition, manipulation of synthesis or layering protocols can generate nanostructures with properties tailored to highly specific applications, for example injectable nanoparticles, or nanofibers for wound dressings [5–8].

Examples of such highly versatile materials include derivates of poly(methyl vinyl ether-*alt*-maleic anhydride) (PMVE/MA), an alternating copolymer of methyl-ether-vinyl and maleic anhydride. This material is marketed by Ashland Inc. as Gantrez® and presents suitable properties for biomedical applications (low toxicity, high biocompatibility, high mucoadhesivity and low cost). In particular, it is reported that PMVE/MA can be structured as loadable nanoparticles [9] and its derivates poly(methyl vinyl ether-alt-maleic acid) (PMVE/MA-Ac) and poly(methyl vinyl ether-alt-maleic monoethyl ester) (PMVE/MA-Es) as nanofibers [10,11]. In addition, these polymers have shown utility as matrix elements, allowing the combination with other materials to generate novel mixtures of chemical properties in the final nanostructures [11–13], such as with homemade fluorescent cationic fluorene-based polyelectrolytes [11,14–16].

In this work, both nanoparticles and nanofibers were synthesized using these polymers, and their relative capacities for encapsulation were compared. There are several possible procedures for the synthesis of each of these nanomaterials. For instance, solvent displacement, emulsion solvent diffusion, interfacial deposition and nanoprecipitation synthesis for nanoparticles [17] and electrospinning, self-assembly, phase separation and template synthesis for nanofibers [18]. The two nanostructuring methodologies utilized here (solvent displacement and electrospinning, for particle and fiber synthesis, respectively), although differing markedly in terms of the processes involved, were selected because of the high degree of uniformity in their products, as well as being scalable processes for industrial production. The range of compounds tested as cargo encompassed four antibiotics of three different classes, differing in their molecular weights, structures and modes of action, i.e., two aminoglycosides (amikacin and neomycin), a cephalosporin (cefotaxime) and a quinolone (ciprofloxacin). Finally, the antimicrobial activity of the encapsulated compounds, as well as their structural stability, were also assessed, to confirm that nanoencapsulation was not an impairment to their antimicrobial potency.

2. Materials and Methods

2.1. Materials

The polymers PMVE/MA-Es (CAS number: 25087-06-3; MW: 130 kg/mol), provided as 50% *w/w* solution in ethanol, and PMVE/MA-Ac (CAS number: 25153-40-6; MW: 216 kg/mol), in powder format, were purchased from Sigma-Aldrich (St. Louis, MO, USA). Amikacin (Mw: 585.6 g/mol; 250 mg/mL) and cefotaxime (MW: 455.5 g/mol; powder) were acquired from Laboratorios Normon (Tres Cantos, Spain). Neomycin (MW: 614.6 g/mol; 10 mg/mL) was obtained from Sigma-Aldrich and ciprofloxacin (MW: 331.3 g/mol; 2 mg/mL) from Genéricos Españoles laboratorios S.A. (Las Rozas, Spain). The chemical structure inputs of all these compounds are shown in Figure 1.

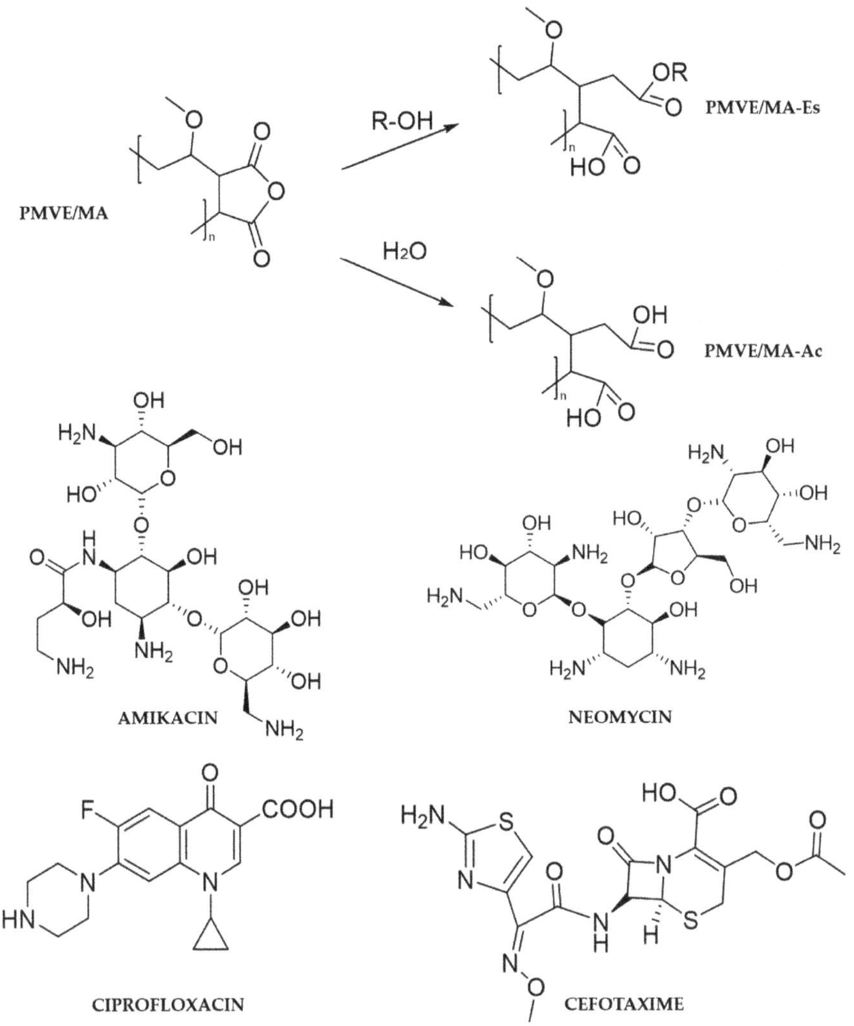

Figure 1. Structures of the principal compounds used. All structures were drawn using ChemBioDraw Ultra v14.0 (CambridgeSoft, Cambridge, MA, USA).

Phosphate buffered saline (PBS; pH 7.4), Mueller-Hinton broth (MHB; powder format), methanol, acetone, phosphoric acid and trifluoroacetic acid with HPLC grade purity, and dimethylsulfoxide (DMSO) and ethanol with purity >95%, were all from Sigma-Aldrich. The water used was double distilled and deionized (DDW) from a Milli-Q Synthesis A10 system (Millipore, Madrid, Spain).

2.2. Preparation of Nanofibers by Electrospinning

From our previous studies, either 20% *w/w* PMVE/MA-Ac in H₂O or 25% *w/w* PMVE/MA-Es in ethanol were selected as optimal polymeric solutions for the creation of electrospun nanofibers [10,11]. Antibiotics were added to the polymeric solutions to reach a final concentration of 1% *w/w* with respect to both PMVE/MA-Ac and PMVE/MA-Es. All solutions were stirred for 1 h and then bath-sonicated for 10 min prior to electrospinning.

As for the electrospinning system, the polymer solutions inside a 2-mL Discardit II syringe (Becton Dickinson, Franklin Lakes, NJ, USA) were pumped through a 20-Gauge blunt-end stainless steel hypodermic needle 316 (Sigma-Aldrich) at a constant flow rate by using a KDS 100 infusion pump (KD Scientific, Holliston, MA, USA). A Series FC high voltage supplier (Glassman High Voltage Inc., Whitehouse Station, NJ, USA) was responsible for the generation of the electrostatic field, which focused the jet onto a collector, made of aluminum foil and located in front of the syringe tip (vertical orientation). Any surface intended to be covered with a mat of electrospun nanofibers would be placed on the collector, for instance common glass slides for fluorescence/optical microscopy analysis or copper grids (diameter 3 mm) for transmission electron microscopy (TEM; Electron Microscopy Sciences, Hatfield, PA, USA). Operational parameters: 15 kV, flow rate 0.5 mL/h, needle-collector distance 15 cm, room temperature (RT) and relative humidity 40%–60%. Finally, in order to evaporate solvent excess, the obtained mats were kept in a fume hood overnight. Synthesized nanofibers were then stored protected from light at RT and dry atmosphere until used.

2.3. Preparation of Nanoparticles by Solvent Displacement

The PMVE/MA-Es nanoparticles were created using the solvent displacement procedure previously described by Arbós et al. (2002) [19] with some modifications. Briefly, different amounts of PMVE/MA-Es (12.5–200 mg) were dissolved in up to 5 g of ethanol by means of magnetic agitation for 10 min at RT. Subsequently, 2 µg of antibiotic in DDW were added (control nanoparticles were prepared only with DDW). After 10 min further magnetic stirring, gradual addition of DDW brought the final volume to 15 g. Then, the ethanol was evaporated under reduced pressure (BUCHI Rotavapor R-230, Flawil, Switzerland). Finally, obtained nanoparticles were purified twice by centrifugation (Sigma 3K30, Sigma Instruments, Osterode, Germany) at 20,000 g for 20 min at 4 °C. The pellet was then resuspended in up to 10 g of DDW and thus the final concentrations were 0.125%–2% w/w (or 1.25–20 mg/g) of PMVE/MA-Es and 200 ng/g of antibiotic. The supernatants were further centrifuged at 30,000 g for 30 min at 4 °C and collected again to quantify the amount of unloaded antibiotics. The purified nanoparticles and supernatants from all batches were stored at −20 °C until use.

2.4. Average Size and Zeta Potential Determination of Nanoparticles

Photon correlation spectroscopy (PCS), also known as dynamic light scattering (DLS), was used to determine the average hydrodynamic diameter (HDD) and polydispersity index (PDI) of each batch of nanoparticles (90 Plus Nanoparticle Size Analyzer; 35 mV red diode laser source, $\lambda = 640$ nm; Brookhaven Instruments Corporations, Holtsville, NY, USA). Each suspension sample was diluted with DDW to yield an appropriate scattering intensity of 100–400 kcps. All measurements were performed three times at 25 °C with angle detection fixed at 90° on 2 mL samples.

The zeta potential (ZP) of the synthesized nanoparticles was determined by electrophoretic laser Doppler anemometry with the 90 Plus Particle Size Analyzer (Brookhaven Instruments Corporation). Suspension samples were prepared as described before and analyzed in triplicate. All results are shown as the mean and standard deviation (s.d.) of the values obtained from three different batches.

2.5. Microscopy

For optical microscopy initial screenings, the nanofibers were electrospun on microscope slides (Deltalab, Barcelona, Spain) and observed by means of a Mycrosystems DMI3000B inverted fluorescence microscope equipped with an EL6000 compact light source and a DFC 3000G digital camera, all from Leica (Bensheim, Germany). All images were taken with a 63× objective in phase contrast and image processing performed manually using the software Leica Application SuiteAF6000 Module Systems.

Detailed observations of selected nanofiber samples were carried out by scanning electron microscopy (SEM), without metal coating, in a JSM-6360 LV device (Jeol, Tokyo, Japan). For size analysis, diameter measurements were performed on 100 nanofibers (minimum three micrographs) per nanofiber synthesis condition using ImageJ software (National Institutes of Health, NIH, Bethesda, MD, USA).

Transmission electron microscopy (TEM) was used to confirm the size and describe the morphology of the synthesized nanoparticles (Jeol 1011 apparatus, at 120 kV). Samples were placed onto Formvar/carbon 300-Mesh, copper grids (Electron Microscopy Science, Hatfield, PA, USA) and then incubated with citrate lead solution (0.03% p/v) to generate contrast in images. Diameter measurements were performed on 50 nanoparticles (minimum three micrographs) per nanoparticle synthesis condition using ImageJ software (National Institutes of Health).

2.6. HPLC Analysis

The quantitation of antibiotics was performed by HPLC in order to determine their encapsulation efficiency (EE, percentage of encapsulated compound relative to the theoretically maximum one) in both nanostructures. For nanoparticles, non-loaded compound, i.e., free in supernatants obtained during synthesis prior to purification, was quantitated then subtracted from the total added antibiotics to determine the amount of loaded compound. Nanoparticle supernatants or 0.1% *w*/*w* nanofiber solutions in DDW were filtered through a 0.4 μm PTFE membrane (Chmlab group, Barcelona, Spain) prior to volume injections (10 μL for amikacin and neomycin, 20 μL for ciprofloxacin and 5 μL for cefotaxime).

A Merck-Hitachi D-7000 HPLC system (Hitachi Instruments, Tokyo, Japan) equipped with an Alltech 3300 evaporative light scattering detector (ELSD; Alltech Associates Inc., Lokeren, Belgium) was used to analyze amikacin and neomycin. Cefotaxime and ciprofloxacin were analyzed by means of an Agilent LC 214 1100 series HPLC system controlled by ChemStation software and equipped with a G1311A quaternary pump, a G1329A ALS automated sample injector, a G1316A thermostat column compartment and a G1316A diode array detector (Agilent Technologies, Inc., Palo Alto, CA, USA).

All methods used were isocratic and their particularities for each antibiotic were as follows. For amikacin and neomycin, the mobile phase was acetone:DDW with 0.15% *v*/*v* TFA at a 1:1 ratio, a flow rate at 1.0 mL/min. ELSD conditions were nitrogen pressure at 3.5 bar and temperature at 45 °C. For cefotaxime, the mobile phase was methanol:DDW at 30:70 *v*/*v* (adjusted to pH 4 with acetic acid), a flow rate at 0.8 mL/min and detection at 235 nm. For ciprofloxacin, the mobile phase was 0.25 M phosphoric acid:acetonitrile at 75:25 *v*/*v*, a flow rate at 0.8 mL/min and detection at 280 nm. Standard curves for each antibiotic were generated using the same concentration range (6.25–300 μg/mL). Within this range, a linear correlation was found between concentration and detected peak area with regression coefficient values (R^2) greater than 0.99 (Figure S1).

2.7. Antibacterial Assays

The antimicrobial activity of the experimental formulations was tested on antibiotic-sensitive strains of Gram-positive *Staphylococcus aureus* (CECT 59) and Gram-negative *Escherichia coli* (CECT 515) obtained from the Spanish Type Culture Collection (Colección Española de Cultivos Tipo, CECT, Universitat de Valencia, Spain). Prior to each assay, a colony of bacteria previously grown in MHB-agar plates was isolated and incubated in MHB for 12 h at 37 °C to prepare bacteria inocula.

Minimal inhibitory concentration (MIC) was determined by the two-fold broth microdilution method according to the Clinical and Laboratory Standards Institute (CLSI) guidelines [20], with some modifications. Briefly, two-fold dilutions in MHB of the experimental formulations and the control antibiotics at twice the final concentration were prepared and 50 μL/well were dispensed in each column (1 column per concentration) of round-bottom 96-well polystyrene plates (Deltalab S.L., Rubí, Spain). Then, bacteria suspension, after adjustment to 0.5 McFarland and then further diluted by 1:100 in MHB, was added to all wells (50 μL/well), except for sterile controls. Plates were then incubated for 24 h at 37 °C and finally MIC was defined as the lowest concentration of antibiotic that visibly inhibited the growth of the bacterium being investigated. All assays were performed in triplicate and results are shown as the mean and s.d.

2.8. Data Analysis and Graphics

Data were analyzed and graphs produced using GraphPad Prism v6 and Microsoft Excel software. Statistical analysis was performed with GraphPad Prism v6.

3. Results

3.1. Electrospinnability of Formulations and Characterization of Obtained Nanofibers

In our previous works [10,11], the optimal conditions for the electrospinning of the polymer solutions for both of these PMVE/MA derivates were determined, and hence taken as starting points for the present study (briefly, 15 kV, flow rate of 0.5 mL/h, needle-collector distance of 15 cm and polymer concentrations of 20% *w/w* PMVE/MA-Ac in H_2O or 25% *w/w* PMVE/MA-Es in ethanol). In this study, such conditions proved suitable for generating morphologically uniform nanofibers when initial polymeric solutions were homogenous. However, amikacin, neomycin and ciprofloxacin were found to be insoluble in ethanol, and therefore insoluble in PMVE/MA-Es solutions. Therefore, nanofibers were obtained from PMVE/MA-Ac solutions in combination with all antibiotics, but PMVE/MA-Es solution was only used with cefotaxime.

From both optical microscopy (data not shown) and SEM images (Figure 2), the morphology of all nanofibers was observed to be uniform (no shape anomalies), continuous (appropriate length, no breaks) and with a smooth surface appearance (no visible pores).

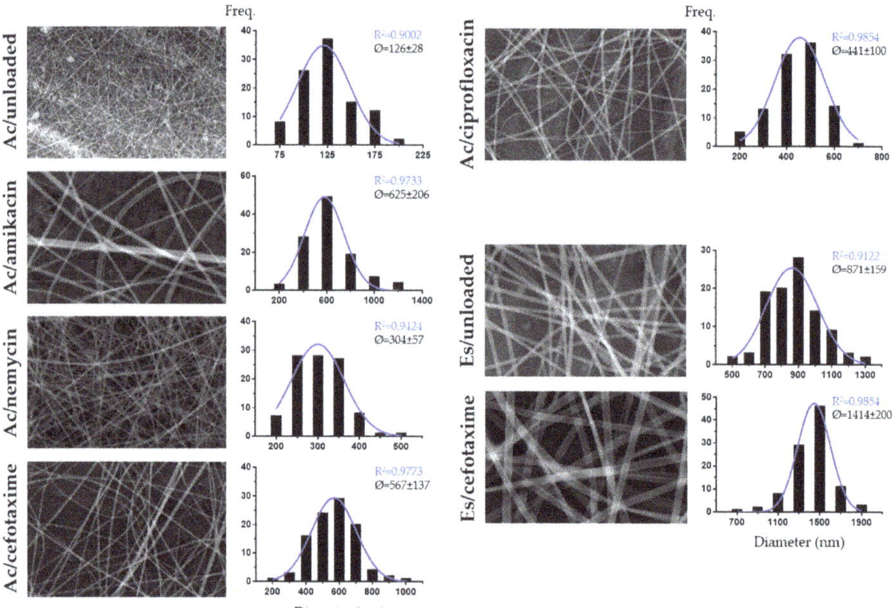

Figure 2. SEM analysis of antibiotic-loaded electrospun nanofibers. Representative SEM micrographs and corresponding diameter frequency histograms for PMVE/MA-Ac, PMVE/MA-Ac/amikacin, PMVE/MA-Ac/neomycin, PMVE/MA-Ac/cefotaxime, PMVE/MA-Ac/ciprofloxacin, PMVE/MA-Es and PMVE/MA-Es/cefotaxime. Histogram data were obtained from multiple micrographs (100 individual measurements). Best-fit adjustments (and their R^2) to a Gaussian distribution are indicated in blue. Average diameter (Ø) ± s.d. is also stated. Scale bar: 5 μm.

As SEM offers higher resolution and contrast imaging, as well as broader depth of field, these images were used for size analysis (Figure 2). All nanofiber types showed average diameter values < 1000 nm, except PMVE/MA-Es/cefotaxime (1414 ± 200 nm). PMVE/MA-Ac nanofibers with no encapsulated compound were the narrowest in diameter (126 ± 28 nm), which contrasts starkly with non-loaded PMVE/MA-Es fibers (871 ± 159 nm). The encapsulation of compounds increased the diameter of both PMVE/MA-Es and PMVE/MA-Ac nanofibers by a minimum of 1.6-fold (PMVE/MA-Es/cefotaxime) and a maximum of 5.0-fold (PMVE/MA-Ac/amikacin). Among PMVE/MA-Ac nanofibers, the smallest diameter increase was found when loading neomycin (2.4-fold), then ciprofloxacin (3.5-fold), with cefotaxime increasing the most (4.5-fold). In terms of size variability, PMVE/MA-Ac/amikacin, PMVE/MA-Ac/cefotaxime, PMVE/MA-Ac/ciprofloxacin and PMVE/MA-Es/cefotaxime nanofibers most closely fitted a Gaussian distribution ($R^2 > 0.95$), although all formulations showed R^2 values higher than 0.9. Of note, EEs reached maximum levels (≥97%) with relatively low s.d. (±2%) in all cases.

3.2. Optimization of the Preparation of PMVE/MA-Es Nanoparticles and Their Characterization

The low solubility in ethanol of most of the antibiotics tested, and the low solubility in DDW of PMVE/MA-Es, severely limited their compatibility for encapsulation in electrospun PMVE/MA-Es nanofibers. In contrast, encapsulation of these antibiotics in PMVE/MA-Es nanoparticles would appear feasible, since their preparation by the solvent displacement method requires, initially, both much lower polymer and ethanol concentrations (32.33% *w/w*). This method continues with the selective evaporation of ethanol to reach final nanoparticle suspensions in just DDW.

Preliminarily, the process of preparation of PMVE/MA-Es nanoparticles was optimized. For this purpose, several concentrations of PMVE/MA-Es were tested, ranging from 0.125% to 2% *w/w* (thus, 1.25–20 mg/g) in the final formulation. The physico-chemical parameters analyzed for these nanoparticles are summarized in Table 1. Overall, within the concentration range tested, greater polymer concentration was associated with larger nanoparticle diameter. No associations were noted between polymer concentration and either PDI or ZP parameters. PDI values ranged 0.06-0.20, indicating suspensions with high quality dispersion in general. In contrast, ZP values were more variable (from −5 to −35 mV), but only nanoparticles made of 10 and 20 mg/g PMVE/MA-Es displayed values yielding fairly good physical stability (≥30 mV approximately). Given these results, nanoparticles made of 10 mg/g PMVE/MA-Es were selected for further studies.

Table 1. Physico-chemical characteristics of nanoparticles made with different PMVE/MA-Es concentrations.

Conc. (mg/g)	HDD (nm)	PDI (a.u.)	ZP (mV)
1.25	84 ± 5	0.11 ± 0.04	−26 ± 10
2.5	139 ± 26	0.19 ± 0.03	−5 ± 4
5	106 ± 4	0.10 ± 0.04	−5 ± 6
10	202 ± 11	0.20 ± 0.06	−35 ± 9
20	230 ± 1	0.06 ± 0.01	−32 ± 1

Conc., concentration in the final formulation. a.u., arbitrary units. Results shown as the mean ± s.d. ($n = 3$).

Such nanoparticle formulations were then analyzed by TEM (Figure 3), which revealed consistent spherical shapes and, in contrast to PCS results, not only a lower average size (114 ± 24 nm), but also a less homogeneous population, as can be observed from the frequency histogram of their diameters.

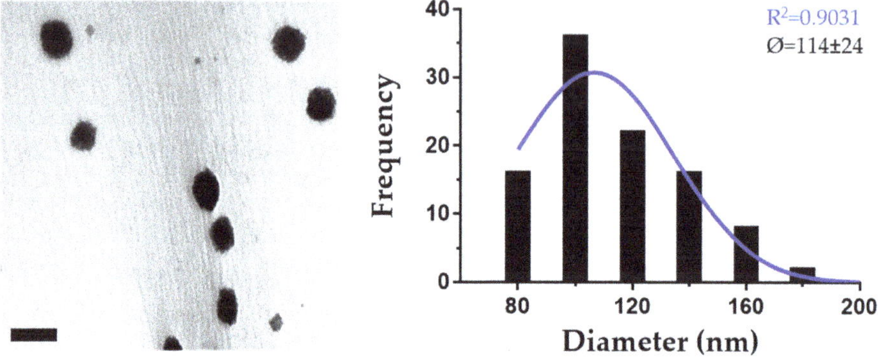

Figure 3. TEM analysis of optimized PMVE/MA-Es nanoparticles. Representative TEM image and diameter frequency histogram of PMVE/MA-Es nanoparticles synthesized by the solvent displacement method. The histogram was generated from data obtained from multiple images, until reaching 50 individual measurements. The best-fit adjustment (and R^2) to a Gaussian distribution is indicated in blue. Average diameter (Ø) and s.d. are also included in the inset. Scale bar: 200 nm.

3.3. Characterization of PMVE/MA-Es Nanoparticles Loaded with Antibiotics

By following the methodology described above, antibiotic-loaded PMVE/MA-Es nanoparticles were prepared containing 10 mg/g PMVE/MA-Es and 200 ng/g antibiotic (if 100% EE). The physico-chemical characteristics and EE displayed by these formulations are summarized in Table 2. Control (non-loaded) nanoparticles showed similar physico-chemical values as observed before (previous section), confirming the high reproducibility of the technique. As for loaded nanoparticles, there were significant compound-dependent variations in these parameters. For instance, both amikacin- and ciprofloxacin-loading was associated with greater particle size, but only the latter showed a notably change in ZP value (-16 ± 3 mV). Nanoparticles loaded with cefotaxime showed lower diameter (156 ± 6 nm) in comparison to control, and only neomycin-loaded particles showed no substantial differences compared to controls. Regarding their EE, only amikacin showed relatively low values ($14\% \pm 4\%$), with the other antibiotics ranging from 40% to 69%.

Table 2. Physico-chemical characteristics and encapsulation efficiency (EE) of PMVE/MA-Es nanoparticles loaded with antibiotics.

Antibiotic	HDD (nm)	PDI (a.u.)	ZP (mV)	EE (%)
None	195 ± 7	0.15 ± 0.04	-38 ± 13	-
Amikacin	253 ± 5	0.17 ± 0.07	-39 ± 2	14 ± 4
Neomycin	209 ± 4	0.20 ± 0.07	-29 ± 7	40 ± 3
Cefotaxime	156 ± 6	0.29 ± 0.03	-31 ± 8	69 ± 7
Ciprofloxacin	256 ± 8	0.17 ± 0.01	-16 ± 3	59 ± 8

a.u., arbitrary units. Results shown as the mean and s.d. ($n = 3$).

3.4. Analysis of the Antibacterial Activity

The antibacterial activity of encapsulated antibiotics was assessed against Gram-positive (*S. aureus*) and Gram-negative (*E. coli*) bacteria by determining their MIC values and comparing to those obtained for the free drug. The MIC values displayed by the encapsulated and free antibiotics were not significantly different (data not shown). As free compounds, MIC values for amikacin, neomycin, ciprofloxacin and cefotaxime were 1.25, 0.31, 2.5 and 0.16 µg/mL against *S. aureus* and 2.5, 1.25, 0.06 and 0.004 µg/mL against *E. coli*, respectively.

4. Discussion

The present work provides evidence of the versatility of polymers for generating nanostructures adapted to the solubility properties of the compounds of interest, for successful encapsulation of these compounds. Our previous studies provided the first demonstration of electrospinning procedures optimized for creating nanofibers of the two PMVE/MA derivates used here [10,11]. These were used to encapsulate 5-aminolevulinic acid (5-ALA) into nanofibers made of each polymer [11], and a combination of salicylic acid, methyl salicylate and capsaicin into PMVE/MA-Es nanofibers [10]. Here, we demonstrated the encapsulation of four structurally different molecules (with a common application: antibiotics) into electrospun nanofibers of either PMVE/MA-Ac or PMVE/MA-Es. All the antibiotics tested were soluble in water, and thus encapsulate into PMVE/MA-Ac nanofibers, but only cefotaxime was also soluble in ethanol, and consequently the only one loadable into PMVE/MA-Es nanofibers. In this regard, as many drugs are hydrophobic, the description of electrospinnable polymers with properties compatible for encapsulating hydrophobic compounds, is of great importance in order to incorporate such molecules into electrospun nanofibers while maintaining their biological functionality. This issue has been revisited by several authors recently [21,22]. Furthermore, these polymers might also work as a shell phase in both co-axial and layer-by-layer electrospinning in order to either manipulate the delivery dynamics of high hydrophilic molecules to the particular requirements or to adapt the delivery system to the needs of the biological context [23,24].

Although the morphology of the nanofibers was not affected by the encapsulation of molecules, their diameter did undergo considerable changes in comparison to non-loaded fibers. The extent of these increases in diameter was related to the compound used (1.6-fold for PMVE/MA-Es/cefotaxime nanofibers and 2.4–5.0-fold for loaded PMVE/MA-Ac nanofibers). These increases were larger than expected, as only 1% *w/w* antibiotic was used in all cases and loading percentages for relatively similar compounds are usually much higher with no, or even reduced, diameter [10,25,26]. In general, within moderate encapsulation levels, the diameter of electrospun nanofibers commonly increases with viscosity and decreases with conductivity [25–29], and since the compounds used here are not salts or polyelectrolytes, they might be increasing the viscosity of the electrospinnable solution [25,26]. In any case, the current study employs the same synthesis parameters for all formulations in order to reliably determine any possible morphological or size changes between the different antibiotics incorporated, and if it is desirable to strictly limit any increase in size, it might be achievable by optimizing such parameters for each particular formulation, as shown in our previous studies [10,11].

This work has shown the flexibility of nanostructuration of some polymers and offers alternative encapsulating nanomaterials for compounds whose solubility limitations hinder their biological/therapeutic applications. Although amikacin, neomycin and ciprofloxacin could not be loaded into PMVE/MA-Es nanofibers by electrospinning, the solubility of all integrating molecules with water:ethanol solutions allowed their encapsulation into PMVE/MA-Es nanoparticles, synthesized by the solvent displacement method. This methodology has been widely reported to create PMVE/MA nanoparticles for drug delivery purposes because of the interesting mucoadhesive properties of this polymer [19,30]; however, to our knowledge, there are no publications describing nanoparticles made of PMVE/MA-Es apart from the patent with reference number US20140161892A1 [31].

The optimized methodology for encapsulating particles (10 mg/g PMVE/MA-Es) produced polymeric nanoparticles with size (200 nm), PDI (0.2) and ZP (−35 mV). Briefly, at the concentrations tested, the greater the polymer concentration used, the larger the nanoparticle diameter produced, while formulations with smaller nanoparticles showed less adequate PDI and ZP values than the selected one. The PDI, which ranges from 0 to 1, reflects the width of the particle size distribution and values lower than 0.3 are commonly considered as optimum [32]. Regarding the ZP, which reflects the level of surface electrostatic potential, values of approximately −30 mV are associated with physical stability and are considered at the limit for colloidal organizations (theoretically, about −60 mV minimum is considered optimal) [33,34]. Comparing to these results, loaded nanoparticles did not differ much in the PDI, ZP and size values obtained, which is in accordance with some studies [35–39].

In contrast, the size of PMVE/MA-Es nanoparticles did change notably (increased by about 25%) for some antibiotics, although not as dramatically as in loaded PMVE/MA-Ac nanofibers. Interestingly, the greatest size increases were found when encapsulating amikacin into both types of nanostructures; however, no further correlations could be established for the remaining antibiotics. These size changes do not appear to be related to their MWs, nor to their EEs. For instance, despite the two aminoglycosides tested, amikacin and neomycin, presenting similar MWs (585.6 and 614.6 g/mol, respectively) and chemical structures (see Figure 1), the size change of their encapsulating nanostructures (625 vs. 304 nm in nanofibers and 253 vs. 209 nm in nanoparticles) and their EEs (14% vs. 40%) are very different. In this regard, the scarcity of nanoencapsulation studies for amikacin, neomycin cefotaxime and, to a lesser extent, ciprofloxacin [40–43], hampers any meaningful assessment or comparison.

Nevertheless, all tested antibiotics, when encapsulated into either nanomaterial, showed similar antibacterial activity as their free forms, indicating that none of the synthesis methodologies used substantially modified their chemical structures. This observation encourages the performance of follow-up experiments focused on assessing any potential advantages conferred by the described encapsulating nanomaterials in terms of controlled release, targeted delivery or other therapeutic impact. Any such advantages may help to overcome the drawbacks restricting the clinical use of some therapeutics (such as toxicity, poor penetration of biological barriers and short circulating half-life) [44].

5. Conclusions

The compatibility in solubility of both the polymer source and the compound of interest is a critical factor limiting the production of loaded polymeric electrospun nanofibers. However, it has been demonstrated here that the high versatility with which some polymers can be assembled into nanomaterials, in this work PMVE/MA-Ac and PMVE/MA-Es, offers novel encapsulating strategies to overcome such constraints. In this context, it has also been shown that the encapsulation efficiency, nanostructuration settings and/or the morphology/size of the final nanostructures are highly dependent upon the chemical properties, rather than the molecular weight, of the compounds to be encapsulated. In contrast, it is confirmed that none of the nanostructuring procedures employed alters the activity of a set molecules with different chemical structures.

Supplementary Materials: The following are available online at http://www.mdpi.com/2079-4991/10/3/486/s1, Figure S1: Representative HPLC chromatograms and calibration curves of tested antibiotics.

Author Contributions: Conceptualization, R.M. and A.F.; methodology, A.M., C.S.-U. and H.C.; software, A.M., C.S.-U. and A.F.; validation, A.M.; formal analysis, S.I.J., R.M. and A.F.; investigation, A.M., R.M. and A.F.; resources, J.C.R.-D., R.M. and A.F.; data curation, A.F.; writing—original draft preparation, A.F.; writing—review and editing, S.I.J.; visualization, A.F.; supervision, R.M. and A.F.; project administration, R.M.; funding acquisition, R.M. and A.F. All authors have read and agreed to the published version of the manuscript.

Funding: This research was funded by the Spanish Ministerio de Economía y Competitividad, grant numbers MAT-2017-86805-R and MAT-2014-53282-R, and Spanish Ministerio de Ciencia e Innovación (MCI)—Agencia Estatal de Investigación (AEI)/Fondo Europeo de Desarrollo Regional (FEDER), grant number RTI2018-101969-J-I00.

Acknowledgments: We thank Elisa Pérez for technical assistance.

Conflicts of Interest: The authors declare no conflict of interest.

References

1. Coates, A.R.; Halls, G.; Hu, Y. Novel classes of antibiotics or more of the same? *Br. J. Pharmacol.* **2011**, *163*, 184–194. [CrossRef] [PubMed]
2. Bassetti, M.; Merelli, M.; Temperoni, C.; Astilean, A. New antibiotics for bad bugs: Where are we? *Ann. Clin. Microbiol. Antimicrob.* **2013**, *12*, 22. [CrossRef] [PubMed]
3. Leucuta, S.E. Nanotechnology for delivery of drugs and biomedical applications. *Curr. Clin. Pharmacol.* **2010**, *5*, 257–280. [CrossRef] [PubMed]
4. Mahapatro, A.; Singh, D.K. Biodegradable nanoparticles are excellent vehicle for site directed in-vivo delivery of drugs and vaccines. *J. Nanobiotechnol.* **2011**, *9*, 55. [CrossRef]

5. Gunn, J.; Zhang, M. Polyblend nanofibers for biomedical applications: Perspectives and challenges. *Trends Biotechnol.* **2010**, *28*, 189–197. [CrossRef]
6. Vasita, R.; Katti, D.S. Nanofibers and their applications in tissue engineering. *Int. J. Nanomed.* **2006**, *1*, 15–30. [CrossRef]
7. Guo, G.; Fu, S.; Zhou, L.; Liang, H.; Fan, M.; Luo, F.; Qian, Z.; Wei, Y. Preparation of curcumin loaded poly(epsilon-caprolactone)-poly(ethylene glycol)-poly(epsilon-caprolactone) nanofibers and their in vitro antitumor activity against glioma 9l cells. *Nanoscale* **2011**, *3*, 3825–3832. [CrossRef]
8. Yoo, J.J.; Kim, C.; Chung, C.W.; Jeong, Y.I.; Kang, D.H. 5-aminolevulinic acid-incorporated poly(vinyl alcohol) nanofiber-coated metal stent for application in photodynamic therapy. *Int. J. Nanomed.* **2012**, *7*, 1997–2005.
9. Lucio, D.; Martinez-Oharriz, M.C.; Gonzalez-Navarro, C.J.; Navarro-Herrera, D.; Gonzalez-Gaitano, G.; Radulescu, A.; Irache, J.M. Coencapsulation of cyclodextrins into poly(anhydride) nanoparticles to improve the oral administration of glibenclamide. A screening on c. Elegans. *Colloids Surf. B Biointerfaces* **2018**, *163*, 64–72. [CrossRef]
10. Martinez-Ortega, L.; Mira, A.; Fernandez-Carvajal, A.; Mateo, C.R.; Mallavia, R.; Falco, A. Development of a new delivery system based on drug-loadable electrospun nanofibers for psoriasis treatment. *Pharmaceutics* **2019**, *11*, 14. [CrossRef]
11. Mira, A.; Mateo, C.R.; Mallavia, R.; Falco, A. Poly (methyl vinyl ether-alt-maleic acid) and ethyl monoester as building polymers for drug-loadable electrospun nanofibers. *Sci. Rep.* **2017**, *7*, 17205. [CrossRef] [PubMed]
12. Iglesias, T.; de Cerain, A.L.; Irache, J.; Martín-Arbella, N.; Wilcox, M.; Pearson, J.; Azqueta, A. Evaluation of the cytotoxicity, genotoxicity and mucus permeation capacity of several surface modified poly (anhydride) nanoparticles designed for oral drug delivery. *Int. J. Pharm.* **2017**, *517*, 67–79. [CrossRef] [PubMed]
13. Ruiz-Gaton, L.; Espuelas, S.; Larraneta, E.; Reviakine, I.; Yate, L.A.; Irache, J.M. Pegylated poly(anhydride) nanoparticles for oral delivery of docetaxel. *Eur. J. Pharm. Sci.* **2018**, *118*, 165–175. [CrossRef] [PubMed]
14. Vazquez-Guillo, R.; Martinez-Tome, M.J.; Kahveci, Z.; Torres, I.; Falco, A.; Mallavia, R.; Mateo, C.R. Synthesis and characterization of a novel green cationic polyfluorene and its potential use as a fluorescent membrane probe. *Polymers* **2018**, *10*, 938. [CrossRef] [PubMed]
15. Vázquez-Guilló, R.; Falco, A.; Martínez-Tomé, M.J.; Mateo, C.R.; Herrero, M.A.; Vázquez, E.; Mallavia, R. Advantageous microwave-assisted suzuki polycondensation for the synthesis of aniline-fluorene alternate copolymers as molecular model with solvent sensing properties. *Polymers* **2018**, *10*, 215. [CrossRef] [PubMed]
16. Kahveci, Z.; Vazquez-Guillo, R.; Martinez-Tome, M.J.; Mallavia, R.; Mateo, C.R. New red-emitting conjugated polyelectrolyte: Stabilization by interaction with biomolecules and potential use as drug carriers and bioimaging probes. *ACS Appl Mater. Interfaces* **2016**, *8*, 1958–1969. [CrossRef]
17. Kumari, A.; Yadav, S.K.; Yadav, S.C. Biodegradable polymeric nanoparticles based drug delivery systems. *Colloids Surf. B Biointerfaces* **2010**, *75*, 1–18. [CrossRef]
18. Sharma, J.; Lizu, M.; Stewart, M.; Zygula, K.; Lu, Y.; Chauhan, R.; Yan, X.; Guo, Z.; Wujcik, E.K.; Wei, S. Multifunctional nanofibers towards active biomedical therapeutics. *Polymers* **2015**, *7*, 186–219. [CrossRef]
19. Arbos, P.; Wirth, M.; Arangoa, M.A.; Gabor, F.; Irache, J.M. Gantrez an as a new polymer for the preparation of ligand-nanoparticle conjugates. *J. Control. Release* **2002**, *83*, 321–330. [CrossRef]
20. Cockerill, F.; Wikler, M.; Alder, J.; Dudley, M.; Eliopoulos, G.; Ferraro, M.; Hardy, D.; Hecht, D.; Hindler, J.; Patel, J. *Methods for Dilution Antimicrobial Susceptibility Tests for Bacteria That Grow Aerobically: Approved Standard*; Clinical and Laboratory Standards Institute: Wayne, PA, USA, 2012.
21. Goke, K.; Lorenz, T.; Repanas, A.; Schneider, F.; Steiner, D.; Baumann, K.; Bunjes, H.; Dietzel, A.; Finke, J.H.; Glasmacher, B.; et al. Novel strategies for the formulation and processing of poorly water-soluble drugs. *Eur. J. Pharm. Biopharm.* **2018**, *126*, 40–56. [CrossRef]
22. Yu, D.G.; Li, J.J.; Williams, G.R.; Zhao, M. Electrospun amorphous solid dispersions of poorly water-soluble drugs: A review. *J. Control. Release* **2018**, *292*, 91–110. [CrossRef] [PubMed]
23. Singh, A.; Rath, G.; Singh, R.; Goyal, A.K. Nanofibers: An effective tool for controlled and sustained drug delivery. *Curr. Drug. Deliv.* **2018**, *15*, 155–166. [CrossRef] [PubMed]
24. Son, Y.J.; Kim, W.J.; Yoo, H.S. Therapeutic applications of electrospun nanofibers for drug delivery systems. *Arch. Pharm. Res.* **2014**, *37*, 69–78. [CrossRef] [PubMed]
25. Shen, X.; Yu, D.; Zhu, L.; Branford-White, C.; White, K.; Chatterton, N.P. Electrospun diclofenac sodium loaded eudragit(r) l 100-55 nanofibers for colon-targeted drug delivery. *Int. J. Pharm.* **2011**, *408*, 200–207. [CrossRef]

26. Samprasit, W.; Akkaramongkolporn, P.; Ngawhirunpat, T.; Rojanarata, T.; Kaomongkolgit, R.; Opanasopit, P. Fast releasing oral electrospun pvp/cd nanofiber mats of taste-masked meloxicam. *Int. J. Pharm.* **2015**, *487*, 213–222. [CrossRef]

27. Nezarati, R.M.; Eifert, M.B.; Cosgriff-Hernandez, E. Effects of humidity and solution viscosity on electrospun fiber morphology. *Tissue Eng. Part C Methods* **2013**, *19*, 810–819. [CrossRef]

28. Reda, R.I.; Wen, M.M.; El-Kamel, A.H. Ketoprofen-loaded eudragit electrospun nanofibers for the treatment of oral mucositis. *Int. J. Nanomed.* **2017**, *12*, 2335. [CrossRef]

29. Canbolat, M.F.; Celebioglu, A.; Uyar, T. Drug delivery system based on cyclodextrin-naproxen inclusion complex incorporated in electrospun polycaprolactone nanofibers. *Colloids Surf. B Biointerfaces* **2014**, *115*, 15–21. [CrossRef]

30. Irache, J.M.; Huici, M.; Konecny, M.; Espuelas, S.; Campanero, M.A.; Arbos, P. Bioadhesive properties of gantrez nanoparticles. *Molecules* **2005**, *10*, 126–145. [CrossRef]

31. Salman, H.H.; Azcarate, I.G. Nanoparticles Comprising Esters of Poly (Methyl Vinyl Ether-Co-Maleic Anhydride) and Uses Thereof. U.S. Patent No. 9,351,940, 31 May 2016.

32. Kaur, I.P.; Bhandari, R.; Bhandari, S.; Kakkar, V. Potential of solid lipid nanoparticles in brain targeting. *J. Control. Release* **2008**, *127*, 97–109. [CrossRef]

33. Hunter, R.J. *Zeta Potential in Colloid Science: Principles and Applications*; Academic Press: Cambridge, MA, USA, 2013; Volume 2.

34. Kovacevic, A.; Savic, S.; Vuleta, G.; Muller, R.H.; Keck, C.M. Polyhydroxy surfactants for the formulation of lipid nanoparticles (sln and nlc): Effects on size, physical stability and particle matrix structure. *Int. J. Pharm.* **2011**, *406*, 163–172. [CrossRef]

35. Agüeros, M.; Zabaleta, V.; Espuelas, S.; Campanero, M.; Irache, J. Increased oral bioavailability of paclitaxel by its encapsulation through complex formation with cyclodextrins in poly (anhydride) nanoparticles. *J. Control. Release* **2010**, *145*, 2–8. [CrossRef] [PubMed]

36. Araujo, R.S.; Garcia, G.M.; Vilela, J.M.C.; Andrade, M.S.; Oliveira, L.A.M.; Kano, E.K.; Lange, C.C.; Brito, M.; Brandao, H.M.; Mosqueira, V.C.F. Cloxacillin benzathine-loaded polymeric nanocapsules: Physicochemical characterization, cell uptake, and intramammary antimicrobial effect. *Mater. Sci. Eng. C Mater. Biol. Appl.* **2019**, *104*, 110006. [CrossRef] [PubMed]

37. Brandhonneur, N.; Hatahet, T.; Amela-Cortes, M.; Molard, Y.; Cordier, S.; Dollo, G. Molybdenum cluster loaded plga nanoparticles: An innovative theranostic approach for the treatment of ovarian cancer. *Eur. J. Pharm. Biopharm.* **2018**, *125*, 95–105. [CrossRef]

38. Lopalco, A.; Ali, H.; Denora, N.; Rytting, E. Oxcarbazepine-loaded polymeric nanoparticles: Development and permeability studies across in vitro models of the blood-brain barrier and human placental trophoblast. *Int. J. Nanomed.* **2015**, *10*, 1985–1996.

39. Calleja, P.; Espuelas, S.; Vauthier, C.; Ponchel, G.; Irache, J.M. Controlled release, intestinal transport, and oral bioavailablity of paclitaxel can be considerably increased using suitably tailored pegylated poly(anhydride) nanoparticles. *J. Pharm. Sci.* **2015**, *104*, 2877–2886. [CrossRef]

40. Sabaeifard, P.; Abdi-Ali, A.; Soudi, M.R.; Gamazo, C.; Irache, J.M. Amikacin loaded plga nanoparticles against pseudomonas aeruginosa. *Eur. J. Pharm. Sci.* **2016**, *93*, 392–398. [CrossRef]

41. Zaki, N.M.; Hafez, M.M. Enhanced antibacterial effect of ceftriaxone sodium-loaded chitosan nanoparticles against intracellular salmonella typhimurium. *AAPS PharmSciTech* **2012**, *13*, 411–421. [CrossRef]

42. Mushtaq, S.; Khan, J.A.; Rabbani, F.; Latif, U.; Arfan, M.; Yameen, M.A. Biocompatible biodegradable polymeric antibacterial nanoparticles for enhancing the effects of a third-generation cephalosporin against resistant bacteria. *J. Med. Microbiol.* **2017**, *66*, 318–327. [CrossRef]

43. Sonam; Chaudhary, H.; Kumar, V. Taguchi design for optimization and development of antibacterial drug-loaded plga nanoparticles. *Int. J. Biol. Macromol.* **2014**, *64*, 99–105. [CrossRef]

44. Natan, M.; Banin, E. From nano to micro: Using nanotechnology to combat microorganisms and their multidrug resistance. *FEMS Microbiol. Rev.* **2017**, *41*, 302–322. [CrossRef] [PubMed]

 *nanomaterials*

Article

Micro- and Nanostructures of Agave Fructans to Stabilize Compounds of High Biological Value via Electrohydrodynamic Processing

Carla N. Cruz-Salas [1], Cristina Prieto [2], Montserrat Calderón-Santoyo [1], José M. Lagarón [2] and Juan A. Ragazzo-Sánchez [1,*]

[1] Laboratorio Integral de Investigación en Alimentos, Tecnológico Nacional de México/Instituto Tecnológico de Tepic, Av. Tecnológico 2595, Tepic C.P. 63175, Nayarit, Mexico; ccruz@ittepic.edu.mx (C.N.C.-S.); montserratcalder@gmail.com (M.C.-S.)

[2] Novel Materials and Nanotechnology Group, IATA-CSIC, Calle Catedrático Agustín Escardino Benlloch 7, 46980 Paterna, Spain; cprieto@iata.csic.es (C.P.); lagaron@iata.csic.es (J.M.L.)

* Correspondence: arturoragazzo@hotmail.com or jragazzo@ittepic.edu.mx; Tel.: +52-311-2119400 (ext. 233)

Received: 24 October 2019; Accepted: 12 November 2019; Published: 21 November 2019

Abstract: This study focuses on the use of high degree of polymerization agave fructans (HDPAF) as a polymer matrix to encapsulate compounds of high biological value within micro- and nanostructures by electrohydrodynamic processing. In this work, β-carotene was selected as a model compound, due to its high sensitivity to temperature, light and oxygen. Ultrafine fibers from HDPAF were obtained via this technology. These fibers showed an increase in fiber diameter when containing β-carotene, an encapsulation efficiency (EE) of 95% and a loading efficiency (LE) of 85%. The thermogravimetric analysis (TGA) showed a 90 °C shift in the β-carotene decomposition temperature with respect to its independent analysis, evidencing the HDPAF thermoprotective effect. Concerning the HDPAF photoprotector effect, only 21% of encapsulated β-carotene was lost after 48 h, while non-encapsulated β-carotene oxidized completely after 24 h. Consequently, fructans could be a feasible alternative to replace synthetic polymers in the encapsulation of compounds of high biological value.

Keywords: HDPAF; electrospinning; micro-nanofibers; β-carotene; thermoprotection; photoprotection

---

1. Introduction

Obtaining micro- and nanostructures in the food industry represents a viable option for the incorporation and stabilization of compounds of high biological value (CHBV). The production of these structures containing CHBV is based on the encapsulation processes in which the bioactive is surrounded by polymeric materials that act as matrices and help to preserve their properties. Currently, there is a huge interest in the use of biopolymers that replace synthetic polymers in the encapsulation and transport processes of compounds of high biological value, as well as expanding their use in biomedical and pharmaceutical technologies [1,2]. Fructans are a new class of heterodispersed biopolymers that are found in various plants, such as the *Agave tequilana* from the Agavaceae family, and serve as an important source of reserve carbohydrates in the plant. Five types of fructans have been identified in nature so far, which are classified into inulins, neo-inulins, levans, neo-levans and mixed fructans, depending on the type of bond [3,4]. Agave fructans are characterized by the presence of fructose units with a terminal glucose connected by bonds β (2-1) and β (2-6) and can present different degrees of polymerization, which are determined by the species [5–7] as well as the environmental conditions in which the agave is produced, stored and processed [8,9].

Toriz et al. (2007) [10] propose a chemical structure for the fructans of *Agave tequilana* Weber var. Azul, based on the combination of permethylation and reductive rupture techniques for identification.

They mainly propose the distribution of two monomers, the β-D fructofuranose terminal type (22%), 1-linked (30.8%), 6-linked (21.2%), 1-6 linked (19.8%) and the α-D terminal glucopiranose (7.3%).

The use of high degree of polymerization agave fructans (HDPAF) has been implemented in recent years as a wall material in encapsulation processes such as spray drying, obtaining remarkable results. Ortiz-Basurto et al. (2017) [11] microencapsulated pitanga juice (*Eugenia uniflora* L.) by spray drying with HDPAF and maltodextrin as encapsulants. These authors reported similar behaviors between both materials. Alternatively, Farías Cervantes et al. (2016) [12] used 50% agave fructans, as an encapsulant to obtain blackberry powder by spray drying, mentioning that agave fructans, in addition to increasing the encapsulation efficiency, could add prebiotic properties and improve the physico-chemical characteristics of the powder.

However, conventional encapsulation processes, such as spray drying, require the use of high temperatures or toxic reactives that could affect the CHBV or compromise its application in a food or pharmaceutical product. The electrohydrodynamic processing, electrospinning or electrospraying, has demonstrated to be a promising alternative to encapsulate and stabilize compounds of high biological value, since it does not require severe conditions of temperature, pressure or aggressive chemicals [13]. This technology allows obtaining structures called fibers or particles at the micron, submicron or nano scale [14]. This technique uses a potential difference for the electro-stretching of a drop of polymer solution from a charged electrode (capillary or free surface) to a collector. Once the drop has gained enough electrical charge to overcome the surface tension and viscosity of the polymer, the drop is stretched, causing ultra-fine fibers to emerge from the polymer drop, forming the so-called Taylor cone [15]. The extent of each phase, whether direct or in drag motion, depends on the operating parameters as well as the physical properties of the polymer, such as surface tension, conductivity and viscosity [16]. The difference between electrospinning and electrospraying (electrohydrodynamic spraying) is based on the degree of molecular cohesion, which can be easily controlled by variation in the concentration of the polymer solution [17]. Ramos-Hernández et al. (2018) [18] managed to obtain spherical structures with a size distribution of 440 to 880 nm, as well as the thermostability and photostability of β-carotene encapsulated by electrospraying, using solutions with concentrations from 10% to 50% of HDPAF.

Electro-spun nanofibers have also been used in CHBV encapsulation processes, such as the encapsulation of bioactive compounds in zein fibers with application in food and pharmaceutical products [19]; however, up to our knowledge, the production of electro-spun micro-nanofibers of HDPAF has not been reported yet. This material could be a feasible alternative to replace synthetic polymers in the encapsulation of compounds of high biological value, as well as expanding their use in biomedical and pharmaceutical products.

Antioxidants are a good example of CHBV due to the high number of health benefits attributed to them and their high sensitivity to physico-chemical factors [20]. Carotenoids are an important member of the antioxidant family, being natural organic pigments present in plants and some photosynthetic organisms. Their consumption is associated with health benefits by reducing the incidence of cancer and heart disease, as well as improving ocular health [21]. β-carotene is one of the most common carotenoids in the functionality of food, supplements and pharmaceuticals due to its high, pro-vitamin A activity and antioxidant capacity. However, its use is sometimes limited due to its sensitivity to oxidation, especially when exposed to high temperatures, light, oxygen, acidic conditions, etc. [19,22].

The aim of this study was to evaluate the feasibility of agave fructans to obtain micro-nanofibers through the electrospinning process, using β-carotene as a model compound, with properties and characteristics, as a first approach, and to provide stability to compounds of high biological value for further applications in the food and pharmaceutical areas.

2. Materials and Methods

2.1. Materials

High degree of polymerization agave fructans (HDPAF) were obtained in the Laboratorio Integral de Investigación en Alimentos (LIIA) of Tecnológico Nacional de México/Instituto Tecnológico de Tepic, Nayarit, México, from native fructans provided by the company Campos Azules Co. (Mexico City, Mexico); β-carotene (>97.0% UV, C40H56) (Sigma Aldrich, Steinheim, Germany), 96% ethyl alcohol (CTR Scientific, Monterrey N.L, Mexico), Chloroform (trichloromethane) HPLC grade (Fermon Episolv, Monterrey N.L, Mexico), Tego® SML (Evonik Industries AG, Essen, Germany).

2.2. Preparation and Characterization of Polymer Solutions

Polymeric solutions were prepared with 70% (*w/w*) of HDPAF in hydroalcoholic solution (ethanol-water, at 10% *w/w*) and 1% (*w/w*) of Tego® SML as surfactant was added. The solution was homogenized under magnetic stirring for 45 min at room temperature. In the alcoholic fraction of the solution, β-carotene was added (1% *w/w*), protecting the mixture from light during the homogenization period with dark paper.

The characterization of the solutions consisted in determining the electrical conductivity, viscosity and surface tension. The electrical conductivity was analyzed with a multiparameter potentiometer Hanna Instruments HI-4521 (Melrose, MA, USA). The viscosity (η) was determined with a Discovery HR-1 Hybrid rheometer (TA Instruments, New Castle, DE, USA), equipped with geometry Smart Swap ™ with automatic detection. The cone and plate geometry option was selected (2°, 60 mm of diameter, 64 μm of gap) and the Peltier system for temperature stabilization was used. The surface tension was measured with the equipment Force tensiometer model K20 EasyDyne (KRÜSS GmbH, Hamburg, Germany), with the Wilhelmy Plate method.

2.3. Obtaining Fibers by Electrospinning

The electrospinning process was performed in a machine LE-10 brand Fluidnatek® from BIOINICIA company (Valencia, Spain), which has a voltage power supply of 19 kV. The injector is based on a syringe pump with a flow from 200 μL/h, and has a distance from the injector to the collector of 20 cm, the collector is cylindrical stainless steel and it has a variable rotation speed from 500 rpm. The process parameters such as voltage, flow, distance and rotation speed of the collector were adjusted in preliminary tests according to the characterization and stability of the solutions.

2.4. Morphology Analysis through SEM

The morphology and size of the fibers obtained were both determined with the scanning electron microscopy (SEM) technique with a Hitachi-S-4800 device (Hitachi High-Technologies Corporation, Tokyo, Japan). Approximately 1.5 mg of sample was fixed with double-sided tape on the sample holder, coated with gold-palladium for 2 min, and an acceleration voltage of 10 kV was used. The determination of the size distribution based on the diameters of the structures was made with the SEM system software (Hitachi High-Technologies Corporation, Tokyo, Japan) with at least 100 measurements per sample.

2.5. Loading and Encapsulation Efficiency

The loading efficiency (LE) of β-carotene in the fibers was determined considering the amount of total extract and final extract by weight difference, obtained with thermogravimetry analysis (TGA) using the TRIOS software (TA Instruments, New Castle, DE, USA) and applying Equation (1). Where the total extract corresponds to the amount of extract added to the fibers, the final extract represents the amount of extract determined by TGA. Regarding the encapsulation efficiency (EE), initially the amount of extract loaded was determined with the application of Equation (2).

Subsequently, the amount of surface was calculated, performing a superficial washing of the fibers with a solvent related to β-carotene and not to the polymer, in order to determine the content of compounds of high biological value (CHBV) on the fiber surface. To this end, 1 mg of fibers was taken and suspended in 1 mL of trichloromethane, centrifuging at 10,000 rpm for 1 min, analyzing the supernatant obtained absorbance at 466 nm in a Varian brand Cary® 50 UV-Vis spectrophotometer (Sydney, Australia). The β-carotene content is calculated according to the calibration curve performed, y = 3.6298x + 0.0059 (R^2 = 0.9998). Finally, knowing the values of loading and surface extract, Equation (3) was applied to determine the encapsulation efficiency (EE).

Loading efficiency (LE)

$$\%LE = \left(1 - \left(\frac{total\ extract - final\ extract}{total\ extract}\right)\right) \times 100 \tag{1}$$

Loading extract

$$Loading\ extract = \frac{\%EC * total\ extract}{100} \tag{2}$$

Encapsulation efficiency (EE)

$$\%EE = \frac{loading\ extract - surface\ extract}{loading} \times 100 \tag{3}$$

2.6. Thermogravimetric Analysis (TGA)

The thermogravimetric analysis to determine the decomposition temperature of each component of the fiber, as well as to determine the LE and to demonstrate the thermoprotective effect of the materials on β-carotene, was performed on a TGA 550 (TA Instruments, New Castle, DE, USA) in a nitrogen atmosphere (N_2), with a heating ramp of 25 to 600 °C at a speed of 5 °C/min. The results were analyzed with the TRIOS software (TA Instruments, New Castle, DE, USA).

2.7. UV Photostability

In order to analyze the oxidation kinetics of β-carotene, the fibers obtained were exposed to a simulator of sunlight. An Osram Ultra-vitalux lamp (300 W) (OSRAM, Munich, Germany) was utilized, which generates a mixture of radiation, using a quartz discharge tube and a tungsten filament [19]. It was placed in a sample holder 1 mg of standard β-carotene, and samples of the fibers with and without β-carotene, which were exposed to ultraviolet (UV) radiation for 48 h at 37 °C. Samples were taken every 6 h. A central segment of the exposed material was cut and dissolved in distilled water at a ratio of 1 mg/mL with magnetic stirring for 1 min. Subsequently, 1 mL of chloroform was added and centrifuged (10,000 rpm, 1 min). The absorbance of the organic phase at 466 nm was measured with the Varian brand Cary® 50 UV-Vis spectrophotometer (Sydney, Australia). Chloroform was used as the target. The results, obtained in relation to oxidation, are reported based on the relative content of β-carotene (% absorbance).

2.8. Statistic Analysis

Data analysis was performed using the least significant digit (LSD) test for comparison of means with STATISTICA version 10 (StatSoft, Inc. (2011), (Tulsa, OK, USA)). All tests were performed in triplicate.

3. Results and Discussion

3.1. Physicochemical Characterization of Polymer Solutions

First, the physicochemical properties of the polymer solutions were evaluated, since the stability of the electrohydrodynamic process and the morphology obtained are highly related to them.

The solutions presented viscosity values of 3.69 ± 0.05 and 3.36 ± 0.03 Pa·s without and with β-carotene, respectively, as shown is Table 1. These results are in the same order of magnitude as those obtained by Kutzli et al. (2019) [23], at high concentrations of maltodextrin combined with whey protein isolated (WPI) or soy protein isolated (SPI). They obtained values of apparent viscosity of 4.85 ± 0.14 Pa·s (WPI 80:5), 5.88 ± 0.18 Pa·s (WPI 80:10), 5.14 ± 0.09 Pa·s (SPI 80:5) and 7.77 ± 0.12 Pa·s (SPI 80:10), respectively. The similarity of fructans with maltodextrin with respect to this property is mainly due to their structural conformation. Concerning the surface tension, similar values in the solutions with and without β-carotene were obtained (29.6 ± 0.2 and 30.1 ± 0.1 mN/m), showing that the β-carotene incorporation does not modify the penetration resistance of the solution ($p \leq 0.05$). Ramos-Hernández et al. (2018) [18] analyzed the surface tension of HDPAF solutions at different concentrations, obtaining a value of 23.46 mN/m for a 50% solution of the polymer, which differs from those obtained in this study. The difference could be associated with the type of surfactant used, but mainly at the concentration of the biopolymer (70%). In contrast, the addition of β-carotene to the polymer solution caused a significant increase in electrical conductivity from 5.54 ± 0.01 to 7.30 ± 0.03 mS/cm, and this can be attributed to the functional group loads of β-carotene (Table 1).

Table 1. Physicochemical characterization of the 70% high degree of polymerization agave fructans (HDPAF) polymer solution with and without β-carotene.

Parameter	70% HDPAF without β-Carotene	70% HDPAF with β-Carotene
Viscosity [1] (Pa·s)	3.69 ± 0.05 [a]	3.36 ± 0.03 [a]
Surface tension (mN/m)	30.1 ± 0.1 [b]	29.6 ± 0.2 [b]
Electrical conductivity (mS/cm)	5.54 ± 0.01 [c]	7.30 ± 0.03 [d]

Different letters within the same row indicate significant differences among samples ($\alpha = 0.05$). The average values were obtained from the analysis of three replicas. [1] Viscosity values were read at a shear rate of 39.8 s^{-1}.

In the electrospinning process, the high concentrations of HDPAF allowed the increase of intermolecular interactions of the polymer with the solvent, and in the same way, the polymer–polymer crosslinking, favoring the stability of the flow [17]. In addition, polymer chain interactions have a close relationship with concentration and viscosity [24]. However, a high viscosity can cause clogging of the injector partially or totally, but it can also affect the morphology of the fibers due to the presence of artifacts in the structures.

3.2. Micro-Nanofiber Formation Process

The stabilization of the electrospinning process was achieved with a voltage of 19 kV, 200 µL/h feed flow of the polymer solution, 20 cm distance from the injector to the collector with 500 rpm rotation. A concentration of the polymer of 70% (*w/w*) was used, which allowed the formation of a network of polymer chains and enough entanglement to electro-stretch the solution, obtaining continuous fibers (Figure 1).

In preliminary studies, the electrospinning process was carried out with a 60% solution (*w/w*) of HDPAF concentration. The presence of structures, with artifacts that cannot be considered as spheres or fibers, was observed (Figure 1a). On the contrary, for a concentration of 70% of HDPAF, it was possible to obtain a homogeneous film on the collector. This is directly related to the increase in concentration, the structure of high-polymerization fructans and the molecular arrangement of the terminal chains and interactions that occur during the electrospinning process.

Despite not finding references in literature about fiber formation with HDPAF, some authors such as Lee et al. (2009) [25]; Kai et al. (2015) [26] reported the use of polysaccharides such as alginate, cellulose, chitosan, starch in the formation of fibers by electrospinning, which could be used as natural encapsulants in the area of medicine. HDPAFs have just been used so far for the production of spherical nanocapsules ([18]), which were obtained using a HDPAF concentration of 30% through the electrospraying process to encapsulate β-carotene. The possibility of obtaining fibers could increase the application field of agave fructans to medicine or biomaterials.

Figure 1. Scanning electron microscopy (SEM) micrographs show structures obtained with (**a**) 60% HDPAF, (**b,c**) fibers obtained with 70% HDPAF without β-carotene at different magnification, (**d,e**) fibers obtained with 70% HDPAF loaded with β-carotene at different magnification, (**f**) distribution of micro-nanofiber diameters obtained with the 70% HDPAF formulation (■) without and (■) with β-carotene. The average values and standard deviation (SD) were obtained from the analysis of three replicas.

3.3. Morphology Analysis

Fibers obtained with solutions without β-carotene presented a smooth and continuous surface (Figure 1d) with a larger distribution of diameters, in the range of 750 to 1000 nm (Figure 1f). Structures obtained with β-carotene showed smooth, continuous surface fibers, with a slight fragmentation of some structures. This is possibly associated with the applied voltage and its effect on the breaking of the cross-links of the polymer chains. An increase in the diameter size distribution was observed, being in the range between 1250 and 1500 nm. The presence of pores and widenings in some segments was observed, which could be due to the lipophilic nature of β-carotene that limits cross-linking with the polymer or the relative humidity and vapor pressure of the solvent used. The high hygroscopicity of the HDPAF could also affect the diameter of the fibers. Bak et al. (2016) [27] analyzed the effect of relative humidity (30% and 60%) in the manufacture of collagen nanofibers. These authors reported that the quantity of humidity affects the morphological characteristics of the fibers obtained, and therefore concluded that at lower humidity the diameter of the fibers decreases.

3.4. Loading and Encapsulation Efficiency (LE and EE)

The encapsulation efficiency of the β-carotene inside the ultrathin fibers of HDPAF was 95%. This indicates that almost all the loaded compound was encapsulated in the center of the fibers. This value differs with the encapsulation efficiencies obtained with other polysaccharides, such as chitosan, where an encapsulation efficiency of β-carotene of 36% was obtained in the encapsulation by nanomicelles. However, it is similar to the encapsulation efficiency reported for the encapsulation of anthocyanins in xanthan gum in combination with starch (96%) by spray drying [28,29]. HDPAFs show similarities in their entrapment capacity compared to some proteins. In this sense, López-Rubio and Lagaron (2012) [30] report a 90% of encapsulation efficiency for the encapsulation of β-carotene in whey protein. Gomez-Estaca et al. (2012) [31] obtain an encapsulation efficiency between 85 and 90% in encapsules of curcumin into zein by electrospraying. The branched structure of the HDPAF favors the entrapment of the bioactive compound due to its available functional groups, forming bonds such as hydrogen bonds that stabilize the nucleus and maintain the compound of interest in the fiber. The loading efficiency was of 85% in the fibers based on thermogravimetric analysis.

3.5. Thermogravimetric Analysis

A thermogravimetric analysis was performed to the fibers as well as to each of the materials used to make the fibers, in order to study the thermal stability of the β-carotene inside the polymeric wall of HDPAF. The β-carotene displayed an initial decomposition temperature of 138 °C and a final temperature of 367 °C (Figure 2b), similar to the temperatures reported by Peinado et al. (2016) [32] when analyzing this powder compound with a maximum peak of decomposition at approximately 120 °C. In addition, there were two important variations in mass, which are related to the breakdown of the basic structure due to thermal energy, generating the formation of volatile organic compounds including 2-methyl-2-hepten-6-yne, 2-methyl-2-hepten-4-yne and β cyclocitral.

The thermal stability of the fibers in the absence of β-carotene (Figure 2b) was characterized by presenting a first weight loss of 5.0% at 100 °C, which is associated with the humidity content. The second variation occurred between 167 to 307 °C, attributed to HDPAF (57.91%), reaching the maximum decomposition temperature at 209 °C. This temperature is similar to that presented by the individually analyzed HDPAFs, which explains that the surfactant used does not modify the thermal properties of the polymer. Espinosa-Andrews and Urias-Silvas (2012) [33] reported decomposition temperatures of 200 to 222 °C for commercial agave fructans, with a similar thermal behavior to that reported in this study. Regarding the fibers with β-carotene, the thermogram showed three mass variations (Figure 2b).

The first corresponds to humidity (4.0%), the second, between 174 to 232 °C, is attributed to the breakdown of HDPAF according to Figure 2a, the third was in the temperature range of 232 to 288 °C, which was related with the breakdown of β-carotene, a higher degree of polymerization fructans and surfactant according to Figure 2b. This can be explained due to the nature of the compounds and their affinity between β-carotene and HDPAF. Likewise, the fibers with β-carotene showed a displacement of 90 °C in the initial decomposition temperature of this compound, demonstrating the thermoprotective effect of HDPAF.

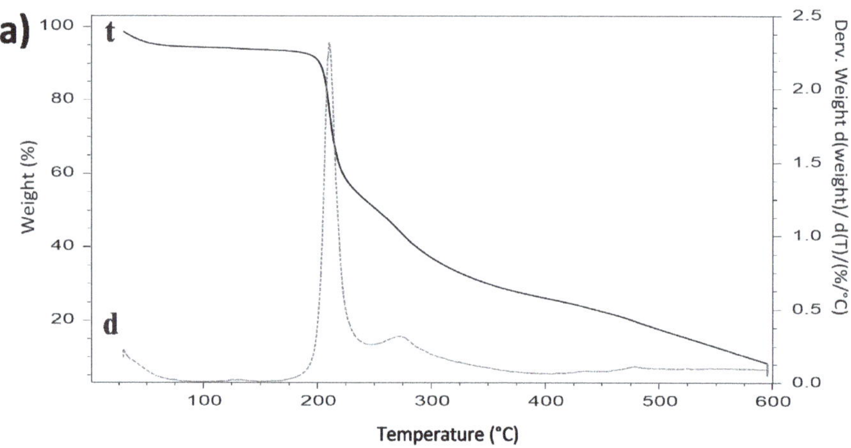
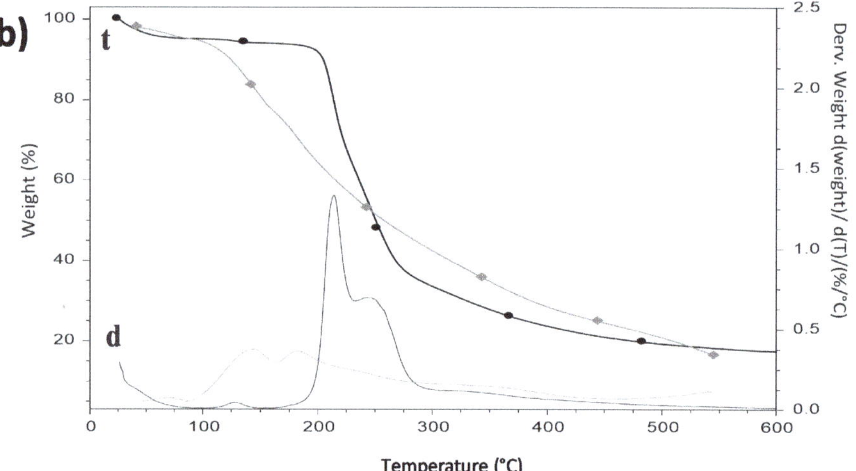

Figure 2. Thermograms of (**a**) HDPAF and (**b**) comparative (♦) β-carotene (●) fibers with β-carotene. (t, indicates thermograms and d, the derivatives of the thermograms).

3.6. Photostability Analysis

The need to evaluate the protection capacity of polymers used in encapsulation processes arises due to the susceptibility of photosensitive compounds to structural modification, such as β-carotene. The photostability of the fibers was evaluated under exposure of the fibers to ultraviolet A (UVA) light. According to the results, the photooxidation of uncapsulated β-carotene was presented from the beginning of the exposure, becoming total at 24 h.

On the contrary, when β-carotene was encapsulated within the fibers, the loss of its relative content was only 21% until 48 h (Figure 3). This behavior shows the ability of HDPAFs to reduce the photooxidation process, thus, prolonging the stability of β-carotene. Ramos-Hernández et al. (2018) [18] prepared capsules with HDPAF using a polymer concentration of 30% and 0.1% of β-carotene, and the results in the oxidation of β-carotene encapsulated with the electrospraying method showed a loss of 10% in capsules within 48 h. If the protective effect of both morphologies is compared, it is possible to observe that the fibers show twice the oxidation than the capsules, although the β-carotene content is 10 times higher in the fibers. The loss of photostability could be attributed to a reduced biopolymer-bioactive ratio, as well as to an effect of morphology, since the fibers have a larger surface exposed to light. For this reason, taking into account the amount of encapsulated β-carotene and the encapsulation efficiency obtained, the photostabilizing effect of the fiber could be considered very adequate.

In comparison with synthetic polymers, the behavior of the fibers with β-carotene is similar to that presented with the use of HDPAF. Peinado et al. (2016) [32] report a 20% loss of the relative content of the bioactive compound when exposed to UVA fibers made with polyethylene oxide (PEO) attributed to the stability of β-carotene once encapsulated, reflecting the polymer ability to limit oxygen diffusion and reduce exposure to light. On the other hand, de Freitas Zômpero et al. (2015) [34] obtain a 20% reduction after 6 h of exposure to UV light in fibers made with polyvinyl alcohol (PVOH) loaded with nanoliposomes, with the aim of analyzing the stability of the β-carotene. Whereas, the fibers in the present study show a reduction of 7% in the same period of time, which proves that HDPAFs have the ability to protect photosensitive compounds.

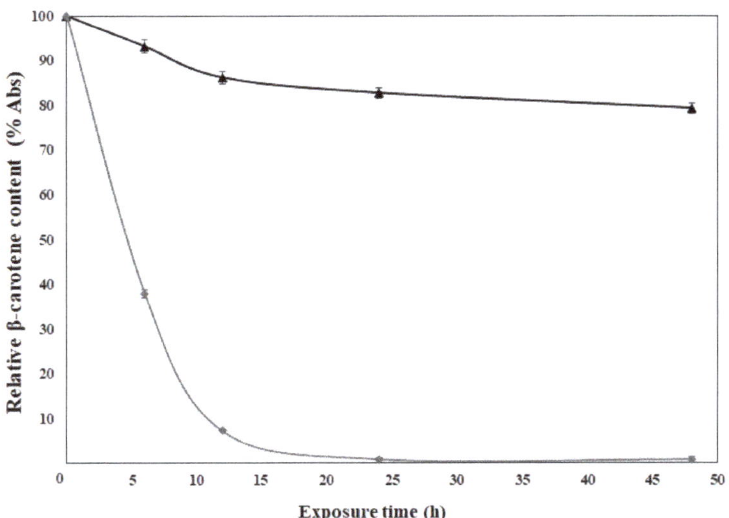

Figure 3. Photostability kinetics of the β-carotene (♦) and 70% HDPAF micro-nanofibers loaded with β-carotene (▲).

4. Conclusions

To the best of our knowledge, this paper reports for the very first time the possibility of generating continuous micro- and nanometric scale fibers from agave fructans of high degree of polymerization in 70% (*w/w*) solutions. This document also reports the possibility to encapsulate a bioactive model compound, such as β-carotene inside the fibers, obtaining a high encapsulation efficiency as well as providing stability to both temperature and UVA radiation.

The good results provided by fructans could suppose the expansion of its use and exploitations, since they could be a feasible alternative to replace synthetic polymers in the encapsulation of compounds of high biological value, as well as expanding their use in biomedical and pharmaceutical products. In this spirit, it is of important interest to consider the study of possible complexes with other polymers that help enhance their characteristics and properties, such as its high hygroscopicity, and structures more homogeneous, as well as their incursion into new applications.

Author Contributions: Conceptualization by J.A.R.-S. and J.M.L.; methodology, validation, and formal analysis was carried out by C.N.C.-S., J.A.R.-S., M.C.-S., C.P., and J.M.L.; investigation, resources, data curation, writing—original draft preparation and writing—review and editing was performed by C.N.C.-S., J.A.R.-S., M.C.-S., C.P., and J.M.L.; visualization, supervision, project administration, funding acquisition was carried out by J.A.R.-S. and J.M.L.

Funding: This study was supported by the Consejo Nacional de Ciencia y Tecnoligía (CONACyT, Mexico) for the scholarship granted (Number 702624) to Carla Norma Cruz Salas; CYTED thematic network code 319RT0576 and Spanish Ministry of Science, Innovation and Universities (proyect code RTI2018-097249-B-C21).

Acknowledgments: We thank Rosa Isela Ortiz-Basurto for providing the high degree of polymerization Agave fructans (HDPAF), Jorge Alberto Ramos-Hernandez and Elda Margarita Gonzalez-Cruz for your technical assistance.

References

1. Apolinário, A.C.; De Lima Damasceno, B.P.G.; De Macêdo Beltrão, N.E.; Pessoa, A.; Converti, A.; Da Silva, J.A. Inulin-Type Fructans: A Review on Different Aspects of Biochemical and Pharmaceutical Technology. *Carbohydr. Polym.* **2014**, *101*, 368–378. [CrossRef]

2. Saénz, C.; Tapia, S.; Chávez, J.; Robert, P. Microencapsulation by Spray Drying of Bioactive Compounds from Cactus Pear (Opuntia Ficus-Indica). *Food Chem.* **2009**, *114*, 616–622. [CrossRef]

3. Moreno-Vilet, L.; Bostyn, S.; Flores-Montaño, J.L.; Camacho-Ruiz, R.M. Size-Exclusion Chromatography (HPLC-SEC) Technique Optimization by Simplex Method to Estimate Molecular Weight Distribution of Agave Fructans. *Food Chem.* **2017**, *237*, 833–840. [CrossRef] [PubMed]

4. Waleckx, E.; Gschaedler, A.; Colonna-Ceccaldi, B.; Monsan, P. Hydrolysis of Fructans from Agave Tequilana Weber Var. Azul during the Cooking Step in a Traditional Tequila Elaboration Process. *Food Chem.* **2008**, *108*, 40–48. [CrossRef]

5. García Gamboa, R.; Ortiz Basurto, R.I.; Calderón Santoyo, M.; Bravo Madrigal, J.; Ruiz Álvarez, B.E.; González Avila, M. In Vitro Evaluation of Prebiotic Activity, Pathogen Inhibition and Enzymatic Metabolism of Intestinal Bacteria in the Presence of Fructans Extracted from Agave: A Comparison Based on Polymerization Degree. *LWT Food Sci. Technol.* **2018**, *92*, 380–387. [CrossRef]

6. Farías-Cervantes, V.S.; Chávez-Rodríguez, A.; Delgado-Licon, E.; Aguilar, J.; Medrano-Roldan, H.; Andrade-González, I. Effect of Spray Drying of Agave Fructans, Nopal Mucilage and Aloe Vera Juice. *J. Food Process. Preserv.* **2017**, *41*, e13027. [CrossRef]

7. Lopez, M.G.; Mancilla-Margalli, N.A.; Mendoza-Diaz, G. Molecular Structures of Fructans from Agave Tequilana Weber Var. Azul. *J. Agric. Food Chem.* **2003**, *51*, 7835–7840. [CrossRef]

8. Da Fonseca Contado, E.W.N.; de Rezende Queiroz, E.; Rocha, D.A.; Fraguas, R.M.; Simao, A.A.; Botelho, L.N.S.; de Fatima Abreu, A.; de Abreu, M.A.B.C.M.P. Extraction, Quantification and Degree of Polymerization of Yacon (Smallanthus Sonchifolia) Fructans. *Afr. J. Biotechnol.* **2015**, *14*, 1783–1789. [CrossRef]

9. Carranza, C.O.; Fernandez, A.Á.; Bustillo Armendáriz, G.R.; López-Munguía, A. *Processing of Fructans and Oligosaccharides from Agave Plants*; Elsevier Inc.: Amsterdam, The Netherlands, 2014. [CrossRef]

10. Toriz, G.; Delgado, E.; Zúñiga, V. A Proposed Chemical Structure for Fructans from Blue Agave Plant (Tequilana Weber Var. Azul). *Rev. Electrónica Y Tecnológica e-Gnosis* **2007**, *5*, 1.

11. Ortiz-Basurto, R.I.; Rubio-Ibarra, M.E.; Ragazzo-Sanchez, J.A.; Beristain, C.I.; Jiménez-Fernández, M. Microencapsulation of Eugenia Uniflora L. Juice by Spray Drying Using Fructans with Different Degrees of Polymerisation. *Carbohydr. Polym.* **2017**, *175*, 603–609. [CrossRef]

12. Farías Cervantes, V.S.; Delgado Lincon, E.; Solís Soto, A.; Medrano Roldan, H.; Andrade González, I. Effect of Spray Drying Temperature and Agave Fructans Concentration as Carrier Agent on the Quality Properties of Blackberry Powder. *Int. J. Food Eng.* **2016**, *12*, 451–459. [CrossRef]

13. Wen, P.; Zong, M.H.; Linhardt, R.J.; Feng, K.; Wu, H. Electrospinning: A Novel Nano-Encapsulation Approach for Bioactive Compounds. *Trends Food Sci. Technol.* **2017**, *70*, 56–68. [CrossRef]

14. Anu Bhushani, J.; Anandharamakrishnan, C. Electrospinning and Electrospraying Techniques: Potential Food Based Applications. *Trends Food Sci. Technol.* **2014**, *38*, 21–33. [CrossRef]

15. Haider, A.; Haider, S.; Kang, I.K. A Comprehensive Review Summarizing the Effect of Electrospinning Parameters and Potential Applications of Nanofibers in Biomedical and Biotechnology. *Arab. J. Chem.* **2018**, *11*, 1165–1188. [CrossRef]

16. Le Corre-Bordes, D.; Hofman, K.; Hall, B. Guide to Electrospinning Denatured Whole Chain Collagen from Hoki Fish Using Benign Solvents. *Int. J. Biol. Macromol.* **2018**, *112*, 1289–1299. [CrossRef] [PubMed]

17. Ghorani, B.; Tucker, N. Fundamentals of Electrospinning as a Novel Delivery Vehicle for Bioactive Compounds in Food Nanotechnology. *Food Hydrocoll.* **2015**, *51*, 227–240. [CrossRef]

18. Ramos-Hernández, J.; Ragazzo-Sánchez, J.; Calderón-Santoyo, M.; Ortiz-Basurto, R.; Prieto, C.; Lagaron, J. Use of Electrosprayed Agave Fructans as Nanoencapsulating Hydrocolloids for Bioactives. *Nanomaterials* **2018**, *8*, 868. [CrossRef]

19. Fernandez, A.; Torres-Giner, S.; Lagaron, J.M. Novel Route to Stabilization of Bioactive Antioxidants by Encapsulation in Electrospun Fibers of Zein Prolamine. *Food Hydrocoll.* **2009**, *23*, 1427–1432. [CrossRef]

20. Ozkan, G.; Franco, P.; De Marco, I.; Xiao, J.; Capanoglu, E. A Review of Microencapsulation Methods for Food Antioxidants: Principles, Advantages, Drawbacks and Applications. *Food Chem.* **2019**, *272*, 494–506. [CrossRef]

21. Fu, D.; Deng, S.; McClements, D.J.; Zhou, L.; Zou, L.; Yi, J.; Liu, C.; Liu, W. Encapsulation of β-Carotene in Wheat Gluten Nanoparticle-Xanthan Gum-Stabilized Pickering Emulsions: Enhancement of Carotenoid Stability and Bioaccessibility. *Food Hydrocoll.* **2019**, *89*, 80–89. [CrossRef]

22. González-Reza, R.M.; Quintanar-Guerrero, D.; Flores-Minutti, J.J.; Gutiérrez-Cortez, E.; Zambrano-Zaragoza, M.L. Nanocapsules of β-Carotene: Thermal Degradation Kinetics in a Scraped Surface Heat Exchanger (SSHE). *LWT Food Sci. Technol.* **2015**, *60*, 124–130. [CrossRef]

23. Kutzli, I.; Gibis, M.; Baier, S.K.; Weiss, J. Electrospinning of Whey and Soy Protein Mixed with Maltodextrin—Influence of Protein Type and Ratio on the Production and Morphology of Fibers. *Food Hydrocoll.* **2019**, *93*, 206–214. [CrossRef]

24. Hulsey, S.; Absar, S.; Choi, H. Comparative Study of Polymer Dissolution Techniques for Electrospinning. *Procedia Manuf.* **2017**, *10*, 652–661. [CrossRef]

25. Lee, K.Y.; Jeong, L.; Kang, Y.O.; Lee, S.J.; Park, W.H. Electrospinning of Polysaccharides for Regenerative Medicine. *Adv. Drug Deliv. Rev.* **2009**, *61*, 1020–1032. [CrossRef] [PubMed]

26. Kai, D.; Liow, S.S.; Loh, X.J. Biodegradable Polymers for Electrospinning: Towards Biomedical Applications. *Mater. Sci. Eng. C* **2015**, *45*, 659–670. [CrossRef] [PubMed]

27. Bak, S.Y.; Yoon, G.J.; Lee, S.W.; Kim, H.W. Effect of Humidity and Benign Solvent Composition on Electrospinning of Collagen Nanofibrous Sheets. *Mater. Lett.* **2016**, *181*, 136–139. [CrossRef]

28. Cai, X.; Du, X.; Cui, D.; Wang, X.; Yang, Z.; Zhu, G. Improvement of Stability of Blueberry Anthocyanins by Carboxymethyl Starch/Xanthan Gum Combinations Microencapsulation. *Food Hydrocoll.* **2019**, *91*, 238–245. [CrossRef]

29. Ge, W.; Li, D.; Chen, M.; Wang, X.; Liu, S.; Sun, R. Characterization and Antioxidant Activity of β-Carotene Loaded Chitosan-Graft-Poly (Lactide) Nanomicelles. *Carbohydr. Polym.* **2015**, *117*, 169–176. [CrossRef]

30. López-Rubio, A.; Lagaron, J.M. Whey Protein Capsules Obtained through Electrospraying for the Encapsulation of Bioactives. *Innov. Food Sci. Emerg. Technol.* **2012**, *13*, 200–206. [CrossRef]

31. Gomez-Estaca, J.; Balaguer, M.P.; Gavara, R.; Hernandez-Munoz, P. Formation of Zein Nanoparticles by Electrohydrodynamic Atomization: Effect of the Main Processing Variables and Suitability for Encapsulating the Food Coloring and Active Ingredient Curcumin. *Food Hydrocoll.* **2012**, *28*, 82–91. [CrossRef]

32. Peinado, I.; Mason, M.; Romano, A.; Biasioli, F.; Scampicchio, M. Stability of β-Carotene in Polyethylene Oxide Electrospun Nanofibers. *Appl. Surf. Sci.* **2016**, *370*, 111–116. [CrossRef]

33. Espinosa-Andrews, H.; Urias-Silvas, J.E. Thermal Properties of Agave Fructans (Agave Tequilana Weber Var. Azul). *Carbohydr. Polym.* **2012**, *87*, 2671–2676. [CrossRef]
34. De Freitas Zômpero, R.H.; López-Rubio, A.; de Pinho, S.C.; Lagaron, J.M.; de la Torre, L.G. Hybrid Encapsulation Structures Based on β-Carotene-Loaded Nanoliposomes within Electrospun Fibers. *Colloids Surf. B Biointerfaces* **2015**, *134*, 475–482. [CrossRef] [PubMed]

 *nanomaterials*

Article

Preparation, Characterization and Properties of Porous PLA/PEG/Curcumin Composite Nanofibers for Antibacterial Application

Feifei Wang [†], Zhaoyang Sun [†], Jing Yin [†] and Lan Xu *

National Engineering Laboratory for Modern Silk, College of Textile and Engineering, Soochow University, 199 Ren-ai Road, Suzhou 215123, China; ffwang@stu.suda.edu.cn (F.W.); zhaoyangsun1991@163.com (Z.S.); 20184215040@stu.suda.edu.cn (J.Y.)
* Correspondence: lanxu@suda.edu.cn; Tel.: +86-512-65884521
† F.W., Z.S. and J.Y. contributed equally to this paper.

Received: 23 February 2019; Accepted: 21 March 2019; Published: 2 April 2019

Abstract: Polylactide/polyethylene glycol/curcumin (PLA/PEG/Cur) composite nanofibers (CNFs) with varying ratios of PEG were successfully fabricated by electrospinning. Characterizations of the samples, such as the porous structure, crystalline structure, pore size, wetting property and Cur release property were investigated by a combination of scanning electron microscopy (SEM), Fourier-transform infrared (FTIR) spectroscopy, X-ray diffraction (XRD) and UV spectrophotometer. The antibacterial properties of the prepared porous CNFs against *Escherichia coli* bacteria were studied. The results showed that with the decrease of PEG in the CNFs, there appeared an evident porous structure on the CNF surface, and the porous structure could enhance the release properties of Cur from the CNFs. When the weight ratio (PEG:PLA) was 1:9, the pore structure of the nanofiber surface became most evident and the amount of Cur released was highest. However, the antibacterial effect of nonporous CNFs was better due to burst release over a short period of time. That meant that the porous structure of the CNFs could reduce the burst release and provide better control over the drug release.

Keywords: electrospinning; curcumin; PLA/PEG/curcumin nanofiber; drug release; porous nanofiber

---

1. Introduction

Electrospinning has been recognized as a simple and efficient technique for the fabrication of composite nanofibers (CNFs). The application of electrospun nanofibers can be improved by increasing the specific surface area and porosity of the nanofibers. Because of high porosity, high specific surface area and high surface activity, porous nanofibers have many existing and potential applications in drug delivery, tissue engineering, electronic engineering, and so on [1–5].

Curcumin (Cur) is a polyphenol compound extracted from the rhizome of the plant *Curcuma longa*. Recently, it was reported that Cur has a wide range of pharmacologic activities, such as anti-inflammation, anti-human immunodeficiency virus (HIV), anti-microbial, anti-oxidant, anti-parasitic, anti-mutagenic and anti-cancer, with low or no intrinsic toxicity [6–8]. Polylactic acid (PLA) has been widely used in biomedical and tissue engineering because of its excellent biocompatibility and biodegradability [9]. Lu et al. [10] investigated the effects of electrospraying a solution system and concentration of polyethylene oxide (PEO) on the morphologies of the single-drug and dual-drug loaded nanocomposites. Zhang et al. [11] reported that the polyethylene glycolation (PEGylation) modification could afford a faster release profile of the encapsulated drug than pure poly(lactic-co-glycolic acid) (PLGA) nanofibers, and the drug-loaded PLGA-PEG nanofibers were able to inhibit the growth of a model bacterium, *Staphylococcus aureus*. Ramírez-Agudelo et al. [12] found that composite nanofibers were easier to obtain an antibacterial property compared to a single polymer system. And the antibacterial composite with bacterial cellulose could prolong the antimicrobial activity against both *E. coli* and *S. aureus* [13]. Phan et al. [14] prepared and characterized a series of polymeric micellar formulations of Cur for targeted cancer therapy. Aytac et al. [15] designed core-shell nanofibers via coaxial-electrospinning using an inclusion complex of curcumin with cyclodextrin in the core and polylactide (PLA) in the shell. Zhong et al. [16] prepared PLA and a PLA-b-PEG composite porous scaffold loaded with a high dose of aspirin, using the solvent casting/particulate leaching technique, and found that the amphiphilic block polymer could efficiently enhance the dispersion property and stabilize the release of hydrophilic drugs. Shaik et al. [17] investigated the preventive role of orally administered *Aloe vera* supplemented probiotic *lassi* (APL) on *Shigella dysenteriae* infection in mice, and found that the immunoprotective effects of APL against *Shigella dysenteriae* induced infection in mice. Feng et al. [18] used tea polyphenol (TP) as an additive to develop to inhibit the spoilage of fish fillet during cold storage.

In this paper, electrospun PLA/PEG/Cur porous CNFs were fabricated by controlling the weight ratio of PEG:PLA. The effects of this ratio on the morphology and porous structure of CNFs were studied by scanning electron microscopy (SEM) and capillary flow porometry. Successful entrapments of Cur in the PEG/PLA CNFs were validated by Fourier-transform infrared (FTIR) spectroscopy and X-ray diffraction (XRD). Then, the properties of these CNFs with various PEG:PLA ratios were investigated by a contact angle (CA) measurement apparatus, UV spectrophotometer and shaking incubator. The results showed the CA and the cumulative release of Cur increased with the decrease of PEG, due to the appearance of a porous structure on the nanofiber surface. However, the antibacterial effect of the nonporous composite nanofiber membranes (CNFMs) was better due to burst release over a short period of time. The initial burst release of Cur was prevented by the porous structure of the porous CNFs compared to nonporous CNFs. That meant that the porous structure of CNFs could reduce the burst release and allow better control of the drug release.

2. Experimental

2.1. Materials

Polylactic acid (PLA), with an average molecular weight of 22,000 g/mol, was purchased from Anqing chemical Co. Ltd. (Anhui, China). The polyethylene glycol (PEG) with an average molecular weight of 400 g/mol was supplied by Sangon Biotech Co. Ltd. (Shanghai, China). The *N,N*-Dimethylformamide (DMF) (Analytical Reagent), chloroform (CF) (Analytical Reagent) and phosphate buffered saline (PBS) (Analytical Reagent, pH 7.4) were purchased from Shanghai Chemical Reagent Co. Ltd. (Shanghai, China). The curcumin (Cur) ($\geq$95.0% purity) was purchased from Shanghai Yuanye Biotechnology Co. Ltd. (Shanghai, China). *E. coli* for antibacterial tests were obtained from Soochow University (Suzhou, China). The nutrient broth medium and nutrient agar medium were purchased from SCAS Ecoscience Technology Inc. (Shanghai, China). All chemicals were used without further purification.

2.2. Preparation of PLA/PEG/Cur CNFs

The concentration was measured by weight. PEG and PLA with various weight ratios of 1:1, 1:3, 1:5, 1:7 and 1:9 were respectively dissolved in a solvent mixture of DMF and CF, with a weight ratio of 1:9. The concentrations of mixed solutions (PEG and PLA weight ratio to solution) were all 8 wt%. Then, a certain amount of Cur was added into the PEG/PLA solutions and the weight ratio of Cur to PEG/PLA was 3:100. The prepared solutions were magnetically stirred for 3 h to ensure homogeneous mixing.

The obtained spinning solution was dropped into a 10 mL syringe. According to our previous work [1,5], the electrospinning parameters were set as follows: The flow rate was 0.6 mL/h, the applied voltage was 15 kV, and the working distance from the needle tip to the collector was 14 cm. All electrospinning experiments were carried out at room temperature (20 °C) and a humidity of 60%.

2.3. Characterizations and Measurements

The electrical conductivity and viscosity of spinning solutions were determined respectively by a conductivity meter (DDS-307, Shanghai instrument & electric Scientific Instrument Co., Ltd., Shanghai, China) and a viscometer (NDJ-5S, Shanghai Nirun Intelligent Technology Co., Ltd., Shanghai, China). The measurement was repeated three times.

The surface morphologies of PLA/PEG/Cur CNFs were carried out using a scanning electron microscopy (SEM, Hitachi S-4800, Tokyo, Japan) at an acceleration voltage of 3 KV. The diameter distribution of the nanofibers was characterized by Image J software (National Institute of Mental Health, Bethesda, MD, USA). Twenty SEM images of 50 nanofibers in each SEM image were chosen at random for diameter distribution analysis.

The mixture, consisting of 1 mg of shredded PLA/PEG/Cur CNFMs and 200 mg of KBr powder, was pressed into a flaky sample, which was used in the Fourier-transform infrared (FTIR) analysis. The sample was characterized using FTIR spectroscopy (Nicolet 5700, Thermo Fisher Scientific, Waltham, WA, USA). The FTIR spectrum of the sample was obtained by the performance of 32 scans with the wavenumber range of 400–4000 cm^{-1} and a resolution of 4 cm^{-1}.

X-ray diffraction (XRD) analyses were performed to illustrate the crystalline structures of Cur powders and PLA/PEG/Cur CNFs using a Philips X'Pert-Pro MPD with a 3 KW ceramic tube as the X-ray source (Cu-Kα) and an X'Celerator detector. Cu-Kα radiation was used with a diffraction angle range of 10–60° at 40 kV and 200 mA at a scanning rate of 10°/min.

The pore size distributions of PLA/PEG/Cur CNFMs were determined using capillary flow porometry (Porometer 3G, Quantachrome Instruments, Boynton, FL, USA), which employed the technique of expelling a wetting liquid, Porofil, from through-pores in the sample. All samples were circular membranes with a diameter of 25 mm and a thickness of 10 μm.

The wettability of the PLA/PEG/Cur CNFMs was characterized using an optical contact angle (CA) measurement instrument (Krüss DSA100, Krüss Company, Hamburg, Germany). The volume of the deionized water droplet used for static CA measurement was 6 μL. The CAs were measured by the sessile drop method [19]. Moreover, the average CAs were determined by measuring six different positions of the same sample.

2.4. In Vitro Release of Cur

Electrospun PLA/PEG/Cur CNFs (10 mg) were dispersed in 3 mL of PBS (PH 7.4) as a release medium and then placed in 25 mL centrifuge tubes (Shanghai Hongsheng Biotech Co. Ltd., Shanghai, China). These centrifuge tubes were incubated in a shaking incubator (FLY-100/200, Shanghai Shenxian Thermostatic Equipment Factory, Shanghai, China) at 37.4 °C with a shaking speed of 60 rpm. At desired intervals, 1 mL of the release medium was withdrawn for Cur release rate analysis, and an equal volume of fresh PBS was replenished for sustaining the incubation. The released amounts of Cur were measured by UV spectrophotometer (Cary 5000, Agilent Technologies, Santa Clara, CA, USA).

2.5. Antimicrobial Tests

The PLA/PEG/Cur CNFMs were cut into 5 mm × 5 mm shreds and were sterilized for 30 min, at a temperature of 125 °C and a pressure of 10^3 KPa. The antibacterial effects of these PLA/PEG/Cur CNFMs on a bacterial strain, *E. coli*, were investigated. Overnight cultures of *E. coli* derived from a single colony and cultivated in nutrient broth medium were used in the study. Each culture solution (1 mL) was inoculated into 9 mL of PBS, resulting in a concentration of 3×10^5–5×10^5 colony forming units (CFUs/mL). These bacterial solutions with a concentration of 3×10^5–5×10^5 CFUs/mL were used for the antibacterial tests, and bacterial solutions incubated in pure PBS only served as controls.

In a conical flask, 375 mg of shredded PLA/PEG/Cur CNFMs were dispersed in 35 mL of PBS, and then 5 mL of bacterial solutions with a concentration of prepared 3×10^5–5×10^5 CFUs/mL were added. The conical flask was shaken in a constant temperature incubator shaker (IS-RDV1,

Crystal Technology & Industries, Inc., Dallas, TX, USA) at 25 °C with a shaking speed of 300 rpm for 20 h. Aliquots (1 mL) of the obtained samples were placed on nutrient agar plates in multiple replicates. Each sample was prepared separately in a triplicate. The growth of the *E. coli* was evaluated by counting CFUs after incubation for 24 h at 37 °C. The inhibition rate was calculated according to the following equation [20]:

$$\text{Inhibition rate } (\%) = \frac{\text{CFUcontrol group} - \text{CFUexperimental group}}{\text{CFUcontrol group}} \times 100\%$$

3. Results and Discussion

3.1. Properties Characterization of Spinning Solutions

The electrical conductivity and viscosity of the spinning solutions with the different weight ratios (PEG:PLA) are shown in Table 1. It was obvious that as the content of the PLA increased, the viscosity of solutions increased due to increased average molecular weight [21], which resulted in a decrease of the electrical conductivity of the solutions [22]. The viscosity played an important role in determining the diameter of the electrospun nanofibers [1].

Table 1. Properties characterization of spinning solutions with the different weight ratio (polyethylene glycol:polylactide, or PEG:PLA).

PEG:PLA	Electrical Conductivity (µs/cm)	Viscosity (mPa·s)
1:1	0.455 ± 0.045	33 ± 2
1:3	0.432 ± 0.038	77 ± 4
1:5	0.421 ± 0.035	112 ± 3
1:7	0.267 ± 0.052	137 ± 3
1:9	0.249 ± 0.048	157 ± 4

3.2. Morphological Characterization of PLA/PEG/Cur CNFs (SEM)

The surface morphologies of the electrospun PLA/PEG/Cur CNFs with different weight ratios (PEG:PLA) were investigated by SEM. Figure 1 shows SEM images of the CNFs and the respective nanofiber diameter distributions. It could be seen that with the decrease in the weight ratio (PEG:PLA), the surface morphology of the CNFs changed from smooth to porous. A further decrease of PEG in the mixtures, with the weight ratio 1:5, resulted in the appearance of evident pores on the CNF surfaces. When the weight ratio (PEG:PLA) was 1:7 and 1:9, the pore structure of the nanofiber surfaces became very noticeable, and the diameters of the porous nanofibers increased due to an increased pore number. This could be explained by the plastic deformation of PLA in water, which led to the formation of the pore structure, due to the solvent volatilization at a suitable humidity in the electrospinning process. Therefore, as the PLA content increased to a certain proportion, evident pore structures appeared on the nanofiber surfaces.

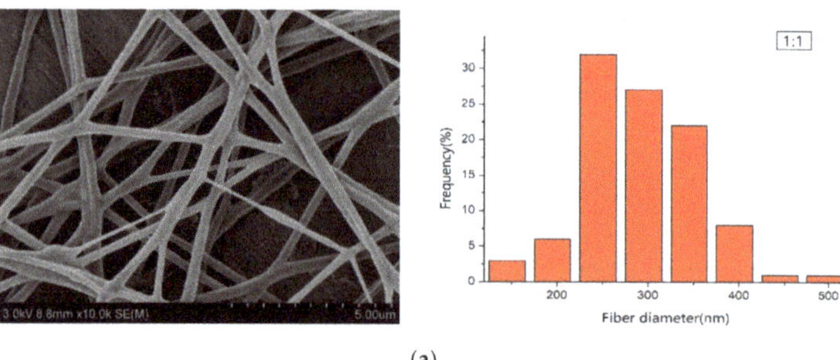

(a)

Figure 1. *Cont.*

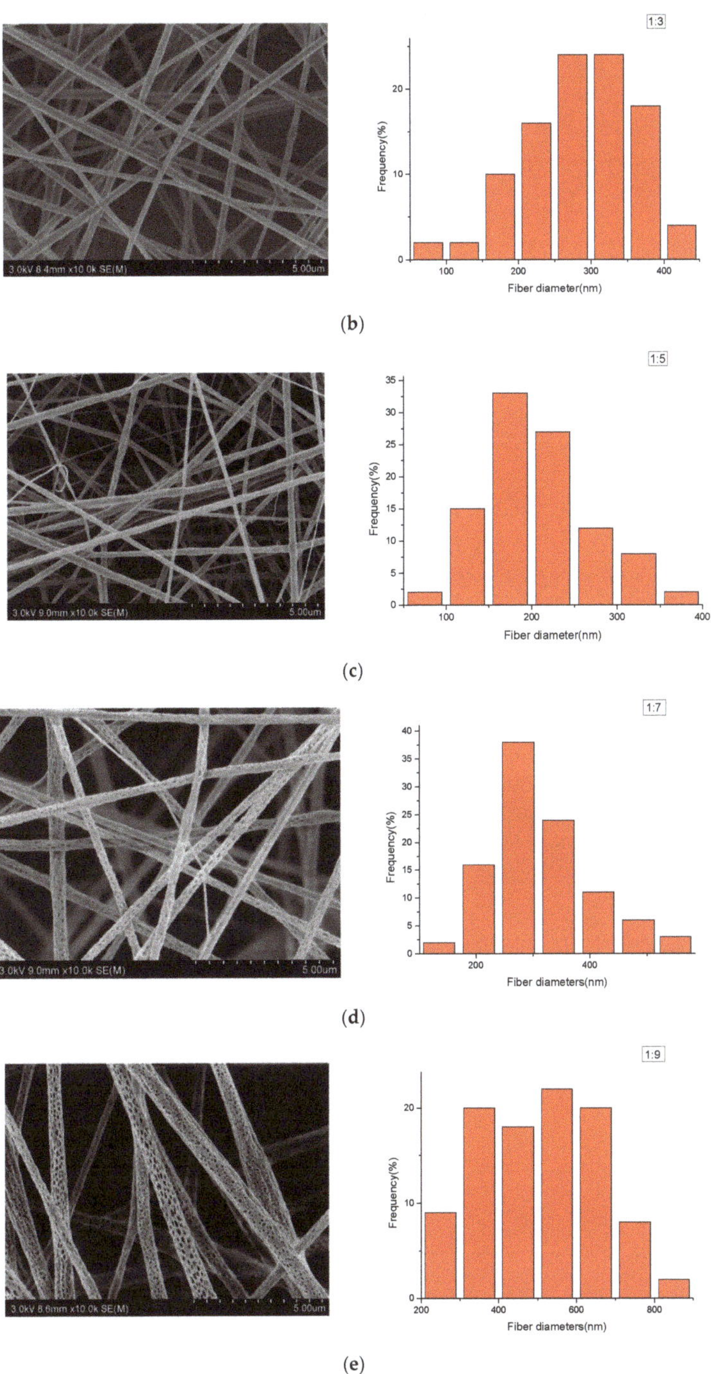

Figure 1. SEM pictures of the electrospun polylactide/polyethylene glycol/curcumin (PLA/PEG/Cur) carbon nanofibers (CNFs) with different weight ratios (PEG:PLA). The right figures were the according diameter distribution: (**a**) 1:1, (**b**) 1:3, (**c**) 1:5, (**d**) 1:7, (**e**) 1:9.

In addition, the relationship between the weight ratio (PEG:PLA) and the average diameters of the electrospun CNFs is shown in Table 2, with the confidence intervals calculated according to Reference [1]. Table 2 illustrates that with the decrease of the weight ratio (PEG:PLA), the average diameters of the CNFs decreased. This was firstly due to better spinnability of the spinning solution, which increased rapidly with a weight ratio of 1:7, due to the formation of pores [21]. When the contents of PEG were too much, such as 1:1 and 1:3 (mass ratios to PLA), the viscosity of the spinning solution was too low, which resulted in poor spinnability of the solution.

Table 2. The relationship between the weight ratio (PEG:PLA) and the average diameters of CNFs.

PEG:PLA	Average Diameter (D) (nm)	Standard Deviation (σ) (nm)	Confidence Interval (nm)
1:1	290.60	64.08	±12.56
1:3	287.70	74.39	±16.91
1:5	205.16	59.43	±11.64
1:7	308.73	91.25	±17.89
1:9	552.81	195.53	±38.32

3.3. Fourier-Transform Infrared (FTIR) and Raman Spectrum Analysis

The presences of Cur in the PEG/PLA CNFs with the different weight ratios (PEG:PLA) were confirmed by Fourier-transform infrared (FTIR) spectroscopy, as shown in Figure 2. The FTIR spectra of these PLA/PEG/Cur CNFs displayed characteristic absorption bands at 1754 cm^{-1} and 1087 cm^{-1}, which represented the backbone ester group of PLA [4], and also showed a CH$_3$ asymmetric bending peak at 1454 cm^{-1}, a C-O stretching peak at 1182 cm^{-1} and a O-H bending peak at 1047 cm^{-1} of PLA. In addition, the absorption peak at 1128 cm^{-1} could be attributed to C-O-C characteristic vibration of PEG [23]. Moreover, it was obvious that a sharp peak appeared at 1272 cm^{-1} due to the C-O stretching vibration in -C-OCH$_3$ of the phenyl ring of Cur [24]. As seen in Figure 2, Cur was encapsulated in the CNFs, and the intensity of the peaks corresponding to PLA in FTIR spectra of samples increased as the PLA contents increased.

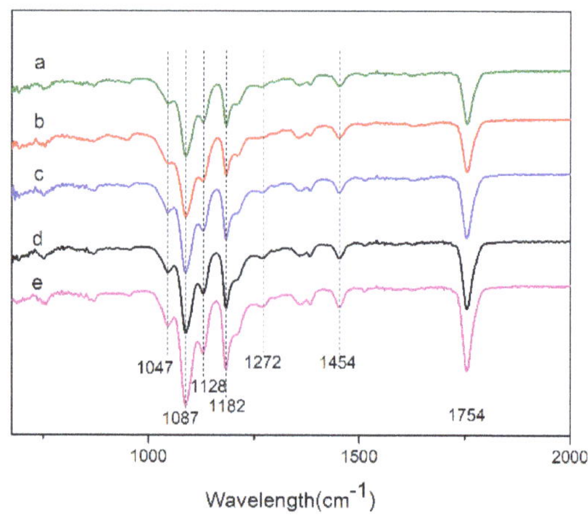

Figure 2. Fourier-transform infrared (FTIR) spectra of PLA/PEG/Cur CNFs with different weight ratios (PEG:PLA): (**a**) 1:1; (**b**) 1:3; (**c**) 1:5; (**d**) 1:7; (**e**) 1:9.

3.4. X-ray Diffraction (XRD) Spectrum Analysis

XRD patterns with distinctive crystalline peaks of Cur and PLA/PEG/Cur CNFs with different weight ratios (PEG:PLA) are shown in Figure 3. As seen in Figure 3A, the XRD spectrum of Cur displayed sharp and intense peaks of crystallinity, which suggested a highly crystalline nature. Figure 3B shows the XRD spectra of the CNFs with different weight ratios (PEG:PLA). It could be observed that the peaks at $2\theta = 16.88°$ and $22.58°$ could be assigned as (010) crystal planes of PEG and (015) crystal planes of PLA, respectively. With the decrease of PEG, the XRD spectra of the CNFs showed a reduction of peak intensity, as compared to the Cur, which indicated decreased crystallinity or changes into an amorphous phase of the drug [6]. The principal peak of Cur at $2\theta = 17.08°$ did not appear due to the low ratio of Cur with respect to the polymer composite.

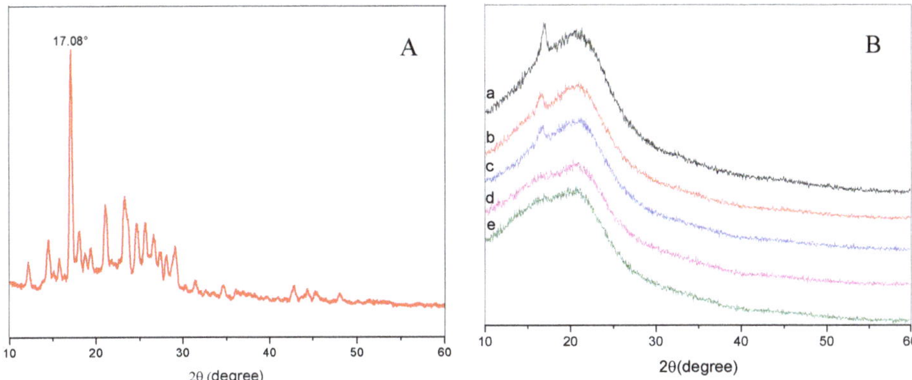

Figure 3. XRD spectra of Cur (**A**). XRD spectra of PLA/PEG/Cur CNFs with the different weight ratio (PEG:PLA): (**a**) 1:1; (**b**) 1:3; (**c**) 1:5; (**d**) 1:7; (**e**) 1:9 (**B**).

3.5. Measuring Pore Size Distribution Composite Membranes

Figure 4 illustrates the pore size distributions of the PLA/PEG/Cur CNFMs with varying PEG weight ratios, measured by a capillary flow porometry, and Table 3 shows the size and number of pores. It could be seen that as the PEG weight ratio decreased, the pore size of the CNFMs increased, and the respective pore number of the CNFMs decreased. Moreover, the pore distributions of the CNFMs with the weight ratios 1:7 and 1:9 (PEG:PLA) demonstrated more uniformity. These results were in agreement with the SEM images, as illustrated in Figure 1.

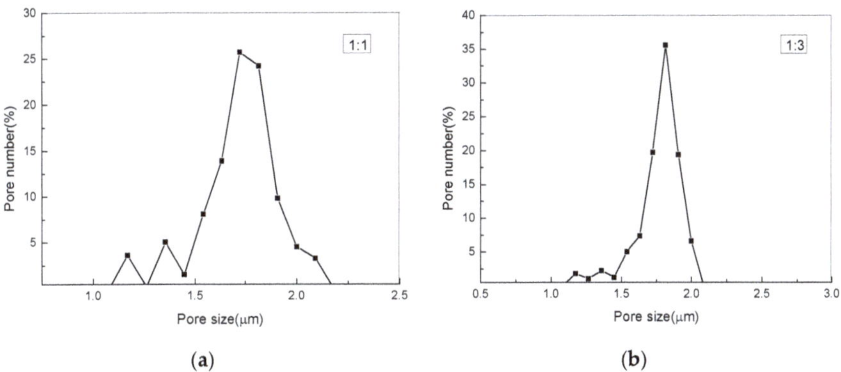

(a) **(b)**

Figure 4. *Cont.*

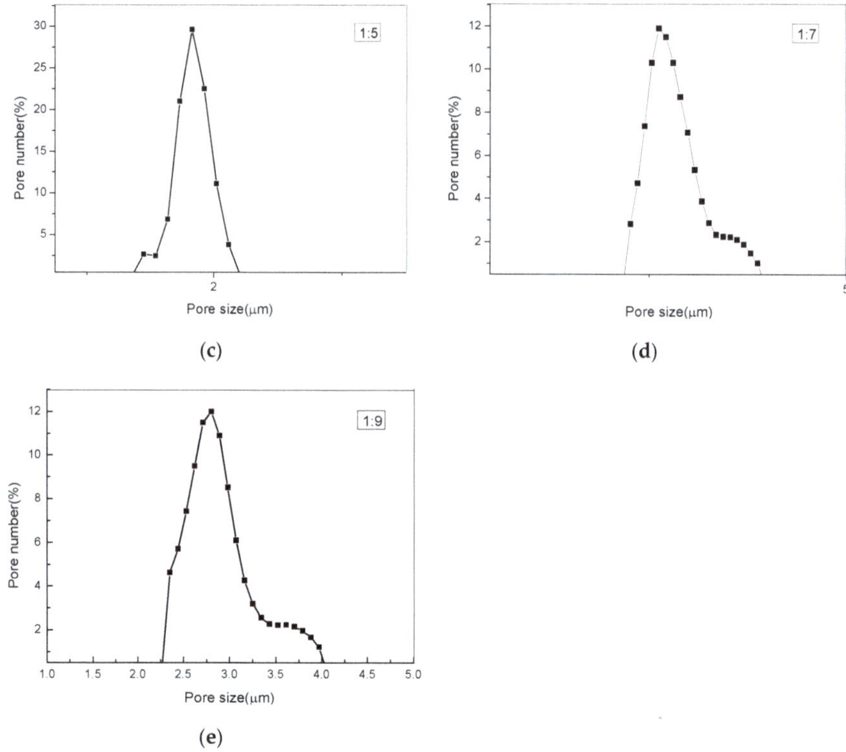

Figure 4. The pore size distributions of PLA/PEG/Cur nonporous composite nanofiber membranes (CNFMs) with different weight ratios (PEG:PLA): (**a**) 1:1; (**b**) 1:3; (**c**) 1:5; (**d**) 1:7; (**e**) 1:9.

Table 3. The analysis of pore size distributions of the CNFMs with different weight ratios (PEG:PLA).

PEG:PLA	Pore Size (μm)	Maximum Pore Number (/cm²) /The According Pore Size (μm)	Pore Number (/cm²)
1:1	1.39–2.11	$4.3 \times 10^7 / 1.769$	1.68×10^8
1:3	1.41–2.18	$4.13 \times 10^7 / 1.861$	1.15×10^8
1:5	1.55–2.39	$2.2 \times 10^7 / 1.879$	7.4×10^7
1:7	2.35–3.93	$5.6 \times 10^6 / 2.668$	5.1×10^7
1:9	2.44–4.04	$5.5 \times 10^6 / 2.848$	4.5×10^7

3.6. Wetting Properties

The CA values of the PLA/PEG/Cur CNFMs with different weight ratios (PEG:PLA) were obtained by an optical (CA) measurement instrument (Table 4). It was obvious that the CNFMs with the weight ratio 1:1 (PEG:PLA) were hydrophilic due to the hydrophilicity of PEG, and with the increase of the PLA weight ratio, the CA increased gradually due to the hydrophobicity of PLA. When the weight ratio was up to 1:3, the CA of the CNFMs increased to 125.4°. With the continuous increase of the PLA weight ratio, the CNFMs were highly hydrophilic. This might be due to the increase of hydrophobic PLA and pores of the CNFs. The pore structure could further enlarge the surface area and enhance the hydrophobicity of the CNFMs [25].

Table 4. Contact angles of PLA/PEG/Cur CNFMs with different weight ratios (PEG:PLA).

Ratio	1:1	1:3	1:5	1:7	1:9
Contact angles	$47.1 \pm 1.1°$	$125.4 \pm 2.6°$	$130.2 \pm 1.1°$	$134.6 \pm 2.0°$	$139.0 \pm 1.7°$

3.7. In Vitro Cur Release

The released amounts of Cur in PBS from the electrospun CNFs were determined spectrophotometrically at 425 nm by a UV spectrophotometer. The standard curve of Cur concentration was calculated according to $y = 14.26087x - 0.06634$ (R = 0.997), where x is the Cur concentration (μg/mL), y is the optical density (OD) value measured. Based on the standard curve, the cumulative release of Cur was calculated. Figure 5 illustrates the cumulative release of Cur from the porous and nonporous CNFs with the different Cur contents, which were prepared with a weight ratio of 1:7 (PEG:PLA), at a humidity of 60% (porous) and 30% (nonporous), respectively. It could be seen that after 240 h the cumulative release rates of Cur from the porous CNFs with varying Cur percentages from 1% to 5%, were 92.17%, 81.23%, 82.89%, 58.36%, and 57.62%, respectively, were higher than those from the according nonporous CNFs, which were 88.51%, 63.83%, 58.10%, 58.35% and 56.66%. This may have been due to the higher specific surface area of the porous nanofibers and the porous structure could promote the release of the drug [4]. In addition, the cumulative release rate of Cur from the CNFs with 1% Cur was highest, and when the Cur percentage was higher than 3%, the cumulative release rate of Cur was at a low level. This may have been due to the fact that Cur could be dissolved less in the spinning solution when the Cur percentage was at a higher level. Therefore, 3% Cur was selected as the experimental parameter in the following study.

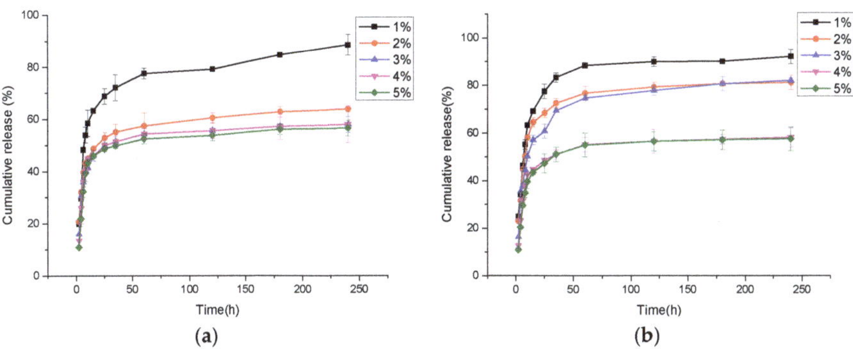

Figure 5. Cumulative release of Cur from porous and nonporous PLA/PEG/Cur CNFs with the different Cur contents: (**a**) Nonporous CNFs; (**b**) Porous CNFs.

Figure 6 shows the cumulative release of Cur from the PLA/PEG/Cur CNFs with 3% Cur and different weight ratios (PEG:PLA). It was evident that after 80 h, the Cur release of CNFs with different weight ratios (PEG:PLA), such as 1:1, 1:3, 1:5, 1:7 and 1:9, reached 48.40%, 53.58%, 55.75%, 66.08% and 69.41%, respectively. That meant the Cur release increased as the PLA weight ratio increased, due to the appearance and increase of pores on the CNFs. The results also demonstrated that the porous structure of the CNFs could absorb more Cur and promote the release of Cur. Therefore, the PLA/PEG/Cur porous CNFs exhibited an improved drug release property.

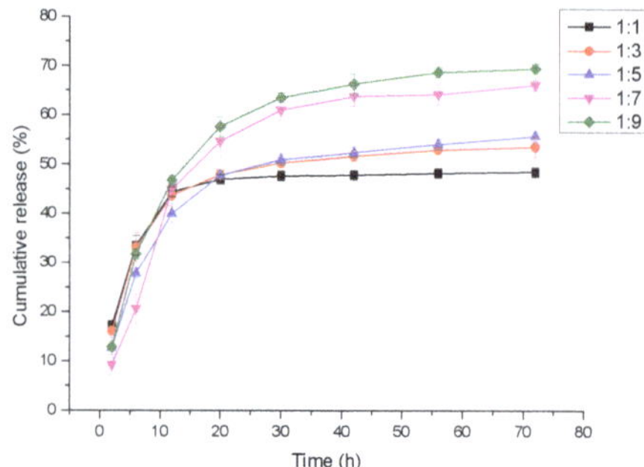

Figure 6. Cumulative release of Cur from PLA/PEG/Cur CNFs with the different weight ratio (PEG:PLA).

3.8. Effects of Porous Structure on Antibacterial Properties

Figure 7 and Table 5 present the antibacterial properties of the PLA/PEG/Cur CNFMs with 3% Cur and different weight ratios (PEG:PLA), and the PEG/PLA CNFMs with a weight ratio (PEG:PLA) of 1:7 used as a comparative sample. It could be seen that the antibacterial effect of the PEG/PLA CNFMs with a weight ratio of 1:7 was poor, and the inhibition rates of the PLA/PEG/Cur CNFMs were all beyond 97%. The antibacterial effect of the PLA/PEG/Cur CNFMs was as good as that of other antibacterial materials in Reference [26]. However, the inhibition rates decreased as the PLA weight ratio increased. To illustrate the results of the antibacterial tests, the initial burst releases of Cur from the PLA/PEG/Cur CNFMs with 3% Cur and different weight ratios (PEG:PLA) for about 20 h were investigated. The standard curve of the Cur concentration was calculated according to $y = 186.428x - 3.3719$ (R = 0.997). According to the standard curve, the cumulative releases of Cur were calculated, as displayed in Figure 8.

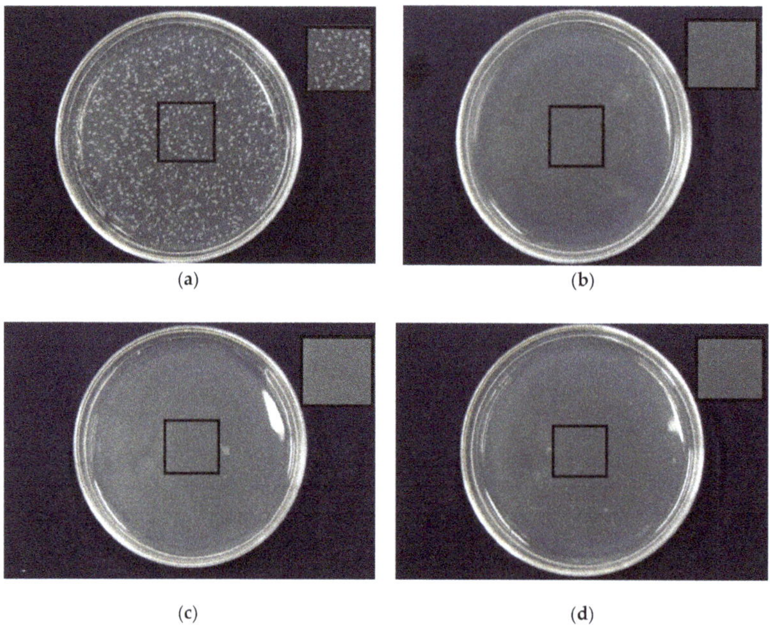

Figure 7. *Cont.*

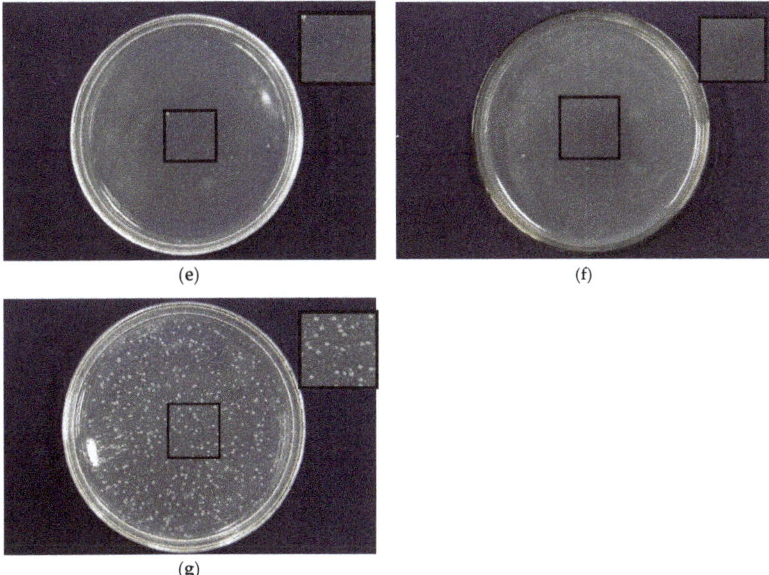

(e)   (f)

(g)

Figure 7. Effects of different composite ratio on antibacterial properties of PLA/PEG/Cur nanofibers membranes: (**a**) PEG:PLA = 1:7 (0% Cur); (**b**) PEG:PLA = 1:1 (3% Cur); (**c**) PEG:PLA = 1:3 (3% Cur); (**d**) PEG:PLA = 1:5 (3% Cur); (**e**) PEG:PLA = 1:7 (3% Cur); (**f**) PEG/PLA = 1:9 (3% Cur); (**g**) cotton.

Table 5. Effects of different composite ratio on antibacterial properties of PLA/PEG/Cur nanofiber membranes.

PEG/PLA	Inhibition Rates (%)
1:1	99.97
1:3	99.96
1:5	99.93
1:7	98.77
1:9	97.61

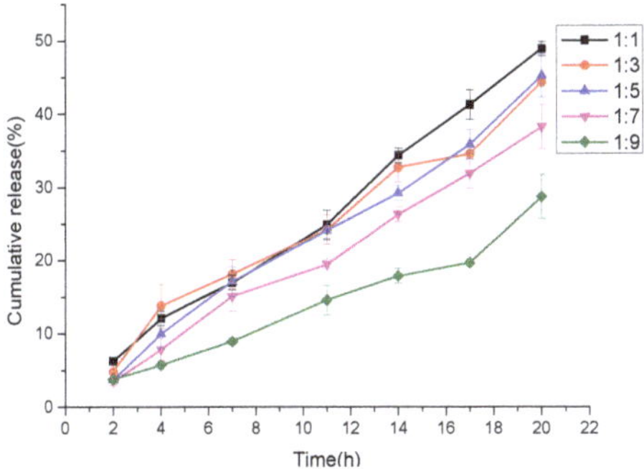

Figure 8. Initial burst release of Cur from PLA/PEG/Cur CNFMs with different weight ratios (PEG:PLA).

As seen in Figure 8, as the PLA weight ratio increased, the initial burst release rate of Cur decreased due to the appearance and increase of pores on the CNFs, which resulted in the decrease of the inhibition rates of the CNFMs. The results showed that the antibacterial effect of the nonporous

CNFs was better due to burst release over a short period of time. This meant that the porous structure of the CNFs could reduce the initial burst release and better control drug release. This could be explained using a modified Fickian diffusion equation, which is presented in our previous work [4]. According to the equation, the porous nanofibers had a longer diffusion path than the nonporous nanofibers, and as the diffusion path increased, the concentration gradient of the drug would become smaller. As a result, the diffusion flux would decrease and lead to the slow release of the drug [4].

4. Conclusions

In the present study, electrospun PLA/PEG/Cur porous CNFs with different weight ratios (PEG:PLA) were successfully prepared by controlling the electrospinning parameters. The electrical conductivity and viscosity of the spinning solutions with different weight ratios (PEG:PLA) were investigated. The results indicated that with the decrease in the weight ratio, the electrical conductivity of spinning solutions decreased, and accordingly, the viscosity of the solutions increased.

The effects of the weight ratio (PEG:PLA) on the morphology and structure of the CNFs were studied systematically. SEM images displayed that as the weight ratio decreased, the morphology of the CNFs changed from smooth to porous. And when the weight ratio increased to 1:7, the pore structure of the nanofiber surfaces became very evident. FTIR and XRD data indicated that Cur was encapsulated successfully in the CNFs. Pore size analysis displayed that with the decrease of the weight ratio, the pore size of the CNFMs increased, and the respective pore number of the CNFMs decreased. CA measurements illustrated that the CA of the CNFMs increased as the weight ratio decreased.

Property characterizations—such as the wetting property, drug release property and antibacterial property—of the PLA/PEG/Cur CNFMs showed that with the decrease of the weight ratio, the CA and the Cur cumulative release of the CNFMs increased due to the appearance of pores on the CNFs. In addition, the antibacterial effect of the CNFMs decreased, as the initial burst release of Cur was prevented by the porous structure of the CNFs. Therefore, the porous structure of the CNFs could improve the drug release property, and the PLA/PEG/Cur porous CNFMs could have great potential for biomedical applications, such as drug delivery, biological scaffold, medical dressing and antibacterial materials.

Author Contributions: L.X. and Z.S. designed the experiments. F.W., Z.S. and J.Y. performed the experiments and the characterizations, and analyzed the data. Z.S. and F.W. drafted the paper. L.X. supervised data analysis and revised the paper.

Funding: This research was funded by National Natural Science Foundation of China: No. 11672198, Six Talent Peaks Project of Jiangsu Province: GDZB-050, and Foundation project of Jiangsu Advanced Textile Engineering Technology Center: No. XJFZ/2018/15.

Acknowledgments: The work was supported financially by National Natural Science Foundation of China (Grant No. 11672198), Six Talent Peaks Project of Jiangsu Province (Grant No. GDZB-050), PAPD (A Project Funded by the Priority Academic Program Development of Jiangsu Higher Education Institutions), and Foundation project of Jiangsu Advanced Textile Engineering Technology Center: No. XJFZ/2018/15.

Conflicts of Interest: The authors declare no conflict of interest.

References

1. Sun, Z.; Fan, C.; Tang, X.; Zhao, J.; Song, Y.; Shao, Z.; Xu, L. Characterization and antibacterial properties of porous fibers containing silver ions. *Appl. Surf. Sci.* **2016**, *387*, 828–838. [CrossRef]
2. Kanehata, M.; Ding, B.; Shiratori, S. Nanoporous ultra-high specific surface inorganic fibres. *Nanotechnology* **2007**, *18*, 1–7. [CrossRef]
3. Liu, P.; Zhu, Y.; Ma, J.; Yang, S.; Gong, J.; Xu, J. Preparation of continuous porous alumina nanofibers with hollow structure by single capillary electrospinning. *Colloids Surf. A-Physicochem. Eng. Asp.* **2013**, *436*, 489–494. [CrossRef]
4. Wang, Y.; Xu, L. Preparation and characterization of porous core-shell fibers for slow release of tea polyphenols. *Polymers* **2018**, *10*, 144. [CrossRef]

5. Zhao, J.; Si, N.; Xu, L. Experimental and theoretical study on the electrospinning nanoporous fibers process. *Mater. Chem. Phys.* **2016**, *170*, 294–302. [CrossRef]

6. Yang, C.; Chen, H.; Zhao, J.; Pang, X.; Xi, Y.; Zhai, J. Development of a folate-modified curcumin loaded micelle delivery system for cancer targeting. *Colloids Surf. B-Biointerfaces* **2014**, *121*, 206–213. [CrossRef]

7. Heni, R.; Yanda, Y.; Rahma, A.; Mase, N. Curcumin-Loaded PLA Nanoparticles: Formulation and Physical Evaluation. *Sci. Pharm.* **2016**, *84*, 191–202. [CrossRef]

8. Lv, L.; Shen, Y.; Li, M.; Xu, X.; Li, M.; Guo, S.; Huang, S. Novel 4-Arm Poly(Ethylene Glycol)-Block-Poly(Anhydride-Esters) Amphiphilic Copolymer Micelles Loading Curcumin: Preparation, Characterization, and In Vitro Evaluation. *Biomed. Res. Int.* **2013**, 1–11. [CrossRef]

9. Bhattarai, N.; Li, Z.; Gunn, J.; Leung, M.; Cooper, A.; Edmondson, D.; Veiseh, O.; Chen, M.; Zhang, Y.; Ellenbogen, R.; et al. Natural-Synthetic Polyblend Nanofibers for Biomedical Applications. *Adv. Mater.* **2009**, *21*, 2792–2797. [CrossRef]

10. Lu, H.; Qiu, Y.; Wang, Q.; Li, G.; Wei, Q. Nanocomposites prepared by electrohydrodynamics and their drug release properties. *Mater. Sci. Eng. C* **2018**, *91*, 26–35. [CrossRef]

11. Zhang, L.; Wang, Z.; Xiao, Y.; Liu, P.; Wang, S.; Zhao, Y.; Shen, M.; Shi, X. Electrospun PEGylated PLGA nanofibers for drug encapsulation and release. *Mater. Sci. Eng. C* **2018**, *91*, 255–262. [CrossRef]

12. Ramírezagudelo, R.; Scheuermann, K.; Galagarcía, A.; Monteiro, A.; Pinzóngarcía, A.D.; Cortés, M.E.; Sinisterra, R.D. Hybrid nanofibers based on poly-caprolactone/gelatin/hydroxyapatite nanoparticles-loaded Doxycycline: Effective anti-tumoral and antibacterial activity. *Mater. Sci. Eng. C* **2018**, *83*, 25–34. [CrossRef]

13. Ma, B.; Huang, Y.; Zhu, C.; Chen, C.; Chen, X.; Fan, M.; Sun, D. Novel Cu@SiO2/bacterial cellulose nanofibers: Preparation and excellent performance in antibacterial activity. *Mater. Sci. Eng. C* **2016**, *62*, 656–661. [CrossRef]

14. Phan, Q.; Mai, H.; Le, T.; Tran, T.; Xuan, P.; Ha, P. Characteristics and cytotoxicity of folate-modified curcumin-loaded PLA-PEG micellar nano systems with various PLA: PEG ratios. *Int. J. Pharm.* **2016**, *507*, 32–40. [CrossRef]

15. Aytac, Z.; Uyar, T. Core-shell nanofibers of curcumin/cyclodextrin inclusion complex and polylactic acid: Enhanced water solubility and slow release of curcumin. *Int. J. Pharm.* **2017**, *518*, 177–184. [CrossRef]

16. Zhong, T.; Jiao, Y.; Guo, L.; Ding, J.; Nie, Z.; Tan, L.; Huang, R. Investigations on porous PLA composite scaffolds with amphiphilic block PLA-b-PEG to enhance the carrying property for hydrophilic drugs of excess dose. *J. Appl. Polym. Sci.* **2017**, *134*, 1–7. [CrossRef]

17. Hussain, S.A.; Patil, G.R.; Reddi, S.; Yadav, V.; Pothuraju, R.; Singh, R.R.B.; Kapila, S. Aloe vera (Aloe barbadensis Miller) supplemented probiotic lassi, prevents Shigella, infiltration from epithelial barrier into systemic blood flow in mice model. *Microb. Pathog.* **2017**, *102*, 143–147. [CrossRef] [PubMed]

18. Feng, X.; Ng, V.K.; Mikš-Krajnik, M.; Yang, H. Effects of Fish Gelatin and Tea Polyphenol Coating on the Spoilage and Degradation of Myofibril in Fish Fillet During Cold Storage. *Food Bioprocess Technol.* **2017**, *10*, 1–14. [CrossRef]

19. Serrano-Aroca, A.; Iskandar, L.; Deb, S. Green synthetic routes to alginate-graphene oxide composite hydrogels with enhanced physical properties for bioengineering applications. *Eur. Polym. J.* **2018**, *103*, 198–206. [CrossRef]

20. Ma, M.; Wan, R.; Gong, H.; Lv, X.; Chu, S.; Li, D.; Peng, C. Study on the In Vitro and In Vivo Antibacterial Activity and Biocompatibility of Novel TiN/Ag Multilayers Immobilized onto Biomedical Titanium. *J. Nanosci. Nanotechnol.* **2019**, *19*, 3777–3791. [CrossRef]

21. Xu, L.; Si, N.; Lee, E.W.M. Effect of humidity on the surface morphology of a charged jet. *Heat Transf. Res.* **2013**, *44*, 441–445. [CrossRef]

22. Achoh, A.; Zabolotsky, V.; Melnikov, S. Conversion of water-organic solution of sodium naphtenates into naphtenic acids and alkali by electrodialysis with bipolar membranes. *Sep. Purif. Technol.* **2019**, *212*, 929–940. [CrossRef]

23. Yan, C.; Yu, Z.; Yang, B. Improvement of thermoregulating performance for outlast/silk fabric by the incorporation of polyurethane microcapsule containing paraffin. *Fibers Polym.* **2013**, *14*, 1290–1294. [CrossRef]

24. Li, J.; Wang, X.; Li, C.; Fan, N.; Wang, J.; He, Z.; Sun, J. Viewing Molecular and Interface Interactions of Curcumin Amorphous Solid Dispersions for Comprehending Dissolution Mechanisms. *Mol. Pharm.* **2017**, *14*, 2781–2792. [CrossRef] [PubMed]

25. Xu, L.; Si, N.; Liu, H.Y. Fabrication and characterization of chinese drug-loaded nanoporous materials. *J. Nano Res.* **2014**, *27*, 103–109. [CrossRef]
26. Frıgols, B.; Martı, M.; Salesa, B.; Herna´ndezOliver, C.; Aarstad, O.; Teialeret Ulset, A.-S.; Inger Sætrom, G.; Lillelund Aachman, F.; Serrano-Aroca, A. Graphene oxide in zinc alginate films: Antibacterial activity, cytotoxicity, zinc release, water sorption/diffusion, wettability and opacity. *PLoS ONE* **2019**, *14*, e0212819. [CrossRef] [PubMed]

 *nanomaterials*

Article

Natural Antibacterial Reagents (*Centella*, Propolis, and Hinokitiol) Loaded into Poly[(*R*)-3-hydroxybutyrate-*co*-(*R*)-3-hydroxyhexanoate] Composite Nanofibers for Biomedical Applications

Rina Afiani Rebia [1], **Nurul Shaheera binti Sadon** [2] **and Toshihisa Tanaka** [2,*]

[1] Interdisciplinary Graduate School of Science and Technology, Shinshu University, 3-15-1 Tokida, Ueda, Nagano 386-8567, Japan; rina_rebia@yahoo.com

[2] Faculty of Textile Science and Technology, Shinshu University, 3-15-1 Tokida, Ueda, Nagano 386-8567, Japan; shera9410@gmail.com

* Correspondence: tanakat@shinshu-u.ac.jp; Tel.: +81-268-21-5531

Received: 19 October 2019; Accepted: 19 November 2019; Published: 22 November 2019

Abstract: *Centella asiatica*, propolis, and hinokitiol, as natural antibacterial reagents, were integrated into the poly[(*R*)-3-hydroxybutyrate-*co*-(*R*)-3-hydroxyhexanoate] (PHBH) polymer to produce antibacterial wound dressings, using electrospinning process. The results showed that the fiber diameters and surface morphology of PHBH composite nanofibers were influenced by the addition of ethanol–*centella* (EC), methanol–*centella* (MC), ethanol–propolis (EP), and ethanol–hinokitiol (EH) at various ratios compared to pristine PHBH nanofibers. From FT-IR, the nanofibrous samples with higher contents of natural antibacterial substances showed the peaks of carboxylic acid, aromatic ring, and tropolone carbon ring from *centella*, propolis, and hinokitiol, respectively. Furthermore, the tensile strength of neat PHBH nanofibers was increased from 8.00 ± 0.71 MPa up to 16.35 ± 1.78 MPa by loading of propolis (EP) 7% into PHBH. X-ray analysis explained that the loading of propolis (EP) was also able to increase the crystallinity in PHBH composite nanofibers from 47.0% to 54.5%. The antibacterial results demonstrated that PHBH composite nanofibers containing natural antibacterial products were potent inhibitors against the growth of *Escherichia coli* and *Staphylococcus aureus*, amongst them hinokitiol and propolis proved to be the most effective. Additionally, the release studies displayed that *centella* and hinokitiol had faster release from PHBH composite nanofibers in comparison to propolis.

Keywords: antibacterial effect; centella; propolis; hinokitiol; biodegradable polymer; PHBH; nanofiber

---

1. Introduction

Electrospinning is a straightforward and efficient method to fabricate nonwoven mats with continuous fibers in the range from micrometer down to nanometer. Nowadays, nanofibers produced by electrospinning have received the considerably high attention to be an excellent candidate for many important applications in medical field such as drug delivery [1], oral cavity [2], tissue engineering [3–6], wound dressing and healing [7,8] due to the ultra-fine diameter, high surface area to volume ratio, and cost-effectiveness. A large number of studies devoted to utilizing nanofibers that could mimic the extracellular matrix (ECM) of the body and cellular activity [9], as well as sustained drug action such as in skin regeneration or healing [1,10,11] have been published over the last few years. Open wounds were reported to be healed in a much faster-recovering pace by covering from infection using antimicrobials. Without covering, the healing process could be hindered by bacterial interference with cell-matrix interactions, followed by the delay of the cell proliferation and tissue regeneration [12].

Recent studies have endeavored to incorporate novel drugs, antibacterial reagents, and other healing enhancers into nanofiber membranes for the sake of facilitating the wound healing process [7,10,11].

Natural antibiotics as one of herb, plant extracts, or pure compounds have been used in pharmaceutical products for centuries. With the advancement of nanotechnology, it provides unlimited opportunities for employing natural antibiotics as additives for engineering novel intriguing materials for wound treatments and healing. One of the advantages of using natural products such as plant extracts, essential oils, aloe vera, honey, and so on instead of synthetic reagents (iodine and silver) is to diminish the risk of sensitization and development of resistance [12]. *Centella asiatica*, a traditional medicine derived from plants, has been used widely in Southeast Asia, India, and other regions for skin treatments. The major biologically active compounds in *centella* are triterpenoids, which include asiaticoside, madecassoside, asiatic acid, and madecassic acid [13–15]. The previous study of gelatin/*centella asiatica* [14] reported that *centella* extract demonstrated a strong inhibition against microbial pathogens such as *S. aureus*, *E. coli*, and *P. aeruginosa*, which are most commonly implicated in wound infection. Polycaprolactone/*centella* biopolymer nanofibers also showed the antibacterial activity against *B. cereus* and *M. luteus* with the improvement of mechanical properties owing to the decrease of the fiber diameters after adding *centella* [5]. *In vitro* test using human dermal fibroblast and human dermal keratinocyte (HaCaT) cells and *in vivo* circular wound excision of rabbits, *centella* extract proved to accelerate wound healing process and the cell migration rates [16]. Furthermore, electrospun gelatin membrane containing *centella* improved cell proliferation *in vitro* for fibroblasts (L-929) and wound recovery *in vivo*, using male Sprague Dawley (SD) rats [14].

Propolis, a resinous substance with dark brown color, was produced via mixing collected from buds and bark of trees with beeswax and saliva by honeybees. The main chemical compounds in propolis are aromatic acids and phenolic compounds, especially flavonoid and phenolic acid [17–19]. A number of studies have demonstrated that propolis exhibited great potential for wound healing. In the study of polyurethane/propolis nanofibers [20], the authors concluded that nanofiber mats containing 30% (*w/w*) propolis had strong inhibitory effects against *E. coli*. In addition, PVA/hydroalcoholic extract propolis nanofiber [21] exhibited the antibacterial activity almost equal to the vancomycin, an effective commercial antibiotic. Propolis was also confirmed to possess antibacterial, antifungal, antiviral, antioxidant, anti-inflammatory, and antitumor characteristics [22].

Hinokitiol is one of the natural components isolated from Japanese cypress and western red cedar. It belongs to the class of tropolones that contain an unsaturated seven-membered carbon ring with β-Thujaplicin and is well-known for its inhibitory effects against the growth of fungi, bacteria, and insects [23]. Research on hinokitiol as a potential medication for the treatment of dental root canal demonstrated strong antibacterial effects against *S. aureus* and anti-inflammation [24]. These natural products (*centella*, propolis, and hinokitiol) show the capability to be used as alternative antibacterial reagents to replace conventional synthetic antiseptics.

Poly(3-hydroxybutyrate-*co*-3-hydroxyhexanoate) (PHBH) is a copolymer, developed from poly[(*R*)-3-hydroxybutyrate] (PHB), the basic homopolymer in Polyhydroxyalkanoates (PHA), which are a family of polyesters produced by bacteria through the fermentation of sugars, lipids, fatty acids, alkanes, alkenes, and alkanoic acids [1,6,25–27]. PHBH, one of copolymers in more than 150 kinds of PHAs, which in turn consist of various co-monomers, has been considered for commercial production [28]. In the biomedical field, PHB and its copolymers have been demonstrated to be fully biodegradable and an excellent biological material with biocompatibility properties. The prominent advantage of these polymers is their ability to be completely degraded by microorganisms under aerobic and anaerobic conditions without forming toxic by-products [26,29]. Recently, these materials have been developed for medical applications such as suture, skin scaffold, tissue engineering, and so on. Moreover, biodegradable nanofibers as a material matrix for wound dressing and healing have received considerable attention because they have been found to be an effective strategy for delivering antibacterial agents or anti-inflammatory substances. Previous studies demonstrated that PHB/caffeic acid electrospun nanofibers coated with polyelectrolyte exerted the antibacterial activity and antitumor [7]. Poly(3-hydroxybutyrate-*co*-3-hydroxyvalerate) (PHBV)/Curcumin nanofibers showed positive effects on wound healing, cell adhesion, and non-toxicity to L929 mouse fibroblast cell line [11]. Besides, PHBH nanofibers promoted the NIH3T3 fibroblast cell attachment and proliferation [4]. This polymer can be regarded as one of the most promising polymers obtained from renewable resources, suitable for drug or antibacterial products carrying.

In this study, *centella*, propolis, and hinokitiol with inherent biological activity were selected as representatives for natural antibacterial products and were integrated in PHBH biodegradable polymer. We focused on developing nanofibrous membranes with antibacterial properties, comprised of natural antibacterial reagents and biodegradable PHBH polymer for combined effects on wound healing. The electrospinning technique was chosen as the mean of fabrication. To our knowledge, the fabrication and characterizations of *centella*, propolis, and hinokitiol loaded in biodegradable PHBH polymer have not been reported hitherto. We attempted to investigate the effects of the addition of *centella*, propolis, and hinokitiol to PHB copolymer with 3-hydroxyhexanoate (3HH) unit as the polymer matrix regarding crystallinity of the PHBH composite nanofibers, antibacterial activities, and sustained release behaviors. The advantages of this research are to promote the use of natural products over conventional synthetic antiseptics by studying three variants of natural antibacterial products, namely *centella*, propolis, and hinokitiol. It is also important to present harmlessness for products made of PHBH and natural antibacterial reagents if they are to be used in medical applications.

The surface morphology, chemical characterization, mechanical properties, crystallinity of PHBH, zone of inhibition, and sustained release behaviors of *centella*, propolis, and hinokitiol loaded into PHBH were analyzed by scanning electron microscopy (SEM), Fourier Transform Infrared (FTIR) Spectroscopy, tensile testing machine, X-ray diffraction, antibacterial experiment, and UV-visible spectrophotometer, respectively. From these results, the influences of natural products and solvents to PHBH structure and the properties of the resultant nanofibrous composites will be revealed.

2. Materials and Methods

2.1. Materials

PHBH containing 5.5 mol% of 3HH (M_w = 5.37 × 10^5 g/mol%) was provided by Kaneka Corp., Osaka, Japan. The natural antibacterial reagents, *centella asiatica* powder, propolis powder, hinokitiol powder were purchased from Dins Natural (pvt) Ltd., Matara, Sri Lanka; Stakich, Inc., Troy, USA; and Wako Pure Chemical Industries, Ltd., Osaka, Japan, respectively. Solvents, used to prepare electrospinning solutions, were 1,1,1,3,3,3-hexafluoro-2-propanol (HFIP) purchased from Wako Pure Chemical Industries, Ltd., Japan. Acetone, ethanol 99.5%, and methanol were obtained from Wako Pure Chemical Industries, Ltd., Japan.

Staphylococcus aureus (NBRC 12732) and *Escherichia coli* (NBRC 3972) for antibacterial activity tests were purchased from the National Institute of Technology and Evaluation, Biological Resource Center (NBRC), Tokyo, Japan.

2.2. Electrospinning Process of PHBH Composite Nanofibers Containing the Natural Antibacterial Reagents

2.2.1. Preparation of Natural Antibacterial Reagent Solutions

Centella solutions were prepared using either ethanol or methanol the concentrations of *centella* to solutions were 15% or 30% (*w/v*), respectively. The propolis solutions were prepared by dissolving propolis in acetone at 10% (*w/v*) or ethanol with the concentrations of 10% or 30% (*w/v*), respectively. Each prepared solution was mixed using a magnetic stirrer for 24 h at room temperature. Afterwards, all solutions were filtered to remove insoluble substances. The hinokitiol solution was prepared by dissolving hinokitiol powder in 99.5% ethanol with a concentration of 30% (*w/v*), also stirred at room temperature for 24 h without filtration.

The concentrations of natural products in ethanol–*centella* (15EC and 30EC), methanol–*centella* (15MC and 30MC), acetone–propolis (10AP), and ethanol–propolis (10EP and 30EP) were determined from the dry weight of each natural product by evaporating solutions under vacuum. Ethanol–hinokitiol (30EH) mentioned the same as 30% because of no insoluble substances in the ethanol solution.

These percentage concentrations of *centella*, propolis, hinokitiol are displayed in Table 1. These solutions were added to PHBH HFIP solutions with different ratios before the electrospinning process (The ratios will be explained in the next step of preparation of PHBH with natural product solutions).

Table 1. The concentrations of natural products in each solution after dry vacuum and ratios of natural product solutions to total weight of spinning poly[(*R*)-3-hydroxybutyrate-*co*-(*R*)-3-hydroxyhexanoate] (PHBH) solutions.

Concentration(*w/v*)_Solution _Natural Product (NP)	Percentage Concentration of NP Inside Solution (*w/w*)	Ratios NP Solutions over Total Weight of Spinning PHBH Solutions (*v/v*) (Code)			
		1%	5%	7%	10%
15% Ethanol–*Centella*	2.16%	15EC (1%)	15EC (5%)	-	15EC (10%)
30% Ethanol–*Centella*	4.14%	30EC (1%)	30EC (5%)	-	30EC (10%)
15% Methanol–*Centella*	3.17%	15MC (1%)	15MC (5%)	-	15MC (10%)
30% Methanol–*Centella*	7.21%	30MC (1%)	30MC (5%)	-	30MC (10%)
10% Acetone–Propolis	8.95%	10AP (1%)	10AP (5%)	10AP (7%)	-
10% Ethanol–Propolis	5.43%	10EP (1%)	10EP (5%)	10EP (7%)	-
30% Ethanol–Propolis	12.25%	30EP (1%)	30EP (5%)	30EP (7%)	-
30% Ethanol–Hinokitiol	30%	30EH (1%)	30EH (5%)	30EH (7%)	-

2.2.2. Preparation of PHBH with Natural Antibacterial Reagent Solutions for Electrospinning

PHBH was dissolved in HFIP at room temperature for 4 h to make solution 2 wt%. The natural antibacterial solutions were added into PHBH HFIP solutions and stirred continuously for 16–20 h to get homogeneous solutions. The ratios of ethanol–*centella* (15EC and 30EC) or methanol–*centella* (15MC and 30MC) were 1%, 5%, and 10% over the total weight of spinning PHBH solutions. Furthermore, ratios of acetone-propolis (10AP), ethanol–propolis (10EP and 30EP), and ethanol–hinokitiol (EH) to total weight of spinning PHBH solutions were 1%, 5%, and 7%, Table 1. These prepared blend solutions were contained into a glass syringe (10 mL) with 0.6 mm diameter of stainless steel needle. The electrospinning process was carried out using a high voltage power supply by Kato Tech, Japan. The solution feeding rate was 0.1–0.3 mm/min, 15–20 kV voltage, and 15 cm tip-to-collector distance. The environmental conditions of the electrospinning chamber were at a humidity level of 20–30 RH% and temperature ranged between 23–26 °C. The electrospun fibers were collected on the surface of the collector followed by drying at room temperature for 24 h to evaporate the solvents.

2.3. Characterization of PHBH Composite Nanofibers with Natural Antibacterial Reagent

2.3.1. Scanning Electron Microscopy (SEM)

The morphology of electrospun nanofibers was investigated by scanning electron microscopy (SEM, JSM-6010LA, JOEL, Tokyo, Japan) at an accelerating voltage of 10 kV with various magnifications. The samples were coated with platinum (Pt) in a sputtering device for 60 s at 30 mA. Image J software was used to measure 50 fibers diameters from each SEM photograph.

2.3.2. Fourier Transform Infrared (FT-IR) Spectroscopy

The presence of *centella*, propolis, and hinokitiol incorporated in electrospun nanofibers were analyzed using attenuated total reflectance (ATR) FT-IR spectroscopy (IR Prestige-21, Shimadzu Corp., Kyoto, Japan). The infrared spectra were recorded from 4000–600 cm^{-1}, with a resolution of 4 cm^{-1}. The infrared beam enters the ATR crystal at an angle of typically 45° and is totally reflected at the crystal to sample interface with a low signal of noise ratio.

2.3.3. X-ray Diffraction

Wide-angle X-ray diffraction (WAXD) was used for crystalline structure analysis of PHBH nanofibers with or without *centella*, propolis, or hinokitiol. The two-dimensional (2D) patterns of nanofibers were recorded by X-ray diffraction equipment (SPring-8 synchrotron radiation facility in Japan) with a wavelength of 0.07085 nm at a 2θ scanning angle between 5–30°. The distance from the sample to the detector (PILATUS 3 × 2 M) was 326.5 mm and the exposure time was 2.0 s. The X-ray diffraction of composite nanofibers was investigated by the MDIP application.

2.3.4. Mechanical Properties

The mechanical properties of neat PHBH, PHBH/*centella*, PHBH/propolis, PHBH/hinokitiol nanofibers were investigated by using a tensile testing machine (EZTest/EZ-S, Shimadzu Corp., Kyoto, Japan) with 50-N load cell at room temperature. Ten specimens were prepared from each composite nanofiber sheet. The specimens with thickness ranging from 0.05 to 0.14 mm were cut into a rectangular shape of 30 mm × 5 mm. The ends of each sample were secured by two clamps with an initial distance of 10 mm and a crosshead speed of 10 mm/min. Tensile strength, Young's modulus, and elongation at break were calculated on the basis of the stress-strain curves.

2.4. Sustained Release Behavior of Natural Antibacterial Reagent from PHBH Composite Nanofibers

The release of *centella*, propolis, and hinokitiol was analyzed by a UV-visible spectrophotometer (UV2550, Shimadzu Corp., Kyoto, Japan). For release experiments, PHBH composite nanofibers with *centella*, propolis, or hinokitiol were cut into 2 × 2 cm and measured specimen weight. For each sample, 2 specimens were placed in a test tube filled with 50 mL of phosphate-buffered saline (PBS) (pH 7.4) incubated at 37 °C and stirred at 150 rpm. During the release, 4 mL of the supernatant was retrieved from the release medium at the time points of 5, 10, 20, 30, 40, 50, 60, 120, 240, 360, 480, 600, 720, 1440, 2880, 4320, and 5760 min and the same amount of fresh PBS was added immediately to maintain the medium volume.

The solutions in various concentrations were used to make a standard calibration curve at wavelength 200–700 nm. In detail, certain weight of extract *centella*, extract propolis, hinokitiol (0.8 mg) was dissolved in 100 mL ethanol/PBS or methanol/PBS (1:99, v/v) and then diluted to 400, 200, 100, 50, 25, 12.5, 6.25, 3.125, and 1.56 µg/mL. UV-vis spectrophotometer at a wavelength of 270 nm, 273 nm, 224 nm was used to measure the amount of released *centella*, propolis, and hinokitiol from PHBH composite nanofibers, respectively. The total amounts of *centella*, propolis, and hinokitiol from the nanofiber mats were determined as the average value of the three tests. These results were presented in the form of a cumulative amount of release [30]:

$$\text{Cumulative amount of release (\%)} = M_t/M_\infty \times 100,$$

where M_t is the amount of *centella*, propolis, or hinokitiol released at the time t, M_∞ is the total amount of *centella*, propolis, and hinokitiol loaded in PHBH composite nanofibers.

2.5. Antibacterial Activity Test

To examine the antibacterial activity of natural antibacterial reagents, *S. aureus* and *E. coli* were chosen as representatives for gram-positive and gram-negative bacteria. The test method was the disk diffusion test, which has been well established by previous reports [2,20,30–33]. The bacterial culture was spread on the Luria Bertani (LB) agar surface by using a sterile cotton bat. The PHBH/*centella*, PHBH/propolis, and PHBH/hinokitiol composite nanofibers were placed on the surface of the Petri dishes then they were incubated at 30 °C (*E. coli*) and 37 °C (*S. aureus*) for 24 h. The diameters of inhibition zone (mm) were measured for 3 specimens each sample and the results were expressed as mean diameters with standard deviations (in millimeters).

2.6. Statistical Analysis

All experiments were conducted in triplicate and the data are presented as mean ± standard deviation (SD). The significant differences were statistically analyzed by one-way analysis of variance (ANOVA) using R free software. Statistical significance was set at $p < 0.05$ to identify which groups were significantly different from other groups.

3. Results and Discussion

3.1. Morphology of PHBH Composite Nanofibers with Natural Antibacterial Reagents

The surface morphology of PHBH nanofibers containing different natural antibacterial products was investigated using SEM. Representative SEM images and the fiber diameter distributions of neat PHBH and PHBH nanofibers loaded with different amounts of *centella* solutions (1%, 5%, and 10%), propolis solutions (1%, 5%, and 7%), and hinokitiol solutions (1%, 5%, and 7%) are exhibited in Figure 1. In general, the different natural product solutions had effects on the surface morphology of composite nanofibers.

Table 2 shows the percentage of each natural product solution used, the conditions of the electrospinning process, the average diameters, standard deviations (SD), maximum, and minimum fiber diameters of each sample. From those data, it can be confirmed that PHBH/ethanol–*centella* (15EC and 30EC) and PHBH/methanol–*centella* (15MC and 30MC) composite nanofibers showed the uniform fiber diameter. These also exhibited the same trend of a decrease in fiber diameter by increasing the ratio of *centella* solutions (EC and MC) in spinning PHBH solution. The fiber diameter, with high concentrations of *centella* solution in PHBH/30EC (10%) and PHBH/30MC (10%), diminished to 349 ± 108 nm and 332 ± 62 nm, respectively. This phenomenon may be due to the decrease in the viscosity of polymer solutions by the addition of ethanol or methanol into HFIP. The viscosity of the mixture solution reduced when the considerable amounts of *centella* solution added into the polymer matrix [5,14]. Interestingly, the high concentration of *centella* in PHBH/30EC (10%) (Figure 1) resulted in fibers merging.

Such a bonded structure was also appeared in PHBH/30EP (7%). The bonded structure in high concentration solutions of EC and EP might be due to using ethanol as the solvent and mixing with HFIP. By contrast, PHBH/30EP (7%) was accompanied by an increase in fiber diameter by loading a high concentration of 30EP. The fiber diameter was increased from 539 ± 99 nm PHBH/30EP (1%) to 739 ± 197 nm PHBH/30EP (7%). A similar behavior was obtained from cellulose acetate nanofibers with a high concentration of honey bee propolis prepared by ethanol solution [33]. Kim et al. [20] also reported that the propolis concentration increased the fiber diameter and provided the linking of PU fibers due to its adhesive properties and demonstrated the optimum utilization of the bonding element for reinforcement of nonwoven fabric. However, different results of fiber diameter were obtained from PHBH composite nanofibers containing acetone-propolis (PHBH/10AP), which are 520 ± 83 nm, 527 ± 161 nm, and 529 ± 109 nm in the cases of 1%, 5%, and 10% (*v/v*) of 10AP, respectively. This is probably because the evaporation of acetone was faster than ethanol during the electrospinning process.

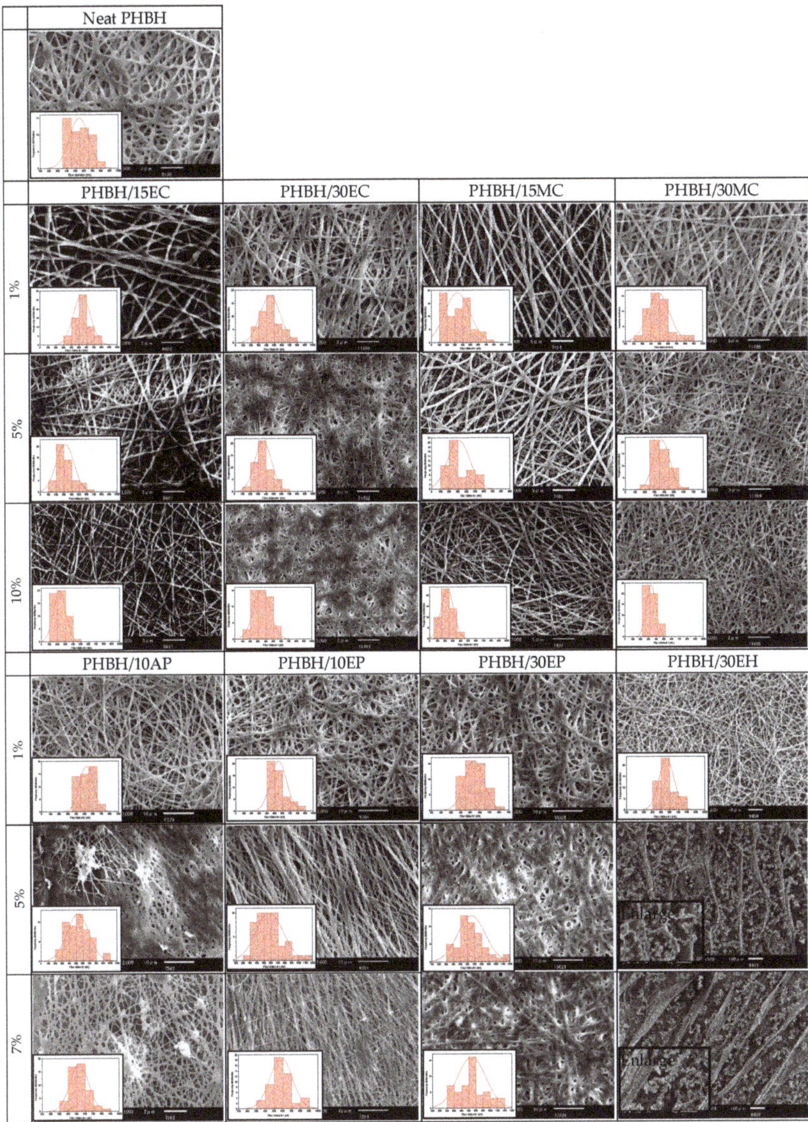

Figure 1. SEM images and diameter distributions of the electrospun neat PHBH 2 wt% and PHBH composite nanofibers containing different natural product solutions. PHBH/15EC, PHBH/30EC, PHBH/15MC, and PHBH/30MC with different concentration ratios of ethanol–*centella* (EC) and methanol–*centella* (MC) (1%, 5%, and 10%). PHBH/10AP, PHBH/10EP, PHBH/30EP, and PHBH/30EH with concentration ratios of 1%, 5%, and 7% of acetone–propolis (AP), ethanol–propolis (EP), and ethanol–hinokitiol (EH).

Table 2. Preparation of PHBH with natural product solutions, electrospinning conditions, and fiber diameter of samples.

No.	Polymer	Ratio Concentration-Solvent-Natural Product-Ratio to PHBH Solutions		Electrospinning Conditions		Fiber Diameters		
		Solvent-Natural Product	Ratios to PHBH Solutions (Code)	Applied Voltage (kV)	Flow Rate (mm/min)	Ave. Dia. (nm) ± SD	Max.	Min.
1	Neat PHBH	-	-	20	0.3	543	785	367
2	PHBH	15% Ethanol–*Centella* (EC)	PHBH/15EC (1%)	15.5	0.13	524 ± 111	760	230
			PHBH/15EC (5%)	15.5	0.13	314 ± 116	631	157
			PHBH/15EC (10%)	15.5	0.13	226 ± 77	413	107
3	PHBH	30% Ethanol–*Centella* (EC)	PHBH/30EC (1%)	15.5	0.15–0.2	447 ± 128	789	255
			PHBH/30EC (5%)	15.5	0.15–0.2	362 ± 107	690	131
			PHBH/30EC (10%)	15.5	0.15–0.2	349 ± 108	593	153
4	PHBH	15% Methanol–*Centella* (MC)	PHBH/15MC (1%)	15.5	0.15	319 ± 107	630	152
			PHBH/15MC (5%)	15.5	0.15	310 ± 126	588	135
			PHBH/15MC (10%)	15.5	0.15	227 ± 81	413	101
5	PHBH	30% Methanol–*Centella* (MC)	PHBH/30MC (1%)	15.5	0.15–0.2	487 ± 127	820	238
			PHBH/30MC (5%)	15.5	0.15–0.2	371 ± 83	571	194
			PHBH/30MC (10%)	15.5	0.15–0.2	332 ± 62	456	236
6	PHBH	10% Acetone–Propolis (AP)	PHBH/10AP (1%)	20	0.1	520 ± 83	717	374
			PHBH/10AP (5%)	20	0.1	527 ± 161	978	286
			PHBH/10AP (7%)	20	0.1	529 ± 109	890	328
7	PHBH	10% Ethanol–Propolis (EP)	PHBH/10EP (1%)	20	0.1	548 ± 104	850	422
			PHBH/10EP (5%)	20	0.1	576 ± 193	1192	289
			PHBH/10EP (7%)	20	0.1	579 ± 126	884	338
8	PHBH	30% Ethanol–Propolis (EP)	PHBH/30EP (1%)	20	0.15–0.2	539 ± 99	752	393
			PHBH/30EP (5%)	20	0.15–0.2	653 ± 194	1180	300
			PHBH/30EP (7%)	20	0.15–0.2	739 ± 197	1185	408
9	PHBH	30% Ethanol–Hinokitiol (EH)	PHBH/30EH (1%)	20	0.15–0.2	562 ± 87	745	438
			PHBH/30EH (5%)	20	0.15–0.2	-	-	-
			PHBH/30EH (7%)	20	0.15–0.2	-	-	-

For PHBH/ethanol–hinokitiol (EH) composite nanofibers, the surface morphology of PHBH/30EH (1%) was smooth, homogeneous without beads. The average diameter was 562 ± 87 nm with a majority of fibers in the range of 438 nm to 745 nm. However, it was difficult to obtain uniform nanofibers with high ratios of hinokitiol solutions (EH), more than 1%. As shown in Figure 1 of PHBH/30EH (5%) and (7%), these resulted in aligned fibers with small grains like beads.

The final surface morphologies of PHBH composite nanofibers were influenced by the kind of used natural products and concentration ratios of natural product solutions to PHBH solutions. This result clearly indicates that the fiber diameters and their distributions can be controlled by concentration ratios of natural products in solutions. In this paper, PHBH composite nanofibers with natural product solutions of PHBH/30EC (10%), PHBH/30MC (10%), PHBH/30EP (7%) and PHBH/30EH (1%) were chosen for further investigation by FTIR, WAXD, mechanical properties, antibacterial test, and release characteristic because these samples had fiber diameter between 300–1000 nm, surface morphology without beads, and high concentration of natural products.

3.2. Characteristic of PHBH Composite Nanofibers with Natural Antibacterial Product

3.2.1. FT-IR Spectral Analysis

The functional groups of natural antibacterial products contained in PHBH composite nanofibers were analyzed by using ATR (FT-IR). The FT-IR spectra of PHBH/30EC (10%), PHBH/30MC (10%), PHBH/30EP (7%), and PHBH/30EH (1%) are shown in Figure 2. The FT-IR spectra of neat PHBH nanofiber and PHBH composite nanofibers with natural products showed similar peaks. The characteristic peaks of PHBH at 2970, 2937, 2870, and 1719 cm^{-1} were attributed to the stretching vibrations of C–CH$_3$, CH$_2$, CH, and C=O, respectively [4].

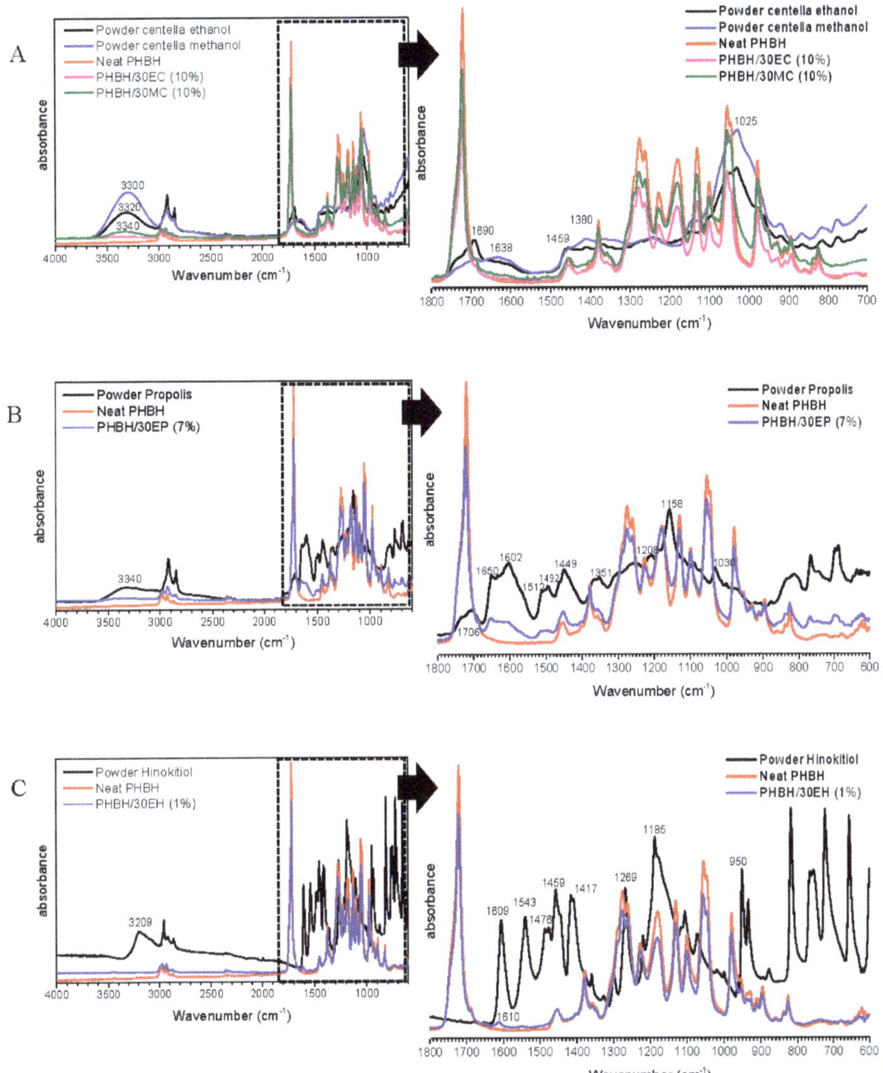

Figure 2. FT-IR spectra of neat PHBH and PHBH composite nanofibers with (**A**) EC and MC, (**B**) EP, (**C**) EH.

FT-IR graphs of *centella* powder from ethanol and methanol solutions after filtering and drying exhibit some bands centered at 3320 or 3300 cm^{-1} corresponding to O–H stretching vibration of carboxylic acid, whereas, 1690 or 1638 cm^{-1} peaks characterizing C–O stretching vibration. The bands at 1459, 1380, and 1025 cm^{-1} shows the presence of alkenes with C–H in-plane bending, the stretching vibration of C–N for aromatic amide group, and C–O stretching, respectively (Figure 2A). These spectra of those compounds were confirmed in the research of *centella asiatica* by Manotham et al. [5], Sondari et al. [34], and Sugunabai et al. [35]. As showed in Figure 2A, it proved that using methanol as solvent yielded higher content of Asiatic acid than using ethanol with the same concentration from raw *centella* (30%EC and 30%MC). In the case of PHBH/30MC (10%) composite nanofibers, the decline of the band at 3340 cm^{-1} suggested the formation of the intermolecular hydrogen bond between *centella* and PHBH polymer.

The FT-IR spectra of propolis powder and PHBH/30EP (7%) nanofibers were exhibited in Figure 2B. In the case of propolis powder, the band at 3340 cm^{-1} was assigned to stretching vibration of the O–H group in the phenolic compound and the band at 2920–2870 cm^{-1} was attributed to C–H aliphatic stretching vibration (stretching vibration of CH$_2$ and CH$_3$) [36]. The bands at 1706 cm^{-1}, 1650–1602 cm^{-1}, and 1190 cm^{-1} were attributed to the C=O group, aromatic ring deformations C=C stretching vibration, and C–O stretching vibration, respectively. In PHBH/30EP (7%) composite nanofibers, some peaks appeared at between 1651 and 1610 cm^{-1}, and 1510 cm^{-1}, confirming the presence of propolis in PHBH nanofiber mats. These bands were assigned to stretching of C=C aromatic ring bands in flavonoids [17,36].

The characteristic absorption peaks of hinokitiol powder were observed at 3200, 1609, 1543, 1476, 1459, 1417, 1269, 1185, and 950 cm^{-1}. These bands represented O–H stretching, C=C and C=O stretching, C=C stretching (in-phase), ring CH bending, C–O–H group, and ring CH bending, respectively [37]. Dyrskov et al. [23] reported that the absorbance bands in the hinokitol spectrum ascribed to C=O stretching in the tropolone carbon ring (1609 and 1543 cm^{-1}) and the C–O–H group (1189 cm^{-1}). The presence of hinokitiol in PHBH composite nanofibers was confirmed by the weak peak appeared at 1610 cm^{-1} that is related to tropolone carbon ring. In general, the peak intensity decreased by loading natural product might reveal that natural product is homogeneously distributed through the polymer matrix [20]. The nanofibrous samples with higher contents of natural antibacterial substances showed the peaks of carboxylic acid, aromatic ring, and tropolone carbon ring from *centella*, propolis, and hinokitiol, respectively.

3.2.2. Crystalline Structure by Wide-Angle X-ray Diffraction (WAXD)

The WAXD measurement was performed in order to confirm the influence of loading *centella*, propolis, and hinokitiol on the crystallinity of PHBH. All reflections of WAXD and intensity profile are displayed in Figure 3. The neat PHBH nanofibers exhibit diffraction peaks at 2θ = 6.1°, 7.7°, 9°, 10°, 11.7°, 12.4°, and 13.8° which were assigned to the (020), (110), (101), (121), (040), and (002) of the orthorhombic unit cell of PHB crystal, respectively [38]. The intensity of diffraction peaks of neat PHBH nanofibers slightly increased by loading *centella* and propolis in PHBH/30EC (10%) and PHBH/30EP (7%) composite nanofibers. Moreover, the calculated crystallinity of neat PHBH (47.0%) was slightly lower than that of PHBH/30EC (55.0%), PHBH/30MC (49.2%), and PHBH/30EP (54.5%) in Table 3. It was previously reported that the crystallinity of PHB nanofibers increased by adding the natural phenolic compound caffeic acid [30]. Kim et al. [39] supposed that intermolecular interaction through a hydrogen bonding of plant polyphenol with polycaprolactone probably increased by slight increment of crystallinity in polycaprolactone. Whereas, the crystallinity of PHBH/30EH (1%) composite nanofibers was declined to 44.1% compared to neat PHBH. These results indicate that an intermolecular interaction between PHBH polymer and *centella* or propolis in nanofibers mat exists.

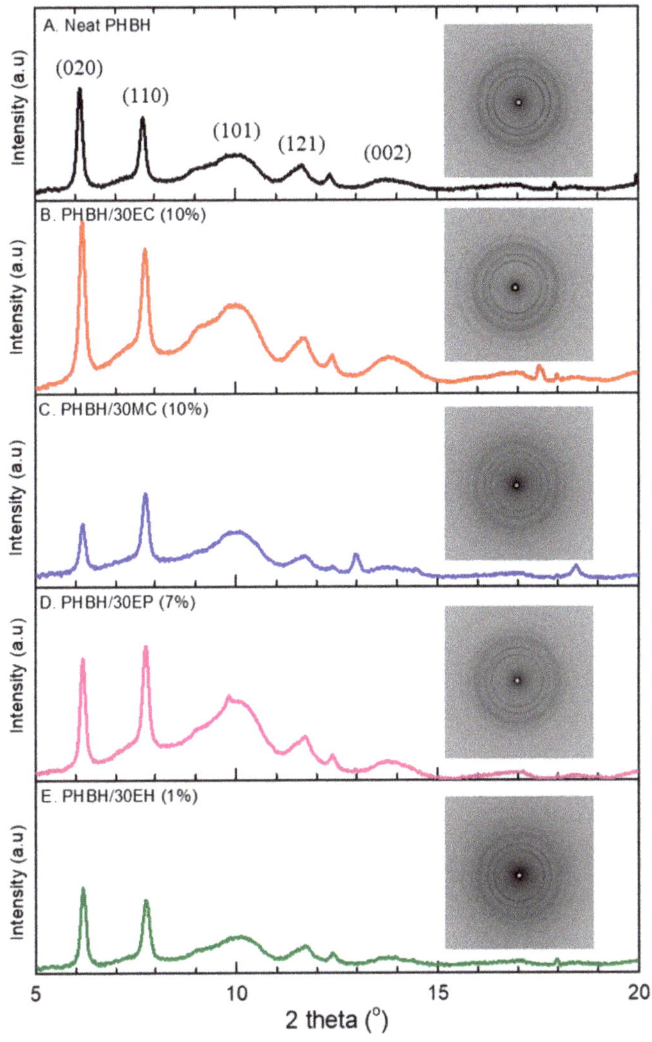

Figure 3. Wide-angle X-ray diffraction (WAXD) intensity profiles and 2D patterns of (**A**) neat PHBH and PHBH composite nanofibers with natural products: (**B**) EC, (**C**) MC, (**D**) EP, (**E**) EH.

Table 3. Mechanical properties and crystallinity of neat PHBH and PHBH composite nanofibers with natural products.

Sample	Tensile Strength	Elongation at Break	Young's Modulus	Crystallinity
	(MPa)	(%)	(MPa)	(%)
Neat PHBH	8.00 ± 0.71	61.49 ± 17.38	291.5 ± 41.8	47.0
PHBH/30EC (10%)	17.79 ± 4.71 *	15.57 ± 4.56 *	420.4 ± 146.1 *	55.0
PHBH/30MC (10%)	8.98 ± 1.47	10 ± 3.79 *	291.7 ± 118.9	49.2
PHBH/30EP (7%)	16.35 ± 1.78 *	9.27 ± 3.32 *	545.6 ± 162.8 *	54.5
PHBH/30EH (1%)	2.14 ± 0.70 *	8.33 ± 1.75 *	78 ± 33.8 *	44.1

* $p < 0.05$ considered statistically significant between PHBH composite nanofibers in each group against neat PHBH nanofibers ($n = 10$) (except crystallinity, $n = 2$).

3.2.3. Mechanical Characteristic of Composite Nanofibers

Tensile test was conducted on neat PHBH and PHBH composite nanofibers with the natural antibacterial products in order to evaluate the influences of natural product loading. The tensile strength, elongation at break, and Young's modulus of neat PHBH, PHBH/30EC (10%), PHBH/30MC (10%), PHBH/30EP (7%), and PHBH/30EH (1%) are shown in Table 3. Figure 4 illustrates the representative stress–strain curves of PHBH composite nanofiber samples. The tensile strength of neat PHBH was 8.0 MPa with high elongation up to 61.5%. Interestingly, the loading of *centella* (EC 10%) and propolis (EP 7%) in PHBH demonstrated the same trend with the tensile strength rose up to 17.8 and 16.4 MPa but it reduced elongation down to 15.6% and 9.3%, respectively. PHBH/30MC (10%) composite nanofibers exhibited no significant differences in tensile strength and Young's modulus compared to neat PHBH. PHBH/30EH (1%) resulted in a decline in both tensile strength and elongation compared to neat PHBH.

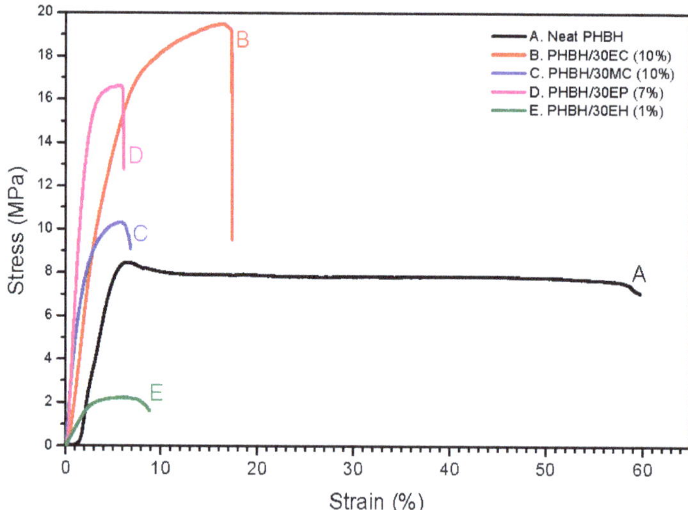

Figure 4. Representative stress-strain curves of (A) neat PHBH and PHBH composite nanofibers with (B) EC, (C) MC, (D) EP, and (E) EH.

PHBH composite nanofibers with *centella* (EC) and propolis (EP) using ethanol as solvent displayed higher tensile strength in comparison to neat PHBH, PHBH/MC, and PHBH/EH. These phenomena may be due to the surface morphology and polymer structure in PHBH composite nanofibers. The surface morphology of PHBH/30EC (10%) and PHBH/30EP (7%) showed bonded-structure between fibers, whereas PHBH/30MC (10%) and PHBH/30EH (1%) surface revealed randomly oriented fibers. Kim et al. [20] reported the similar behaviors of merged structure, the adhesive properties of propolis might be useful to increase the mechanical strength of polyurethane fibers. The mechanical properties of nonwoven nanofibers also depended on parameters such as surface interaction among the fibers, average fiber diameter, fiber defects (bead formation) during electrospinning [10,11].

Additionally, the enhancement of physical properties in PHBH/30EC (10%) and PHBH/30EP (7%) might be related to the rise of the crystallinity of PHBH polymer in PHBH composite nanofibers. It is probably due to the interaction of natural products (EC and EP) to PHBH polymer that was explained by FT-IR and X-ray analysis.

3.3. In Vitro Antibacterial Activity

The antibacterial activity of neat PHBH nanofibers, PHBH composite nanofibers with *centella*, propolis, and hinokitol against the gram-positive bacteria *S. aureus* and gram-negative bacteria *E. coli* was evaluated using inhibition zone method [20,33,40,41]. The antimicrobial effect of samples was evaluated at 37 °C (*S. aureus*) and 30 °C (*E. coli*) for 24 h. Table 4 presents the diameters of the inhibition zone for PHBH composite nanofibers with natural products. The effects of natural antibacterial products via the agar diffusion test are clearly shown in Figure 5. PHBH/30EC (10%) was observed to have no effect against gram-positive and gram-negative bacteria. Whereas, *centella* in PHBH/30MC (10%) demonstrated low antibacterial activity against *S. aureus* with inhibitory zone of 7.7 mm and no effect against *E. coli*. The diameter of the inhibition zone of PHBH/30EP (7%) was 18.3 mm for *S. aureus* and 17.3 mm for *E. coli*. PHBH/30EH (1%) composite nanofibers have extensive inhibition zone against *S. aureus* (25.7 mm) and *E. coli* (29.7 mm), respectively. The zones of inhibition for PHBH/30EP (7%) against *S. aureus* were slightly broader than that against *E. coli*. Zeighampour et al. [21] explained that hydroalchoholic extract of propolis showed better antibacterial activities against gram-negative bacteria than gram-positive bacteria due to the different cell wall chemical structures of bacteria. However, the opposite results for PHBH/30EH (1%) might be due to the different modes of action and the bactericidal effects of natural products.

Table 4. Zones of inhibition for PHBH composite nanofibers with different natural products against *S. aureus* and *E. coli*.

Sample	Inhibition Zones (mm)	
	S. aureus	*E. coli*
Neat PHBH	0	0
PHBH/30EC (10%)	0	0
PHBH/30MC (10%)	7.7 ± 6.7	0
PHBH/30EP (7%)	18.3 ± 1.5 *	17.3 ± 5.1 *
PHBH/30EH (1%)	25.7 ± 4.0 *	29.7 ± 1.5 *

* $p < 0.05$ considered statistically significant, PHBH composite nanofibers in each group were compared with PHBH nanofibers ($n = 3$).

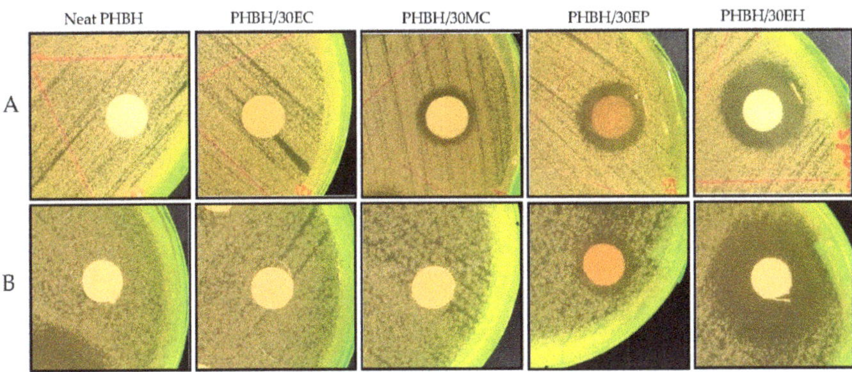

Figure 5. Representative inhibition zones of neat PHBH and PHBH composite nanofibers with *centella* (30EC) and (30MC), propolis (30EP), and hinokitiol (30EH) on gram-positive bacteria (*S. aureus*) (**A**) and gram-negative bacteria (*E. coli*) (**B**).

Regarding these results, the antimicrobial effects of PHBH composite nanofibers with propolis and hinokitiol were observed to be better than these of *centella*. However, the low effect antibacterial activity of *centella* (EC or MC) might be due to the low chemical content in PHBH composite nanofiber as analysed by FT-IR, which PHBH/30EC (10%) showed no clear detection of asiatic acid. For further work, raising the *centella* concentrations in ethanol or methanol solutions might be one of the solutions to increase the antibacterial activity, if those will be used for wound healing application. Overall, the natural products (especially propolis and hinokitiol) loaded into PHBH composite nanofibers inhibited the growth of *S. aureus* and *E. coli* powerfully.

3.4. Release Behavior of Natural Antibacterial Product

The release profiles of *centella*, propolis, and hinokitiol from electrospun PHBH composite nanofibers were plotted as a function of time in PBS with pH 7.4 at 37 °C. The release curves of PHBH/30EC (10%), PHBH/30MC (10%), PHBH/30EP (7%), and PHBH/EH (1%) composite nanofibers demonstrated different release behaviors in each sample as shown in Figure 6. *Centella* from PHBH/30EC started to release within the 10 min and completely released in 20 min. In the case of PHBH/30MC, Figure 6B, the similar behaviors were also observed with complete release in 20 min. The maximum releases of *centella* from PHBH/30EC and PHBH/30MC were 13.9 % and 31.5%, respectively. While the significant release amounts of propolis from PHBH/30EP started to be detected after 10 min and gradually increased up to 2880 min (48 h) with the release amount of 9.6%. The release of propolis from PHBH/30EP sustained longer than others but the release amount of propolis was less. The fast release of hinokitiol from PHBH/30EH was noticeable in the first 5 min and gradually increased up to 20 min with a maximum release of 46.1%. Afterwards, the release of hinokitiol gradually declined over 240 min (4 h). This decline might be due to the degradation of hinokitiol in PBS solution by light, heat, or solvent (ethanol). In the study of ciprofloxacin hydrochloride (CpHCl) release from electrospun alginate [42], at around 24% release of total loaded drug, the cross-linking process between CpHCl molecules in the nanofibers happened and affected the release behaviors. This result also suggests the relation between the release capacity and the characteristics of PHBH, which is hydrophobic polymer. Ignatova et al. [43] reported that the release of caffeic acid phenethyl ester (CAPE) in material was influenced by hydrophilic–hydrophobic features of the fibrous mats, the release of CAPE increased markedly when CAPE was incorporated in PVP (hydrophilic) matrix than when incorporated in PHB (hydrophobic) matrix.

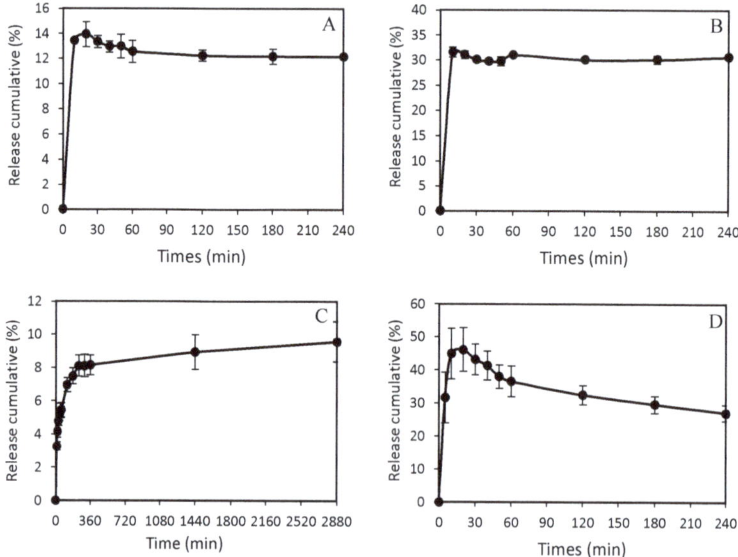

Figure 6. Cumulative release profiles of natural antibacterial products in PHBH composite nanofibers (**A**) PHBH/30EC (10%), (**B**) PHBH/30MC (10%), (**C**) PHBH/30EP (7%), and (**D**) PHBH/30EH (1%).

Especially for PHBH/30EC (10%) and PHBH/30EP (7%), those have similar release behaviors, when only small amounts of natural products were released to PBS. It can be considered due to the high crystallinity of those PHBH composite nanofibers that affected release characteristics. In comparison, PHBH/30EH (1%) with low crystallinity showed fast and high release amounts. Study of drug release of ampicillin incorporated poly(methyl methacrylate)–nylon6 core/shell nanofibers proved that the increase of released drug, most probably due to decreased crystallinity in the polymer matrix [44].

4. Conclusions

The natural antibacterial products (*centella*, propolis, and hinokitiol) loaded into PHBH composite nanofibers were successfully fabricated by the electrospinning process. The fiber diameter and surface morphology of PHBH composite nanofibers with antibacterial reagents can be controlled by different kinds of natural antibacterial products, solvent systems for dissolution, and component concentrations in the PHBH-HFIP polymer solutions. The presence of carboxylic acid and aromatic amide groups in PHBH/*centella*, aromatic ring bands due to flavonoid in PHBH/propolis, and tropolone carbon ring in PHBH/hinokitiol were confirmed by FT-IR. The loading of *centella* and propolis led to an increase in the crystallinity of the PHBH polymer. Furthermore, the loading of *centella* and propolis improved the tensile strength, compared to neat PHBH nanofibers. Hinokitiol and propolis were proved to be potent antibiotics by large inhibition zones against both *E. coli* and *S. aureus*. The release of *centella* and hinokitiol from PHBH nanofibers was fast and finished in 20 min with maximum release of 13.9% (PHBH/30EC), 31% (PHBH/30MC), and 46% (PHBH/30EH). Whereas, the release of propolis was continuously to 48 h and maximal release was 9.5%. These results in our study indicated that natural antibacterial products loaded into PHBH composite nanofibers can improve the mechanical properties (PHBH/30EP and PHBH/30EC) and prove the composite nanofibers antibacterial effects against the gram-negative and gram-positive bacteria (PHBH/30EP and PHBH/30EH), which are important characteristics as wound healing materials. Interestingly, PHBH/30EC (10%) showcased good mechanical properties and PHBH/30MC (10%) exhibited good morphology in nanofiber form, however, it may necessary to increase the concentration of *centella* in *centella* solution (ethanol or methanol) if it is to be used as antibacterial reagents. PHBH/30EH (1%) might be used as drug delivery with rapid release but possesses low mechanical properties. PHBH/30EP (7%) could be used in wound healing with needed mechanical properties and slow release for long-time effects.

Author Contributions: Conceptualization, R.A.R., N.S.b.S., and T.T.; formal analysis, R.A.R. and N.S.b.S.; investigation, R.A.R. and N.S.b.S.; data curation, R.A.R. and T.T.; writing—original draft, R.A.R.; writing—review and editing, R.A.R. and T.T.; supervision, T.T.; funding acquisition, R.A.R. and T.T.

Funding: This research was supported by Shinshu University Advanced Leading Graduate Program for Fiber Renaissance from the Ministry of Education, Culture, Sport, Science, and Technology (MEXT), Japan.

Acknowledgments: The authors are grateful to Kaneka Corp. for providing the PHBH as materials. The synchrotron radiation experiments were performed at the BL45XU of SPring-8 with the approval of RIKEN (proposal No. 20180079) and at the BL40B2 of SPring-8 with the approval of the Japanese Synchrotron Radiation Research Institute (JASRI) (proposal No. 2019A1213).

Conflicts of Interest: The authors declare no conflict of interest.

References

1. Sh Asran, A.; Razghandi, K.; Aggarwal, N.; Michler, G.H.; Groth, T. Nanofibers from blends of polyvinyl alcohol and polyhydroxy butyrate as potential scaffold material for tissue engineering of skin. *Biomacromolecules* **2011**, *11*, 3413–3421. [CrossRef] [PubMed]
2. Asawahame, C.; Sutjarittangtham, K.; Eitssayeam, S.; Tragoolpua, Y.; Sirithunyalung, B.; Sirithunyalug, J. Antibacterial activity and inhibition of adherence of Streptococcus mutans by propolis electrospun fibers. *AAPS Pharm. Sci. Tech.* **2015**, *16*, 182–191. [CrossRef] [PubMed]

3. Lee, H.S.; Lee, S.Y.; Park, S.H.; Lee, J.H.; Ahn, S.K.; Choi, Y.M.; Choi, D.J.; Chang, J.H. Antimicrobial medical sutures with caffeic acid phenethyl ester and their in vitro/in vivo biological assessment. *Med. Chem. Commun.* **2013**, *4*, 777–782. [CrossRef]

4. Rebia, R.A.; Rozet, S.; Tamada, Y.; Tanaka, T. Biodegradable PHBH/PVA blend nanofibers: Fabrication, characterization, in vitro degradation, and in vitro biocompatibility. *Polym. Degrad. Stab.* **2018**, *154*, 124–136. [CrossRef]

5. Manotham, S.; Pengpat, K.; Eitssayeam, S.; Rujijanagul, G.; Sweatmam, D.R.; Tunkasiri, T. Fabrication of Polycaprolactone/Centella asiatica Extract Biopolymer Nanofiber by Electrospinning. *Appl. Mech. Mater.* **2015**, *804*, 151–154. [CrossRef]

6. Philip, S.; Keshavarz, T.; Roy, I. Polyhydroxyalkanoates: Biodegradable polymers with a range of applications. *J. Chem. Technol. Biotechnol* **2007**, *82*, 233–247. [CrossRef]

7. Ignatova, M.G.; Manolova, N.E.; Rashkov, L.B.; Markova, N.D.; Toshkova, R.A.; Georgieva, A.K.; Nikolova, E.B. Poly(3-hydroxybutyrate)/caffeic acid electrospun fibrous material coated with polyelectrolyte complex and their antibacterial activity and in vitro antitumor effect against HeLa cells. *Mater. Sci. Eng. C Mater.* **2016**, *65*, 379–392. [CrossRef]

8. Bulman, S.E.L.; Goswami, P.; Tronci, G.; Russell, S.J.; Carr, C. Investigation into the potential use of poly (vinyl alcohol)/methylglyoxal fibres as antibacterial wound dressing components. *J. Biomater. Appl.* **2015**, *8*, 1198–1200. [CrossRef]

9. Gun, J.; Zhang, M. Polyblend nanofibers for biomedical applications: Perspectives and challenges. *Cell Press* **2010**, *28*, 189–197. [CrossRef]

10. Scaffaro, R.; Lopresti, F. Processing, structure, property relationships and release kinetics of electrospun PLA/Carvacrol membranes. *Eur. Polym. J.* **2018**, *100*, 165–171. [CrossRef]

11. Mutlu, G.; Calamak, S.; Ulubayram, K.; Guven, E. Curcumin-loaded electrospun PHBV nanofibers as potential wound-dressing material. *J. Drug Deliv. Sci. Technol.* **2018**, *43*, 185–193. [CrossRef]

12. Andreau, V.; Mendoza, G.; Arruebo, M.; Irusta, S. Review: Smart dressings based on nanostructured fibers containing natural original antimicrobial, anti-inflammatory, and regenerative compounds. *Material* **2015**, *8*, 5154–5193. [CrossRef] [PubMed]

13. Kwon, M.C.; Choi, W.Y.; Seo, Y.C.; Kim, J.S.; Yoon, C.S.; Lim, H.W.; Kim, H.S.; Ahn, J.H.; Lee, H.Y. Enhancement of the skin-protective activities of Centella asiatica L. Urban by a nano-encapsulation process. *J. Biotechnol.* **2012**, *157*, 100–106. [CrossRef] [PubMed]

14. Yao, C.H.; Yeh, J.Y.; Chen, Y.S.; Li, M.H.; Huang, C.H. Wound-healing effect of electrospun gelatin nanofibres containing *Centella asiatica* extract in a rat model. *J. Tissue Eng. Regen. Med.* **2017**, *11*, 905–915. [CrossRef]

15. Sikareepaisan, P.; Suksamrarn, A.; Supaphol, P. Electrospun gelatin fiber mats containing a herbal-Centella asiatica-extract and release characteristic of asiatocoside. *Nanotechnology* **2008**, *19*, 015102. [CrossRef]

16. Azis, H.A.; Taher, M.; Ahmed, A.S.; Sulaiman, W.M.A.W.; Susanti, D.; Chowdhury, S.R.; Zakaria, Z.A. In vitro and In vivo wound healing studies of methanolic fraction of Centella asiatica extract. *S. Afr. J. Bot.* **2017**, *108*, 163–174. [CrossRef]

17. Lima, G.G.; Souza, R.O.; Bozzi, A.D.; Poplawska, M.A.; Devine, D.M.; Nugent, M.J.D. Extraction Method Plays Critical Role in Antibacterial Activity of Propolis-Loaded Hydrogels. *J. Pharm. Sci.* **2016**, *105*, 1–10.

18. Adomavičiūtė, E.; Stanys, S.; Žilius, M.; Juškaitė, V.; Pavilonis, A.; Briedis, V. Formation and biopharmaceutical characterization of electrospun PVP mats with propolis and silver nanoparticles for fast releasing wound dressing. *Biomed. Res. Int.* **2016**. [CrossRef]

19. Wang, J.; Vermerris, W. Antimicrobial nanomaterials derived from natural products—A review. *Material* **2016**, *9*, 255. [CrossRef]

20. Kim, J.I.; Pant, H.R.; Sim, H.J.; Lee, K.M.; Kim, C.S. Electrospun propolis/polyurethane composite nanofibers for biomedical applications. *Mater. Sci. Eng. C Mater.* **2014**, *44*, 52–57. [CrossRef]

21. Zeighampour, F.; Alihosseini, F.; Morshed, M.; Rahimi, A.A. Comparison of prolonged antibacterial activity and release profile of propolis-incorporated PVA nanofibrous mat, microfibrous mat, and film. *J. Appl. Polym. Sci.* **2018**, *135*, 45794. [CrossRef]

22. Paul, S.; Emmanuel, T.; Matchawe, C.; Alembert, T.T.; Elisabeth, Z.M.; Sophie, L.; Luce, V.E.; Maurice, T.F.; Joel, Y.G.A.; Alex, A.D.T.; et al. Pentacyclic triterpenes and crude extracts with antimicrobial activity from Cameroonian brown propolis samples. *J. Appl. Pharm. Sci.* **2014**, *4*, 1–9.

23. Dyrskov, L.; Strobel, B.W.; Svensmark, B.; Hansen, H.C.B. β-Thujaplicin: New Quantitative CZE Method and Adsorption to Goethite. *J. Agric. Food Chem.* **2004**, *52*, 1452–1457. [CrossRef] [PubMed]

24. Shih, Y.-H.; Lin, D.-J.; Chang, K.-W.; Hsia, S.-M.; Ko, S.-Y.; Lee, S.-Y.; Hsue, S.-S.; Wang, T.-H.; Chen, Y.-L.; Shieh, T.-M. Evaluation physical characteristics and comparison antimicrobial and anti-inflammation potentials of dental root canal sealers containing hinokitiol in vitro. *PLoS ONE* **2014**, *9*, e94941. [CrossRef] [PubMed]

25. Hosoda, N.; Tsujimoto, T.; Uyama, H. Green composite of poly (3-hydroxybutyrate-co-3-hydroxyhexanoate) reinforced with porous cellulose. *ACS Sustain. Chem. Eng.* **2014**, *2*, 248–253. [CrossRef]

26. Bugnicourt, E.; Cinelli, P.; Lazzeri, A.; Alvare, V. Polyhydroxyalkanoate (PHA): Review of synthesis, characteristics, processing and potential applications in packaging. *Express Polym. Lett.* **2014**, *8*, 791–808. [CrossRef]

27. Iwata, T.; Tsuge, T.; Taguchi, S.; Abe, H.; Tanaka, T. Bio-Polyesters. In *Bio-Based Polymers*, 1st ed.; Kimura, Y., Ed.; CMC Publishing Co., Ltd.: Tokyo, Japan, 2013; pp. 71–85. ISBN 978-4-7813-0271-3.

28. Lu, X.; Wang, L.; Yang, Z.; Lu, H. Strategies of polyhydroxyalkanoates modification for the medical application in neural regeneration/nerve tissue engineering. *Adv. Biosci. Biotechnol.* **2013**, *4*, 731–740. [CrossRef]

29. Lee, S.Y. Bacterial polyhydroxyalkanoates. *Biotechnol. Bioeng.* **1996**, *49*, 1–14. [CrossRef]

30. Ignatova, M.; Manolova, N.; Rashkov, I.; Markova, N. Quaternized chitosan/κ-carrageenan/caffeic acid-coated poly(3-hydroxybutyrate) fibrous materials: Preparation, antibacterial and antioxidant activity. *Int. J. Pharm.* **2016**, *513*, 528–537. [CrossRef]

31. Phan, D.-N.; Dorjjugder, N.; Khan, M.Q.; Saito, Y.; Taguchi, G.; Lee, H.; Mukai, Y.; Kim, I.-S. Synthesis and attachment of silver and copper nanoparticles on cellulose nanofibers and comparative antibacterial study. *Cellulose* **2019**, *26*, 6629–6640. [CrossRef]

32. Phan, D.-N.; Dorjjugder, N.; Saito, Y.; Taguchi, G.; Lee, H.; Lee, J.S.; Kim, I.-S. The mechanistic actions of different silver species at the surfaces of polyacrylonitrile nanofibers regarding antibacterial activities. *Mater. Today Commun.* **2019**, in press. [CrossRef]

33. Sharaf, S.; El-Naggar, M.E. Eco-friendly technology for preparation, characterization and promotion of honey bee propolis extract loaded cellulose acetate nanofibers in medical domains. *Cellulose* **2018**, *25*, 5195–5204. [CrossRef]

34. Sondari, D.; Harmami, S.B.; Ghozali, M.; Randy, A.; Amanda-S, A.; Irawan, Y. Determination of the active asiaticoside content in *centella asiatica* as anti-cellulite agent. *Indones. J. Cancer Chemoprevent.* **2011**, *2*, 222–227. [CrossRef]

35. Sugunabai, J.; Jeyaraj, M.; Karpagam, T. Analysis of functional compounds and antioxidant activity of centella asiatica. *World J. Pharm. Pharm. Sci.* **2015**, *4*, 1982–1993.

36. Oliveira, R.N.; Mancini, M.C.; Oliveira, F.C.S.; Passos, T.M.; Quilty, B.; Thiré, R.M.S.; McGuinness, G.B. FTIR analysis and quantification of phenols and flavonoids of five commercially available plants extracts used in wound healing. *Rev. Mater.* **2016**, *21*, 767–779. [CrossRef]

37. Ikegami, Y. The infrared spectra of troponoid compounds. VI. The infrared and Raman spectra of tropolone, 3- and 4-isopropyltropolones. *Bull. Chem. Soc. Jpn.* **1963**, *36*, 1118–1125. [CrossRef]

38. Ying, T.H.; Ishii, D.; Mahara, A.; Murakami, S.; Yamaoka, T.; Sudesh, K.; Samian, R.; Fujita, M.; Maeda, M.; Iwata, T. Scaffolds from electrospun polyhydroxyalkanoate copolymers: Fabrication, characterization, bioabsorption and tissue response. *Biomaterials* **2008**, *29*, 1307–1317. [CrossRef]

39. Kim, Y.-J.; Park, M.R.; Kim, M.S.; Kwon, O.H. Polyphenol-loaded polycaprolactone nanofibers for effective growth inhibition of human cancer cells. *Mater. Chem. Phys.* **2012**, *133*, 674–680. [CrossRef]

40. Moradkhannejhad, L.; Abdouss, M.; Nikfarjam, N.; Mazinani, S.; Heydari, V. Electrospinning of zein/propolis nanofibers; antimicrobial properties and morphology investigation. *J. Mater. Sci. Mater. Med.* **2018**, *29*, 165–175. [CrossRef]

41. Klančnik, A.; Piskernik, S.; Jeršek, B.; Možina, S.S. Evaluation of diffusion and dilution methods to determine the antibacterial activity of plant extracts. *J. Microbiol. Methods* **2010**, *81*, 121–126. [CrossRef]

42. Kyziol, A.; Michna, J.; Moreno, I.; gemez, E.; Irusta, S. Preparation and characterization of electrospun alginate nanofibers loaded with ciprofloxacin hydrochloride. *Eur. Polym. J.* **2017**, *96*, 350–360. [CrossRef]

43. Sohrabi, A.; Shaibani, P.M.; Etayash, H.; Kaur, K.; Thundat, T. Sustained drug release and antibacterial activity of ampicillin incorporated poly (methyl methacrylate)enylon6 core/shell nanofibers. *Polymer* **2013**, *54*, 2699–2705. [CrossRef]

44. Ignatova, M.; Manolova, N.; Rashkov, I.; Markova, N.D. Antibacterial and antioxidant electrospun materials from Poly(3-hydroxybutyrate) and polyvinylpyrrolidone containing caffeic acid phenethyl ester-"in" and "on" strategies for enhanced solubility. *Int. J. Pharm.* **2018**, *545*, 342–356. [CrossRef] [PubMed]

 *nanomaterials*

Article

Influence of the Introduced Chitin Nanofibrils on Biomedical Properties of Chitosan-Based Materials

Ekaterina N. Maevskaia [1,*], Anton S. Shabunin [1,2], Elena N. Dresvyanina [1,3], Irina P. Dobrovol'skaya [1,4], Vladimir E. Yudin [1,4], Moisey B. Paneyah [5], Andrey M. Fediuk [5], Petr L. Sushchinskii [5], Gerald P. Smirnov [3], Evgeniy V. Zinoviev [5,6] and Pierfrancesco Morganti [7]

[1] Department of Medical Physics, Peter the Great St. Petersburg Polytechnic University, Polytechnicheskaya Street 29, 195251 Saint Petersburg, Russia; anton-shab@yandex.ru (A.S.S.); elenadresvyanina@gmail.com (E.N.D.); zair2@mail.ru (I.P.D.); yudinve@gmail.com (V.E.Y.)

[2] H.Turner National Medical Research Center for Children's Orthopedics and Trauma Surgery, Parkovaya Street 64-68, 196603 Pushkin, Saint-Petersburg, Russia

[3] Institute of Textile and Fashion, Saint Petersburg State University of Industrial Technologies and Design, Bolshaya Morskaya Street 18, 191186 Saint Petersburg, Russia; tpnm@sutd.ru

[4] Laboratory of Mechanics of Polymers and Composites, Institute of Macromolecular Compounds, Bolshoy pr. V.O. 31, 199004 Saint Petersburg, Russia

[5] Laboratory of Experimental Surgery of Scientific Research Center, Saint Petersburg State Pediatrical Medical University, Litovskaya Street 2, 194100 Saint Petersburg, Russia; moisey031190@gmail.com (M.B.P.); andrej.fedyuk@gmail.com (A.M.F.); petersuchinsky@mail.ru (P.L.S.); evz@list.ru (E.V.Z.)

[6] Saint-Petersburg I. I. Dzhanelidze Research Institute of Emergency Medicine, Budapeshtskaya Street 3, 192242 Saint Petersburg, Russia

[7] Department of Experimental Medicine, University of Campania Luigi Vanvitelli, via L. De Crecchio 7, 80138 Naples, Italy; pierfrancesco.morganti@mavicosmetics.it

* Correspondence: ma.eka@yandex.ru

Received: 30 March 2020; Accepted: 12 May 2020; Published: 15 May 2020

Abstract: Hemorrhage occurring during and after surgery still remains one of the biggest problems in medicine. Although a large number of hemostatic products have been created, there is no universal preparation; thus, the development of new materials is an urgent task. The aim of this research is to increase hemostatic properties of chitosan by introducing chitin nanofibrils (ChNF). The blood absorbance by ChNF-containing chitosan sponges and time-until-arrest of bleeding were studied. Non-woven materials containing 0.5% of ChNF and materials without chitin were obtained. The studies of ζ-potential showed that the material containing 0.5% ChNF had relatively a high positive charge, but efficiencies of both materials for hemorrhage arrest were comparable to those of commercial hemostatic products (Surgicel and TachoComb). To investigate the interaction between the materials and living organism, histological studies and optical microscopy studies were conducted after implantation of fibers. Despite bioinertness of fibers, implantation of non-woven materials led to formation of significant granulomas.

Keywords: chitosan; chitin nanofibrils; hemostatic material; hemorrhage

---

1. Introduction

Uncontrolled hemorrhage remains one of the most significant causes of mortality [1] in patients with traumas of various etiologies and medical conditions. Moreover, a subsequent blood loss leads to numerous possible complications and side effects in these patients, up to lethal hemorrhagic shock and multi-organ failure. Thereby, rapid and effective arrest of bleeding is an extremely important task. Nowadays, there is a huge variety of methods intended to reduce bleeding time and amount of blood loss, but the use of topical hemostatic materials remains the most popular and common way due to its simplicity, efficiency and ease of application. The ideal hemostatic material should be able to stop massive bleeding from large vessels and parenchymatous organs quickly and effectively to prevent repeated hemorrhage. Such properties as biocompatibility and ability for biodegradation are also quite important, as well as the absence of any negative effects in the adjacent tissues and major systems, ease of use and low cost [2,3]. Although numerous hemostatic materials are presented on the market, there is still no universal and ideal material; thus, development of new hemostatic products is an important task of medical materials science.

Hemostatic agents can be produced in various forms, such as films, sponges, hydrogels, granules, powders, fibers, as well as woven and non-woven materials based on these forms. Each form has its own limitations and advantages, which ensure high efficiency when a material is used in a certain specific case. Modern hemostatic materials are divided into several groups depending on the mechanism of action [4]: (1) physical agents, which provide occlusion of bleeding; (2) absorbing agents, whose main characteristic is swelling; (3) biologic agents that take part in biologic coagulation cascade; (4) synthetic agents (which, for example, glue the adjacent surfaces together). According to the literature, most polymers used to produce hemostatic materials are capable of realizing these mechanisms.

Thrombin and fibrin are widely used to create hemostatic agents, since they are directly involved in the process of blood clotting; therefore, they are highly effective. The same is true for microfibrillar collagen, which activates the intrinsic pathway of hemostasis reactions cascade. However, the use of human blood clotting factors significantly increases the risk of developing an immune response [5,6] as well as the cost of the product [7].

The materials which provide hemostasis due to blood absorbance and swelling are used much more frequently. This group of materials includes gelatin and starch, whose mechanisms of action are related to absorbance of the liquid blood phase; thus, the concentration of macromolecular blood clotting factors increases, and hemostasis is achieved [8]. Moreover, the material can expand and thereby limit the blood flow [9]. At the same time, the ability to expand is the main disadvantage of these materials, which restricts their use in the limited space [10].

Another example of the polymer commonly used in development of hemostatic materials is oxidized cellulose. Its mechanism of action is based on depolymerization with formation of cellulosic acid upon contact with wet environment. In this case, the local decrease in pH provokes lysis of red blood cells and release of hemoglobin, which turns into acidic hematin upon contact with cellulosic acid [11]. Decrease in the pH values contributes to the appearance of additional antimicrobial action [12]; however, in this case, possibility of inflammation and delayed wound healing increases [13]. In addition, this factor limits the joint use of hemostatic materials based on oxidized cellulose with other hemostatic agents [14].

Chitosan is one of the most promising polymers used in biomedicine, due to its biocompatibility and biodegradation properties as well as antibacterial effect and absence of cytotoxicity. As for hemostatic action, chitosan can serve as a promising base for hemostatic materials. In particular, such commercial products as Celox, HemCon and QuikClot demonstrate increased efficiency of bleeding arrest [15]. When these chitosan-based materials are used, hemostasis is achieved in a short period of time (from 53 s [16] to 6 min [17]), and this time depends directly on the nature of a damaged vessel and the species used in an experiment.

Various fillers are introduced to increase hemostatic properties of chitosan-based materials. For example, introduction of gelatin increases swelling degree of a material, and the use of oxidized cellulose reduces the period of resorption of a material in the body [18]. The additives can improve antibacterial properties of the product, reduce its cytotoxicity, promote faster wound healing and accelerate bleeding arrest, the latter being the most important parameter of a hemostatic material.

Since chitosan and chitin have similar natures, use of their combinations may enhance some properties necessary for biomedical applications. The materials containing chitin additives are widely used in tissue engineering, in development of drug delivery agents, wound dressings, anticancer preparations, antimicrobial agents [19]. There is information about the improved cell adhesion in the case of using low ratios of chitin nanoform [20] and enhancing of mechanical characteristics of materials [21]. The goal of this research was to obtain a chitosan-based material containing chitin nanofibrils (ChNF), evaluation of its hemostatic properties and study of interaction with living organism.

To obtain complete and reliable results, in vitro and in vivo studies should be used in combination, since each of these two methods has certain limitations. When in vitro methods are used, the experiment only approximately reproduces real conditions; all reactions that occur in the body cannot be taken into account; thus, reliability of this experiment is limited. Animal tests are associated with ethical difficulties, as well as with the error occurring due to individual characteristics; on the other hand, conditions of such experiments are significantly closer to the real conditions of clinical use [22].

The preliminary in vitro experiments involving chitosan-based composite materials were conducted in the previous works. It was shown [23] that upon the contact between blood plasma and composite chitosan-based ChNF-containing fibers, optical density of blood plasma decreases. This result is associated with sorption of hemoglobin by the fibers. The material containing 0.5% of ChNF demonstrated the least hemocompatibility due to the associated hemolysis processes. The behavior of composite fibers in wet environment was also investigated. This research was mostly dedicated to the in vivo studies of the behavior of composite materials. Chronic experiments provide an opportunity to study long-term interactions of a sample with living organism, tissue reactions, changes in cellular composition, thickness of the connective tissue capsule; it is also possible to reveal the presence or absence of the material's bioresorption ability. It is believed that the optimal duration of the experiment is 14 days [22].

2. Materials and Methods

2.1. Materials

Chitosan, with molecular mass of 164 kDa and deacetylation degree of 92% (BiologHeppe, Landsberg, Germany) and chitin nanofibrils (ChNF) (SRL Mavi Sud, Aprilia, Italy) were used in our experiments. The length and diameter of ChNF were approximately 600 nm and 25 nm, respectively.

The following two commercial products with different mechanisms of action were used as control samples in hemostatic efficiency tests: Surgicel Nu-Knit consisting of oxidized cellulose (Ethicon, Somerville, NJ, USA) and TachoComb (Takeda Austria GmbH, Linz, Austria), which contains fibrinogen and thrombin as active substances.

2.2. Preparation of Composites

Chitin nanofibrils (ChNF) were preliminarily dispersed in water using an IL10-0.63 ultrasonic generator (Saint Petersburg, Russia) for 7 min at a frequency of 23.4 kHz. The mixture of chitosan and ChNF was dissolved in 2 wt% water solution of acetic acid. Concentrations of ChNF were 0.5 wt%, 5 wt% and 50 wt% with respect to chitosan amounts; total concentrations of polymers in the solution were 3 wt% (for sponges) and 4 wt% (for fibers) [24]. The prepared solutions were filtered and deaerated at a pressure of 0.1 atm.

The sponges were obtained by lyophilization. The solutions were preliminary frozen at −20 °C, then lyophilized at −2 °C and at a pressure of 1.6 Pa with the aid of a LABCONCO Triad Freeze Dry System (Labconco Corporation, Kansas City, MO, USA). In order to transform the sponges into the insoluble basic form, they were treated with 10% water solution of NaOH and then washed with distilled water.

The fibers were obtained by wet spinning [24]. The solutions were fed through a spinneret die into coagulation bath (10% water solution of NaOH and ethanol, 1:1), then the obtained fibers were washed with distilled water and dried at 40 °C. In order to produce monofilaments, the die with the hole diameter of 0.6 mm was used; polyfilaments were spun using a die with 100 holes, diameter of each hole was 100 μm. The feed rate during monofilament preparation was 0.2 mm/min; in the case of polyfilaments, feed rate was 0.3 mm/min. During spinning, oriented fibers were obtained due to 50% drawing. The non-woven material was obtained by needle-punching from polyfilament fibers.

2.3. Measurements of Blood Absorbance by Chitosan Sponges

The sponges were cut into pieces of the same area (10×10 mm^2) and different thicknesses (3 and 5 mm). After pre-weighing (W1), the sponges were brought into contact with rat blood for 1 min and weighed again (W2). The absorbance ability of sponges (W) was calculated as a difference between the weights: W = W2 − W1. The obtained parameters were compared taking into account sponge thicknesses and ChNF concentrations.

2.4. Measurements of Z-Potential

The measurements of ζ-potential of non-woven materials were carried out using a SurPASS 3 instrument (Anton Paar, Graz, Austria) according to the method described in [25]. Briefly, the samples (size: 10×20 mm^2) were mounted on sample holders in such a fashion that the gap between them was approximately 100 μm. The pressure was varied from 600 mbar to 200 mbar; the temperature was 23 °C. Solution of KCl (0.001 mol/L) was used as the background electrolyte; the pH value was varied from 5 to 9 by addition of 0.05 mol/L KOH.

2.5. In Vivo Experiments

The use of experimental animals in the studies of the obtained materials was permitted by the Local Ethics Committee of Saint-Petersburg State Pediatric Medical University (Saint-Petersburg, Russia).

All manipulations were performed under full anesthesia (mixture of tiletamine hydrochloride and zolazepam hydrochloride) in strict compliance with the European Convention for the Protection of Vertebrate Animals used for Experimental and other Scientific Purposes (ETS 123). Euthanasia was performed in strict compliance with the Recommendations for Euthanasia of Experimental Animals of European Commission.

Sixty-five animals were used in the studies (Wistar Kyoto male rats, body weight 230–250 g). The animals were divided into 3 groups for different experiments.

2.5.1. Experiments Related to Arrest of Arterial and Vein Bleeding

The first group of 54 experimental animals was used in the experiments involving stop of bleedings from femoral artery and vein. The mixture of tiletamine hydrochloride and zolazepam hydrochloride was used for anesthesia (Zoletil 100, Vibrac, France). The dosage was calculated individually for each animal (15 mg per 1 kg of the animal's mass). Intramuscular injections of the anesthetic mixture were administered.

To simulate femoral bleeding under conditions of complete anesthesia, the inner side of the animal thigh was shaved, and the animal was fixed to the operating table with ligatures. An incision 1 cm long was made on the inner side of the thigh, and a frame retractor was introduced into the incision. The fascia was dissected using a scalpel and microsurgical scissors, access to the neurovascular bundle was ensured, then the femoral artery (vein) was isolated. When the selected blood vessel was intersected, the studied material ($7 \times 7 \times 4$ mm^3) was applied to the damaged area. The time until the complete arrest of bleeding was measured.

Eight samples were investigated in 2 phases. The first phase included tests of chitosan sponges without ChNF and the sponges containing 0.5 wt%, 5 wt% and 50 wt% of ChNF as materials for bleeding arrest. At the second stage, pure chitosan non-woven materials and chitosan-based non-woven materials containing 0.5% of ChNF were compared with commercial hemostatic materials and the control group. The materials that demonstrated the best properties in the above experiments were selected for further studies. The reference groups included commercial materials (TachoComb and Surgicel) of similar sizes. In the control experiments, hemostatic agents were not used.

2.5.2. Studies of Resorption and Implantation

To investigate the potential tissue response to the studied materials and to make preliminary assessments of applicability, implantation tests were performed in the second group of animals. All investigated materials were introduced into subfascial plane in the projection of both m. latissimus dorsi of 3 intact rats for a period of 14 days. The distance between the samples was at least 1 cm to avoid cross-exposure. In 14 days after the beginning of the experiment, biopsy samples were taken.

Chitin–chitosan fibers were introduced into subfascial plane of m. latissimus dorsi of 8 intact rats for 14 and 91 days. The introduced chitosan and chitin–chitosan (0.5%) fibers were used as a basis for preparation of the non-woven materials mentioned above. The experiments involving long-term implantation of singular elements allowed us to assess the consequences of destruction of materials in the wound when these materials are removed after use.

After processing the skin on the back in the region of m. latissimus dorsi, a 1 cm incision was made up to the fascia. A frame retractor was introduced into the section, and the edges of the section were fixed to facilitate the ongoing manipulations. The fascia was incised and moved apart with the help of pointed four-tooth hooks. The test samples were placed under the fascia. After a specimen was placed into the desired area, the fascia was sutured with the Vicryl 4-0 suture. The distance between the suturing area and the area containing the specimen was at least 1 cm to exclude the influence of suture material. The skin incision was sutured using the Vicryl 2-0 suture.

2.6. Histological Studies

Tissue samples were fixed in 10% solution of neutral formalin in phosphate buffer (pH = 7.4) for 24 h, then treated with ethanol solutions of increasing concentrations (10%, 30%, 50%, 80% and 99%) and embedded in paraffin. Paraffin sections (5 μm thick) were dyed with hematoxylin and eosin (Bio-Optica, Milan, Italy). Microscopic analysis and registration of images were performed with the use of a Leica DM750 light microscope (Wetzlar, Germany).

2.7. Statistical Processing of Data

Statistical analysis of the results was carried out using the Origin software (OriginLab Corporation, USA) and the R software (R Foundation for Statistical Computing, Vienna, Austria). The Kruskal–Wallis test and Mann–Whitney test were used to compare the independent samples. The differences were considered statistically significant when the p-value was less than 0.05.

3. Results

3.1. Preparation of Non-Woven Materials

Addition of ChNF to chitosan solutions facilitates wet spinning of fibers, which, in turn, leads to enhancement of their mechanical properties [24] and improvement of the structure of the resulting filaments and the non-woven material produced on their basis (Figure 1). This phenomenon can be explained by the fact that introducing a low amount of ChNF (0.3–0.5%) into the chitosan matrix ensures additional orientation of the chitosan macromolecules on their surface [24]. Orientation of chitosan macromolecules on the surface of ChNF was also confirmed [24] by energy minimization and molecular dynamics simulation of the systems containing one chitosan molecule on the surface of chitin nanocrystallite. Mutual ordering of the components of the mixture contributes to the formation of hydrogen bonds between them, this resulting in the formation of a more stable and energetically favorable structure.

Figure 1. Non-woven material without chitin nanofibrils (ChNF) (left) and the material containing 0.5% ChNF (right).

3.2. Measurement of Blood Absorbance by Chitosan Sponges

For all samples, an increase in sample thickness led to higher blood absorption (Figure 2). Regardless of the size, the highest value of blood absorption was shown by the sponges containing 0.5% ChNF; this effect becomes more pronounced with increasing sample size.

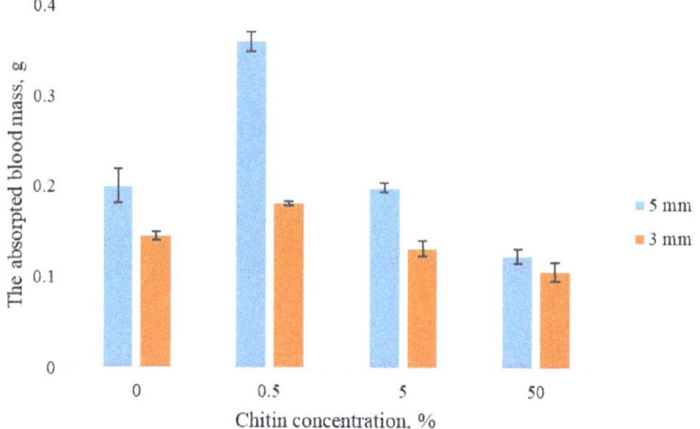

Figure 2. Blood absorbance by sponges vs. ChNF content in samples.

3.3. Arrest of Bleeding

The results of studies of hemostatic properties of the chitosan sponges containing different amounts of ChNF are shown in Figure 3. It is seen that all materials demonstrate hemostatic properties, i.e., the bleeding from femoral artery is arrested faster than that in the case when the materials were not applied (the control group). The fastest bleeding arrest was observed in the experiment involving the material with 0.5% ChNF. Therefore, this concentration was used in preparation of non-woven materials.

The results of studies of bleeding arrest by non-woven materials are shown in Figure 4. It can be noted that for venous bleeding, both non-woven materials (the "NW" group in the picture) and the commercial samples showed higher effectiveness in comparison with that of a sponge. At the same time, in the case of arterial bleeding, only non-woven material containing ChNF demonstrated the hemostatic properties comparable to those of the sponge material with the same ChNF concentration. In both cases, the non-woven material showed the efficiency comparable with those of commercial products, and arterial bleeding stopped even faster than in the experiments with TachoComb.

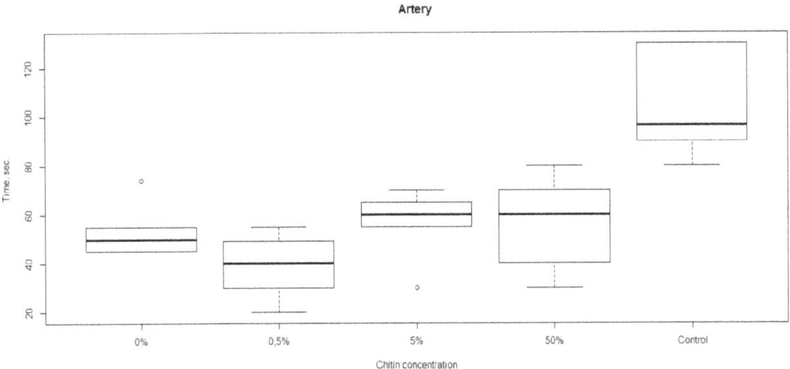

Figure 3. Rates of bleeding arrest (time until cessation of bleeding) in the experiments with chitosan sponges containing different amounts of ChNF.

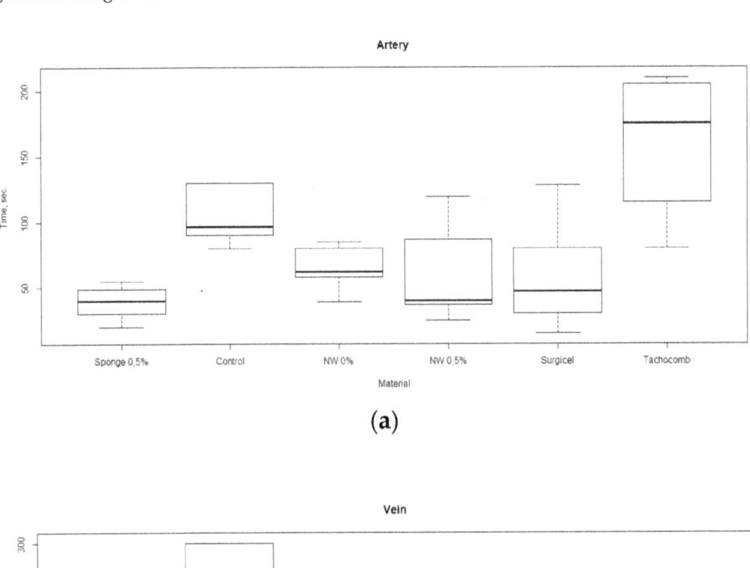

(**a**)

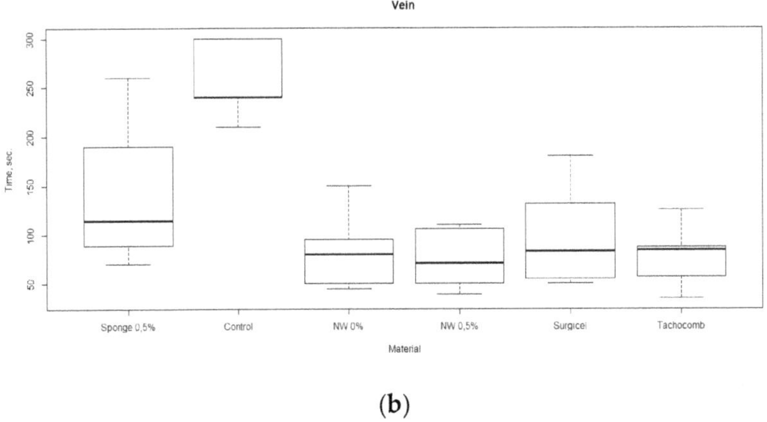

(**b**)

Figure 4. Rates of bleeding arrest (time until cessation of bleeding) in the experiments with different materials, in which femoral (**a**) and vein (**b**) artery hemorrhages were modeled.

3.4. Measurements of Z-Potential

Hemostatic properties of the obtained materials described above can be explained by the value of the charge present on the composite surface. The pH dependences of the ζ-potential of the non-woven materials based on chitosan fibers and ChNF-containing composite fibers are shown in Figure 5. It is seen that both fibers have positive surface charges. Introducing ChNF in small amounts (0.5 wt%) leads to an increase in the surface potential of the composite fiber.

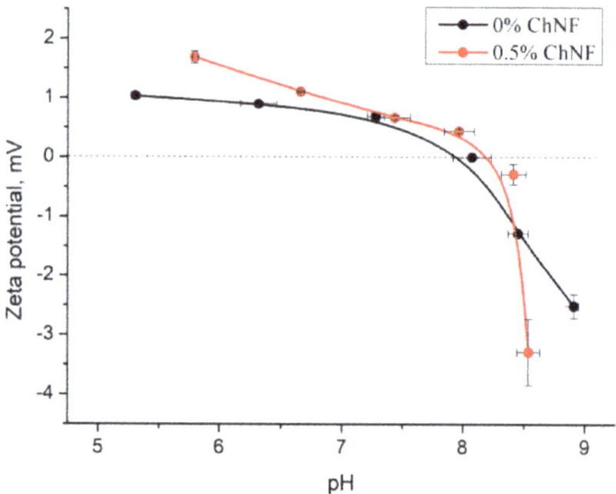

Figure 5. The pH dependences of ζ-potential of non-woven materials.

It is known [26] that chitosan possesses polyelectrolytic properties. In solutions of weak acids, amino group (NH_2) is transformed into positively charged NH_3^+ group; hydroxyl and amino groups of chitosan form internal and intermolecular hydrogen bonds. It can be assumed that regular arrangement of chitosan macromolecules in a composite fiber (in particular, their mutual orientation), as well as their orientation relative to ChNF, will have a certain influence on the number of hydrogen bonds and, as a result, on the value of surface charge of the fiber. If macromolecules have stretched conformation, probability of formation of chitosan intramolecular bonds will be reduced, thereby increasing the free surface charge of the fiber.

It has been shown [24] that introducing ChNF into chitosan solution facilitates orientation of chitosan macromolecules on the surface of nanoparticles. In addition to the hydrogen bonds described above, chitosan amino groups can interact with carbonyl $-C = O-$ groups of chitin. As indicated by the data shown in Figure 5, introduction of a small amount of ChNF (0.5 wt%) increases the value of the ζ-potential. A sharp decrease in the potential occurs when pH of the medium reaches the isoelectric point.

3.5. Studies of Degradation of Fibers in Organism

In 2 weeks after introducing the fibers into organism, weak signs of fiber fragmentation are observed (Figure 6). Small cracks appear on fiber surface. In 91 days after implantation, a dense connective tissue capsule appears (which is confirmed by histological studies). Despite partial fragmentation, the resulting pieces of material are held together by the capsule. Resorption of chitosan fibers may be associated with mechanical loads appearing during the movements of rats due to contact between fibers and muscles.

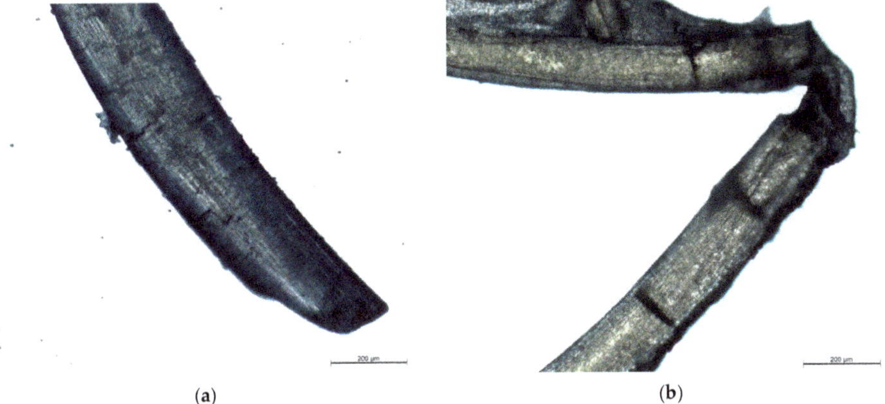

(**a**)   (**b**)

Figure 6. Images of the chitosan fiber (0.5% ChNF) located under m. latissimus dorsi that underwent bioresorption for (**a**) 14 days and (**b**) 91 days.

3.6. Histological Studies

3.6.1. Histological Studies of Tissues around Implanted Fibers

In the biopsy samples of the ChNF-chitosan filaments (0.5% of chitin) used in subfascial implantation, we observed a circle of coarse, fibrous connective tissue containing numerous fibroblasts with large bright oval-shaped nuclei. It was formed around the material in 14 days after beginning of the experiment (Figure 7a). In addition, focal lymphomacrophagous infiltration with an admixture of neutrophilic leukocytes (1–3 in the field of view, magnification 400×) and foreign body giant cells (1–2 in the field of view, magnification 400×) were found. On the 91st day of observation (Figure 7b), no signs of fiber resorption were observed. A narrow rim of scar tissue with small foci of lymphomacrophagous infiltration and an admixture of single neutrophilic leukocytes were present in the field of view (magnification 400×) around the fibers.

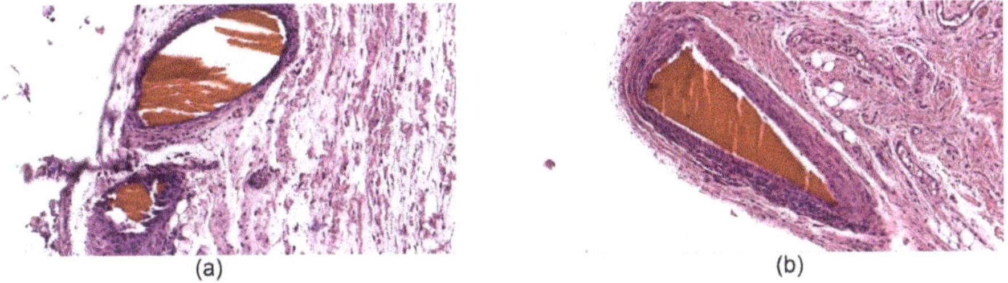

(a)   (b)

Figure 7. Histological microphotographs of biopsy samples of [chitosan + 0.5% ChNF] fibers, magnification 50×. (**a**) subfascial implantation, day 14; (**b**) subfascial implantation, day 91.

3.6.2. Histological Studies of Tissues around Implanted Hemostatic Materials

Histological examination of tissues with implanted chitosan hemostatic non-woven materials performed on the 14th day of observation revealed a large macrophage granuloma surrounding brightly eosinophilically stained chitosan fibers (Figure 8a). In addition, numerous foreign body giant cells were detected in the granuloma (Figure 9a). In the group of animals with implanted chitosan hemostatic sponges containing 0.5% of ChNF, a large granuloma with numerous foreign body giant cells and a small number of infiltrated neutrophilic leukocytes were detected among brightly stained eosinophilic filaments (Figure 8b). The group of samples obtained from the animals with implanted

"Surgicel" hemostatic sponge shows a small sclerosis focus with small focal lymphomacrophagic infiltration and low amount of foreign body giant multinuclear cells in the field of view (Figure 8c).

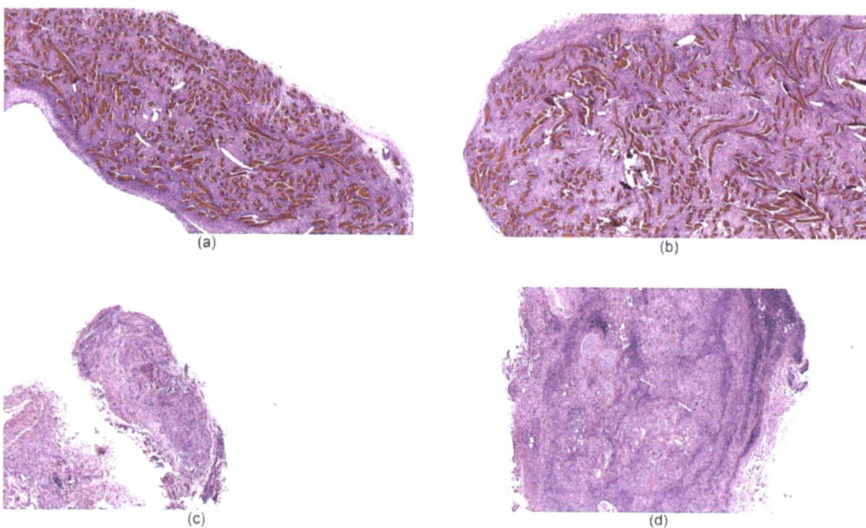

Figure 8. Histological microphotographs of biopsy samples of hemostatic materials taken on the 14th day of the experiment, magnification 50×. (**a**) non-woven chitosan material; (**b**) [chitosan + 0.5% ChNF] non-woven material; (**c**) "Surgicel" hemostatic material; (**d**) "TachoComb" hemostatic material.

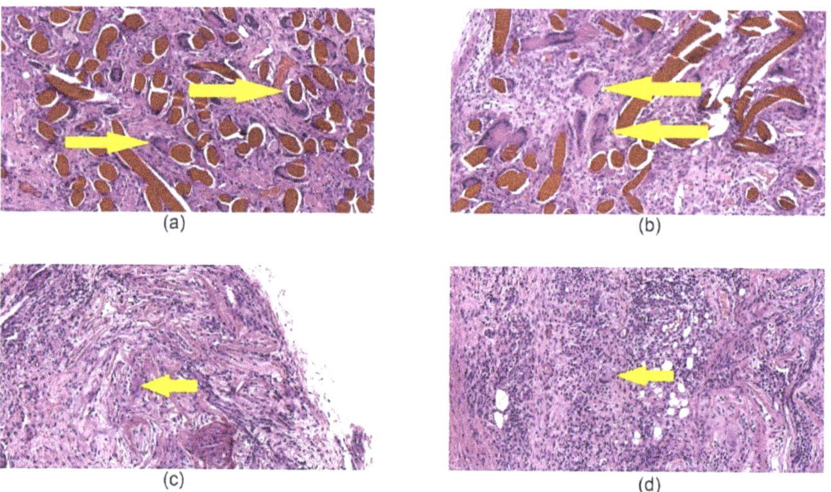

Figure 9. Histological microphotographs of biopsy samples of tissues with implanted hemostatic materials taken on the 14th day after beginning of the experiment, magnification 200×. (**a**) chitosan non-woven material, the arrow points to giant cells in granuloma; (**b**) [chitosan + 0.5% ChNF] non-woven material, the arrow points to giant cells in granuloma; (**c**) "Surgicel" hemostatic material, the arrow points to giant cells; (**d**) "TachoComb" hemostatic material, the arrow points to giant cells between infiltration.

4. Discussion

The best results in the blood sorption experiments and the best hemostatic properties were revealed for the chitosan sponges containing 0.5% of ChNF. Therefore, the non-woven material containing this amount of ChNF was prepared and used in the experiments involving arrest of venous and arterial bleeding. Efficiency of this material was comparable to those of commercial hemostatic agents (Surgicel and TachoComb). TachoComb demonstrated somewhat deteriorated properties; this result may be explained by the fact that it contains human fibrinogen and thrombin, while the experiments were conducted in rats. It also should be noted that the Surgicel application instruction forbids its use in the cases of massive artery hemorrhage; thus, chitosan-based materials can be used as substitutes.

In order to reveal possible causes of different hemostatic effect of non-woven materials, the measurement of ζ-potential was carried. It was shown that addition of 0.5% of ChNF leads to the increment of positive charge at pH value from 5 to 8; after the isoelectric point is achieved, sharp decline occurs. It can be assumed that this phenomenon is related to higher ordering of the surface structure of composite fibers, more regular arrangement of positively charged groups. In [24], it was shown that introduction of 0.5 wt% of ChNF increases the orientation of chitosan macromolecules due to adhesion of chitosan macromolecules on the surface of ChNF. Model calculations performed by molecular dynamics have shown that the most energy-efficient location of chitosan macromolecules is along the ChNF axes. The orientation of chitosan macromolecules, including their localization on fiber surface, may lead to regular arrangement of positive charges and thus may minimize their interaction with other groups of chitosan and chitin macromolecules (primarily carboxylic and carbonyl fragments). Moreover, the sharp drop in the ζ-potential at the isoelectric point may indicate higher ordering of the charged groups of composite fibers and, as a result, narrower distribution of the charge value.

The existence of higher positive charge on the surface of non-woven material containing 0.5% of ChNF can explain better interaction of this material with blood cells (which are negatively charged). Moreover, finer structure of the composite non-woven material may be related to increased plasma sorption. To study the subsequent interaction of materials with organism, the samples were placed under the fascia of m. latissimus dorsi for 14 days.

According to the results of histological examination, the implanted chitosan filaments containing 0.5% of ChNF demonstrate bioinertness. An extremely low number of foreign body giant cells and leukocyte infiltration were observed on the 91st day in the case of subfascial implantation. In addition, a connective tissue capsule was formed around the fiber. Taking these results into account, we studied implantation of non-woven materials based on fibers of similar composition. Nevertheless, implantation of such material led to the formation of a significant granuloma, while low degree of resorption of the tested material was observed. At the same time, the commercial analogs (Surgicel and TachoComb) demonstrated high degrees of resorption on the 14th day of the experiment with the formation of separate foci of sclerosis.

The long-term study of behavior of fibers in organism showed that biodegradation signs appear in 3 months after implantation into abdomen and m. latissimus dorsi. This result may be connected to the mechanical stress during the rats' movement and to chitosan ability for bioresorption.

5. Conclusions

In this work, we studied the influence of ChNF additives present in chitosan-based materials on hemostatic properties of samples. The sponges containing 0.5% of ChNF demonstrated the best results in the experiments involving blood absorbance in vitro, and the highest rate of arterial bleeding arrest in vivo. Therefore, this concentration was chosen for preparation of the non-woven material. The quality of ChNF-containing material was higher than that of pure chitosan samples, but efficiencies of both materials were comparable with those of commercial hemostatic agents Surgicel and TachoComb. The studies of ζ-potential indicate better interaction between the chitin-containing material and blood cells due to higher surface positive charge of this material. This high surface charge appears due to orientation of chitosan macromolecules along the ChNF. To study the long-term interaction of the materials with organism, ChNF-containing chitosan fibers were implanted into rats. The histological studies demonstrated bioinertness of the fibers, and light microscopy studies showed the signs of biodegradation in 91 days after implantation due to their fragmentation. Meanwhile, histological studies of non-woven materials indicated formation of granulomas with foreign body giant cells; therefore, the detailed further research of the process is required.

Author Contributions: Data curation, E.N.M., M.B.P., A.M.F., P.L.S. and G.P.S.; formal analysis, E.N.M.; investigation, E.N.M., A.S.S.; E.N.D. and I.P.D.; methodology, A.S.S.; resources, P.M.; supervision, V.E.Y. and E.V.Z.; writing—original draft, E.N.M., A.S.S.; writing—review & editing, E.N.D. All authors have read and agreed to the published version of the manuscript.

Funding: This research was funded by RSF, Grant Number 19-73-30003.

Conflicts of Interest: The authors declare no conflict of interest.

References

1. Kauvar, D.S.; Lefering, R.; Wade, C.E. Impact of Hemorrhage on Trauma Outcome: An Overview of Epidemiology, Clinical Presentations, and Therapeutic Considerations. *J. Trauma Inj. Infect. Crit. Care* **2006**, *60*, S3–S11. [CrossRef] [PubMed]
2. Johnson, D.; Bates, S.; Nukalo, S.; Staub, A.; Hines, A.; Leishman, T.; Michel, J.; Sikes, D.; Gegel, B.; Burgert, J. The effects of QuikClot Combat Gauze on hemorrhage control in the presence of hemodilution and hypothermia. *Ann. Med. Surg.* **2014**, *3*, 21–25. [CrossRef] [PubMed]
3. Sun, X.; Tang, Z.; Pan, M.; Wang, Z.; Yang, H.; Liu, H. Chitosan/kaolin composite porous microspheres with high hemostatic efficacy. *Carbohydr. Polym.* **2017**, *177*, 135–143. [CrossRef] [PubMed]
4. Jensen, D.M.; Silpa, M.L.; Tapia, J.I.; Beilin, D.B.; Machicado, G.A. Comparison of different methods for endoscopic hemostasis of bleeding canine esophageal varices. *Gastroenterology* **1983**, *84*, 1455–1461. [CrossRef]
5. Vadasz, Z.; Toubi, E. Hemostasis in Allergy. *Semin. Thromb. Hemost.* **2018**, *44*, 669–675.
6. Göbel, K.; Eichler, S.; Wiendl, H.; Chavakis, T.; Kleinschnitz, C.; Meuth, S.G. The Coagulation Factors Fibrinogen, Thrombin, and Factor XII in Inflammatory Disorders—A Systematic Review. *Front. Immunol.* **2018**, *9*, 1731. [CrossRef]
7. Lee, S.; Pham, A.M.; Pryor, S.G.; Tollefson, T.; Sykes, J.M. Efficacy of Crosseal Fibrin Sealant (Human) in Rhytidectomy. *Arch. Facial Plast. Surg.* **2009**, *11*, 29–33. [CrossRef]
8. Murat, F.-J.L.; Ereth, M.H.; Dong, Y.; Piedra, M.P.; Gettman, M.T. Evaluation of Microporous Polysaccharide Hemospheres As A Novel Hemostatic Agent In Open Partial Nephrectomy: Favorable Experimental Results In The Porcine Model. *J. Urol.* **2004**, *172*, 1119–1122. [CrossRef]
9. Larson, P.O. Review: Topical Hemostatic Agents for Dermatologic Surgery. *J. Dermatol. Surg. Oncol.* **1988**, *14*, 623–632. [CrossRef]
10. Lewis, K.M.; Atlee, H.; Mannone, A.; Lin, L.; Goppelt, A. Efficacy of hemostatic matrix and microporous polysaccharide hemospheres. *J. Surg. Res.* **2015**, *193*, 825–830. [CrossRef]
11. Achneck, H.E.; Sileshi, B.; Jamiolkowski, R.M.; Albala, D.M.; Shapiro, M.L.; Lawson, J.H. A Comprehensive Review of Topical Hemostatic Agents. *Ann. Surg.* **2010**, *251*, 217–228. [CrossRef] [PubMed]
12. Palm, M.D.; Altman, J.S. Topical Hemostatic Agents: A Review. *Dermatol. Surg.* **2008**, *34*, 431–445. [PubMed]
13. Tomizawa, Y. Clinical benefits and risk analysis of topical hemostats: A review. *J. Artif. Organs* **2005**, *8*, 137–142. [CrossRef] [PubMed]
14. Howe, N.; Cherpelis, B. Obtaining rapid and effective hemostasis. *J. Am. Acad. Dermatol.* **2013**, *69*, e1–e659. [CrossRef] [PubMed]
15. Kozen, B.G.; Kircher, S.J.; Henao, J.; Godinez, F.S.; Johnson, A.S. An Alternative Hemostatic Dressing: Comparison of CELOX, HemCon, and QuikClot. *Acad. Emerg. Med.* **2008**, *15*, 74–81. [CrossRef]
16. Kale, T.P.; Singh, A.K.; Kotrashetti, S.M.; Kapoor, A. Effectiveness of Hemcon Dental Dressing Versus Conventional Method of Haemostasis in 40 Patients on Oral Antiplatelet Drugs. *Sultan Qaboos Univ. Med J.* **2012**, *12*, 330–335. [CrossRef]
17. Hejazi, F.; Hosseinzadeh, F.; Irani Rad, L.; Bagheri, A.; Vahedian, M.; Damanpak, V. Effectiveness of Celox Powder and Standard Dressing in Control of Angiography Location Bleeding. *JBUMS* **2013**, *15*, 30–36.

18. Hu, Z.; Zhang, D.-Y.; Lu, S.-T.; Li, P.-W.; Li, S.-D. Chitosan-Based Composite Materials for Prospective Hemostatic Applications. *Mar. Drugs* **2018**, *16*, 273. [CrossRef]

19. Ling, S.; Chen, W.; Fan, Y.; Zheng, K.; Jin, K.; Yu, H.; Buehler, M.J.; Kaplan, D.L. Biopolymer nanofibrils: Structure, modeling, preparation, and applications. *Prog. Polym. Sci.* **2018**, *85*, 1–56. [CrossRef] [PubMed]

20. Zubillaga, V.; Salaberria, A.M.; Palomares, T.; Alonso-Varona, A.; Kootala, S.; Labidi, J.; Fernandes, S.C.M. Chitin Nanoforms Provide Mechanical and Topological Cues to Support Growth of Human Adipose Stem Cells in Chitosan Matrices. *Biomacromolecules* **2018**, *19*, 3000–3012. [CrossRef]

21. Torres, F.G.; Troncoso, O.P.; Pisani, A.; Gatto, F.; Bardi, G. Bardi Natural Polysaccharide Nanomaterials: An Overview of Their Immunological Properties. *Int. J. Mol. Sci.* **2019**, *20*, 5092. [CrossRef] [PubMed]

22. Dutton, R.P. Haemostatic resuscitation. *Br. J. Anaesth.* **2012**, *109*, i39–i46. [CrossRef] [PubMed]

23. Maevskaia, E.; Kirichuk, O.; Kuznetzov, S.; Dresvyanina, E.; Yudin, V.; Morganti, P. Hemocompatible chitin-chitosan composite fibers. *Cosmetics* **2020**, *7*, 28. [CrossRef]

24. Yudin, V.E.; Dobrovolskaya, I.P.; Neelov, I.M.; Dresvyanina, E.N.; Popryadukhin, P.V.; Ivan'kova, E.M.; Elokhovskii, V.Y.; Kasatkin, I.A.; Okrugin, B.M.; Morganti, P. Wet spinning of fibers made of chitosan and chitin nanofibrils. *Carbohydr. Polym.* **2014**, *108*, 176–182. [CrossRef] [PubMed]

25. Smirnova, N.V.; Kolbe, K.A.; Dresvyanina, E.N.; Grebennikov, S.F.; Dobrovolskaya, I.P.; Yudin, V.E.; Luxbacher, T.; Morganti, P. Effect of Chitin Nanofibrils on Biocompatibility and Bioactivity of the Chitosan-Based Composite Film Matrix Intended for Tissue Engineering. *Materials* **2019**, *12*, 1874. [CrossRef]

26. Verma, A.; Nagarwal, R.C.; Sharma, S.D.; Pandit, J.K. Preparation and characterization of floating gellan-chitosan polyelectrolyte complex beads. *Lat. Am. J. Pharm.* **2012**, *31*, 138–146.

Article

Plasma-Coated Polycaprolactone Nanofibers with Covalently Bonded Platelet-Rich Plasma Enhance Adhesion and Growth of Human Fibroblasts

Svetlana Miroshnichenko [1,2], Valeriia Timofeeva [1], Elizaveta Permyakova [1,3], Sergey Ershov [4], Philip Kiryukhantsev-Korneev [3], Eva Dvořaková [5], Dmitry V. Shtansky [3], Lenka Zajíčková [5], Anastasiya Solovieva [1,*] and Anton Manakhov [1,*]

[1] Scientific Institute of Clinical and Experimental Lymphology—Branch of the ICG SB RAS, 2 Timakova str., 630060 Novosibirsk, Russia; svmiro@yandex.ru (S.M.); leravalera0204@mail.ru (V.T.); permyakova.elizaveta@gmail.com (E.P.)

[2] Institute of Biochemistry – subdivision of the FRC FTM, 2 Timakova str., 630117 Novosibirsk, Russia

[3] Laboratory of Inorganic Nanomaterials, National University of Science and Technology "MISiS", Leninsky pr. 4, 119049 Moscow, Russia; kiruhancev-korneev@yandex.ru (P.K.-K.); shtansky@shs.misis.ru (D.V.S.)

[4] Physics and Materials Science Research Unit, Laboratory for the Physics of Advanced Materials, University of Luxembourg, 162a, avenue de la Faïencerie, L-1511 Luxembourg, Luxembourg; sergey.ershov@uni.lu

[5] CEITEC—Central European Institute of Technology—Masaryk University, Kamenice 5, 625 00 Brno, Czech Republic; evke.dvorakova@gmail.com (E.D.); lenkaz@physics.muni.cz (L.Z.)

* Correspondence: solovey_ao@mail.ru (A.S.); ant-manahov@ya.ru (A.M.); Tel.: +7-913-906-0098 (A.S.); +7-915-849-4059 (A.M.)

Received: 11 March 2019; Accepted: 15 April 2019; Published: 19 April 2019

Abstract: Biodegradable nanofibers are extensively employed in different areas of biology and medicine, particularly in tissue engineering. The electrospun polycaprolactone (PCL) nanofibers are attracting growing interest due to their good mechanical properties and a low-cost structure similar to the extracellular matrix. However, the unmodified PCL nanofibers exhibit an inert surface, hindering cell adhesion and negatively affecting their further fate. The employment of PCL nanofibrous scaffolds for wound healing requires a certain modification of the PCL surface. In this work, the morphology of PCL nanofibers is optimized by the careful tuning of electrospinning parameters. It is shown that the modification of the PCL nanofibers with the COOH plasma polymers and the subsequent binding of NH_2 groups of protein molecules is a rather simple and technologically accessible procedure allowing the adhesion, early spreading, and growth of human fibroblasts to be boosted. The behavior of fibroblasts on the modified PCL surface was found to be very different when compared to the previously studied cultivation of mesenchymal stem cells on the PCL nanofibrous meshes. It is demonstrated by X-ray photoelectron spectroscopy (XPS) that the freeze–thawed platelet-rich plasma (PRP) immobilization can be performed via covalent and non-covalent bonding and that it does not affect biological activity. The covalently bound components of PRP considerably reduce the fibroblast apoptosis and increase the cell proliferation in comparison to the unmodified PCL nanofibers or the PCL nanofibers with non-covalent bonding of PRP. The reported research findings reveal the potential of PCL matrices for application in tissue engineering, while the plasma modification with COOH groups and their subsequent covalent binding with proteins expand this potential even further. The use of such matrices with covalently immobilized PRP for wound healing leads to prolonged biological activity of the immobilized molecules and protects these biomolecules from the aggressive media of the wound.

Keywords: polycaprolactone; nanofibers; COOH plasma; cell adhesion and spreading; cell viability; freeze–thawed platelet-rich plasma immobilization

---

1. Introduction

During the last decade, a lot of effort was focused on the research and development of biodegradable polymer materials for medical applications. The advantage of synthetic biodegradable polymer matrices is the lack of immunogenicity, the numerous possibilities for their modification, the controllability of their mechanical properties, and the rate of decomposition. This allows the creation of materials with the necessary properties for a wide range of applications in regenerative medicine. Both cellular and cell-free materials are being developed. Cell matrices demonstrate greater efficacy for the treatment of deep and chronic wounds [1–4]. Matrices for cell colonization must have the structure of an extracellular matrix (ECM), which provides structural support and intercellular contact, serves as a reservoir for signaling molecules, and thereby regulates cell migration, proliferation, and angiogenesis [1,2].

The fibrillary, porous structure of the polycaprolactone nanofibers obtained by the electrospinning method mimics the ECM structure [5,6]. However, their surface is hydrophobic and unsuitable for cell viability. In order to increase the bioavailability of polycaprolactone (PCL) nanofibers, plasma treatment [7,8], the deposition of plasma coatings [9], biomineralization [10,11], and the grafting of silk fibroin [12] are used. It is worth noting that the electrospinning of PCL nanofibers is very versatile and the addition of nanoparticles into the electrospinning solution can induce antibacterial properties [13].

The effectiveness of cell matrices depends on the functional activity of the cells used, which in turn depends on the initial processes of adhesion and spreading on the matrix. Cell adhesion to a functionally well-founded substrate is characterized by a well-defined time, as well as mechanical and energy-dependent processes, determining further cell proliferation, cell migration, cell differentiation, and gene expression [14–17]. It should be noted that the spreading of cells on the surface of substrates is a complex process and it is controlled by the matrix–integrin interactions that form adhesion sites [15,18–21]. The co-spinning of PCL with collagen [22,23], gelatin [24], chitosan [25], or other bioactive substances [26,27] enriches matrices by natural adhesion sites, thereby improving the cell adhesion properties of nanofibers and consequently contributing to the normal cell functioning [17,18]. Platelet-rich plasma is often used in regenerative medicine because it is a naturally balanced ensemble of growth factors, components of the extracellular matrix including fibronectin (FN) necessary for dynamic connection with substrate [14,28–32].

Fibroblasts are most often used in tissue engineering [33]. These cells are characterized by high plasticity and, depending on the conditions, their functionality changes. The secretion of ECM components (fibronectin, collagens, elastin, laminin, hyaluronic acid, proteoglycans) is one of the main properties of fibroblasts, and their nature depends on the quality of the substrate.

In this work, the morphology of PCL nanofibers was optimized by tuning the needle-less electrospinning process parameters, whereas the chemistry of the PCL nanofibers was activated by the plasma deposition of a COOH layer. This study determined the effect of the modification of PCL nanofibers by the COOH plasma polymerization and freeze–thawed platelet-rich plasma immobilization on the adhesion and spreading of fibroblasts, as well as their proliferative potential and viability. It was shown that the effective immobilization of COOH groups, followed by the covalent bonding of freeze–thawed platelet-rich plasma, led to a physiological course of adhesion and spreading processes, resulting in good viability, proliferation, and contact inhibition of proliferation when the monolayer is reached by the cells.

2. Materials and Methods

2.1. Preparation of Freeze–Thawed Platelet-Rich Plasma

The methodology of preparation of freeze–thawed platelet-rich plasma (PRP) was described in [34]. In brief, platelet-rich plasma was taken from healthy donors and, subsequently, it was activated using three freeze–thaw cycles. Then, PRP was centrifuged at 12,000× g for 10 min at 4 °C. The supernatant containing growth factors was recovered and kept frozen at −70 °C until later steps. The study was approved by the Ethics Committee of the RICEL—branch of ICG SB RAS (Identifier: N115 from 24 December 2015).

2.2. Electrospining of Nanofibers

The electrospun nanofibers were prepared by the electrospinning of solution with different concentrations of PCL. A description of the processing of samples can be found elsewhere [5]. The granulated polycaprolactone (80,000 molecular weight (mw)) was dissolved in a mixture of acetic acid (99%) and formic acid (98%). All compounds were purchased from Sigma Aldrich (Darmstadt, Germany. The weight ratio of acetic acid (AA) to formic acid (FA) was 2:1. The concentration of PCL was varied from 7 to 12 wt.%. The PCL solutions in AA and FA were stirred for 24 h at 25 °C. The PCL solution was electrospun with a 20-cm-long wired electrode using a Nanospider™ NSLAB 500 machine (ELMARCO, Liberec, Czech Republic). The applied voltage was in the range from 30 to 60 kV. The distance between the electrodes was set at 100 mm. The as-prepared PCL nanofibers are referred to as PCL-ref throughout the text.

2.3. Deposition and Characterization of Plasma Polymer Layers

The COOH plasma polymer layers were deposited using the vacuum system UVN-2M equipped with a rotary and oil diffusion pumps. The residual pressure of reactor was below 10^{-3} Pa. The plasma was ignited using a radio frequency (RF) power supply Cito 1310-ACNA-N37A-FF (Comet, San Jose, CA, USA) connected to the RFPG-128 disc generator (Beams & Plasmas, Zelenograd, Russia) installed in the vacuum chamber. The duty cycle and the RF power were set to 5% and 500 W, respectively.

CO_2 (99.995%), Ar (99.998%), and C_2H_4 (99.95%) were fed into the vacuum chamber. The flows of the gases were controlled using a Multi Gas Controller 647C (MKS, Andover, MA, USA). The flow rates of Ar, CO_2, and C_2H_4 were set to 50, 16.2, and 6.2 sccm, respectively. The pressure in the chamber was measured by a VMB-14 unit (Tokamak Company, Oxfordshire, UK) and D395-90-000 BOC (Edwards, Heinenoord, The Netherlands) controllers. The distance between the RF electrode and the substrate was set to 8 cm. The deposition time was 15 min and it led to the growth of ~100-nm-thick plasma coatings. The plasma-coated PCL nanofibers are referred to as PCL-COOH throughout the text.

The surface of the samples was imaged by scanning electron microscopy (SEM) Tescan LYRA3 (Tescan, Brno, Czech Republic) in secondary emission mode (5–10 kV acceleration voltage, working distance 8–10 mm). The samples were coated with a 10-nm-thick gold film deposited by RF magnetron sputtering prior to the imaging in order to avoid charging of surface. The Gwyddion 2.29 software (Brno, Czech Republic) was used for the determination of the diameter of polymer nanofibers from SEM micrographs. The diameters were determined by averaging of values obtained for 30 single independent filaments present in the micrographs taken at the magnification of 10,000×.

The chemical composition of sample surfaces was determined by X-ray photoelectron spectroscopy (XPS) using an Axis Supra spectrometer (Kratos Analytical, Manchester, UK) equipped with a monochromatic Al Kα X-ray source. The maximum lateral resolution of the analyzed area was 0.7 mm. The spectra were fitted using the CasaXPS software after subtracting the Shirley-type background. The binding energies (BE) for all carbon and nitrogen environments were taken from literature [5,9,35]

The water contact angle (WCA) of the nanofibers surface was determined by a contact angle analyzer (KRÜSS EasyDrop DSA20, Hamburg, Germany). Plasma-treated and untreated samples were cut into 1-cm^2 mats. Then, 5-μL droplets of deionized water were deposited in random locations on the surface and the water contact angle was measured. All the measurements were completed in triplicate. The infrared spectrum in the range from 370 to 4000 cm^{-1} was measured using a Fourier-transform infrared (FTIR) spectrophotometer (Bruker Vertex 80v, Billerica, MA, USA) using attenuated total reflectance mode (ATR-FTIR).

2.4. Coating of Scaffolds with PRP

Prior to immobilization of PRP, all samples were sterilized under ultraviolet (UV) light for 45 min. At first, the adsorption of PRP by PCL-ref was investigated. The PCL-ref was immersed in the PRP solution at 25 °C for 15 min and then it was thoroughly washed with phosphate-buffered saline (PBS). The sample was denoted as PCL-PRP. Then, the ionic bonding of PRP to the plasma-coated PCL nanofibers (PCL-COOH) was investigated. The PCL-COOH was immersed in the solution of PRP for 15 min and then it was washed with PBS. This sample was denoted as PCL-COOH-PRP1. In order to achieve covalent bonding of PRP to the PCL-COOH surface, the latter was immersed in a 1-ethyl-3-(3-dimethylaminopropyl) carbodiimide (EDC) (Sigma Aldrich, St. Louis, MO, USA, 98%) solution in water (2 mg/mL) for 15 min. The sample was carefully washed with PBS and then incubated with PRP for 15 min at room temperature. After the reaction, the sample was thoroughly washed with PBS. The samples were denoted as PCL-COOH-PRP2.

2.5. Cell Tests

Human fibroblast cells (MRC-5 line) were purchased from the State Research Center of Virology and Biotechnology VECTOR (Novosibirsk, Russian). Cells were cultured in Dulbecco's modified Eagle medium (DMEM, Sigma Aldrich, St. Louis, MO, USA) supplemented with 10% fetal bovine serum (Gibco) under standard culture conditions (humidified atmosphere, 5% CO_2 and 95% air, at 37 °C).

2.5.1. Adhesion Assay

Round-shaped scaffolds (PCL-ref, PCL-CCOH, and PCL-COOH-PRP2) with a diameter of 6 mm were placed in a 96-well plate. The cells were seeded on the scaffolds at a concentration of 10×10^3 cells per scaffold. After 20 min (short-time adhesion), 2 and 24 h, and three and seven days, the scaffolds with fibroblasts were fixed using 4% paraformaldehyde solution for 10 min. Then, 0.1% Triton X-100 was used for the cell membrane permeabilization. For visualization of cytoskeleton actin filaments, we used the Alexa Fluor 532 phalloidin (Thermo Fisher Scientific, Waltham, MA, USA). The fibroblast-seeded scaffolds were stained with Hoechst 33342 for 15 min at room temperature. The cell adhesion on the PCL nanofibers was studied by fluorescence microscopy (Zeiss, Axio observer Z1, Oberkochen, Germany).

2.5.2. Cell Proliferation

The Click-iT™ EdUAlexa Fluor™ 488 Imaging Kit (Thermo Fisher Scientific, Waltham, MA, USA) was used for detection of the cell proliferation rate. Each round-shaped scaffold with a diameter of 6 mm was seeded with the fibroblasts at a concentration of 7×10^3 cells per well in the 96-well plates. The cell-seeded scaffolds were incubated under standard condition (5% CO_2 and 95% air) at 37 °C. On the third and seventh day after seeding the cells, EdU was added to each sample for 2 h. The cells were studied by fluorescence microscopy (Zeiss, Axio observer Z1, Oberkochen, Germany). The cell quantification was performed on 50 microscopic images per studied scaffold. The level of proliferation rate was estimated by determining the ratio of total cells (Hoechst-positive) to the EdU-positive cells using the Cell Activision software (Yokogawa Electric Corporation, Tokyo Japan).

2.5.3. Cell Apoptosis

The fluorescent dye Hoechst 33342 was used to stain cell nuclei. The percentage of apoptotic cells was evaluated as the ratio of normal cell nuclei (middle bright and uniformity of Hoechst) to fragmented and brighter nuclei.

2.5.4. Immunofluorescent Staining of Scaffold Cells

Fibroblasts were grown on scaffolds for three and seven days. Afterward, they were fixed with 4% paraformaldehyde, permeabilized with 0.1% Triton X-100, and blocked with 1% BSA (bovine serum albumin). The cells seeded on the scaffold were stained with primary antibodies for 4 h at room temperature, washed with PBS, and incubated with secondary antibodies for 1 h at room temperature. The stained cells were analyzed with the fluorescence microscope (Zeiss, Axio observer Z1, Oberkochen, Germany).

The following primary antibodies were used: anti-fibronectin (Abcam, Cambridge, UK, ab6328, 1:200) and anti-collagen IV (Life Span, Hamilton, OH, USA, 1:200). The following secondary antibodies were used: Alexa Fluor 594 goat anti-mouse immunoglobulin G1 (IgG1; Life Technologies, Carlsbad, CA, USA).

All results were displayed as mean values and standard errors. The statistical significance was assessed by the nonparametric Mann–Whitney U test using the Statistica 10 software (StatSoft. Inc., Tulsa, Ok, USA).

3. Results

3.1. Optimization of Electrospinning of the PCL Nanofibers

The effect of the electrospinning process on the resulting polymer nanofibers was studied in order to obtain homogeneous substrates for subsequent immobilization and biological tests. The nanofibers' topography analyzed by SEM is shown in Figure 1. The comparison of the micrographs reveals a highly porous network of fibers with considerable differences in morphology and homogeneity depending on the PCL concentration in solution and the applied voltage.

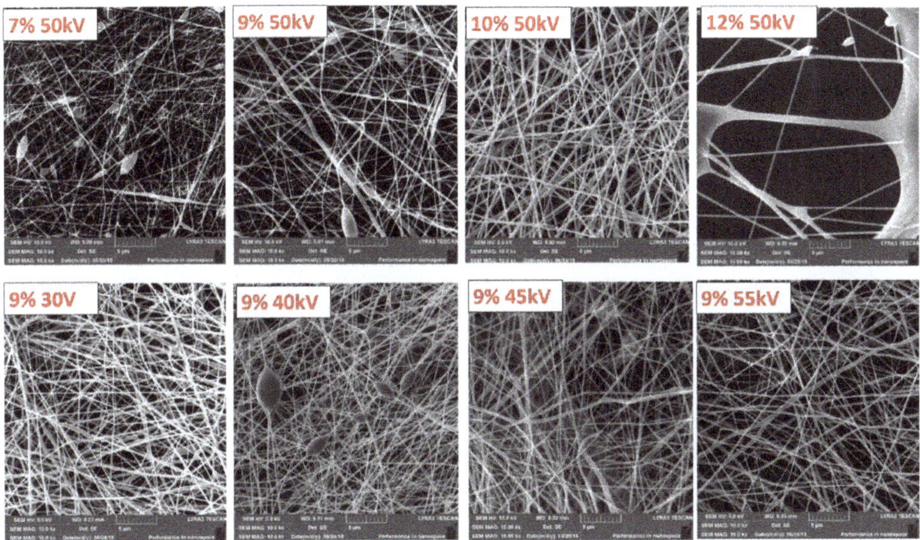

Figure 1. SEM micrographs of the polycaprolactone (PCL-ref) nanofibers obtained from the PCL solutions with the different PCL concentrations and at different voltages. The PCL concentration and voltage are reported on the images. The size of the bar corresponds to 1 μm.

The effect of the PCL concentration on the electrospinning solution is shown in the upper row of images in Figure 1, captured at the magnification of 10,000×. The applied voltage was consistent for all substrates at 50 kV. The electrospun nanofibers prepared from the solution with 7 wt.% of the PCL polymer were thin with a calculated thickness of 132 ± 42 nm, and they exhibited a significant beaded structure. Beads and beaded fibers are more likely formed from less concentrated solutions, as their viscosities are low, making the jet formation unstable [36]. Increasing the PCL concentration led to an increase in fiber uniformity and a more regular morphology with a considerable decrease in beaded fibers [36,37].

The nanofibrous meshes prepared from solutions with PCL concentrations of 9 and 11 wt.% gave fiber thicknesses of 263 ± 57 nm and 290 ± 112 nm, respectively. A higher thickness of nanofibers and their lower homogeneity prepared from more concentrated solutions might be caused by the viscoelastic force in the nanofiber jet resisting the stretching repulsive forces of opposing charges [38]. The effect of PCL concentration on the thickness of nanofibers is summarized in Figure 2a.

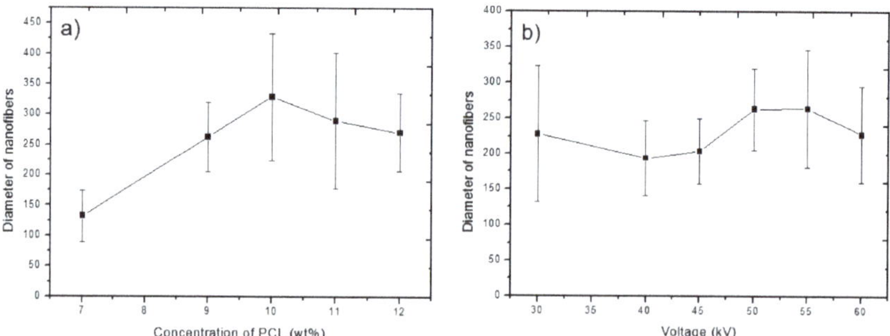

Figure 2. Thickness of resulting PCL nanofibers as a function of the PCL concentration (**a**) and applied voltage (**b**) during the electrospinning process. Error bars represent standard deviation of the nanofibers' thickness distribution.

For further experiments, a PCL concentration of 9 wt.% was selected, and the effect of the applied voltage was studied. The SEM micrographs of the electrospun PCL nanofibers obtained with the applied voltages of 30, 40, 45, and 55 kV are shown in the lower row of images in Figure 1. The effect of applied voltage is summarized in Figure 2b. It can be seen that the effect of the applied voltage on the fiber diameter is not linear. It is worth noting that the extremely low values of applied voltage led to a higher dispersion of fiber diameters. Depending on the combination of process parameters, it is possible to obtain a substrate with fine homogeneous nanofibers.

The observed effect of polymer concentration and applied voltage on fiber thickness and homogeneity is summarized in Figure 2a,b. The error bars represent the calculated standard deviation of nanofiber thickness on selected substrates and, thus, illustrate their homogeneity. As visible in the graphs, a lower concentration of PCL polymer in solution (7 wt.%) led to thinner fibers. Nevertheless, because of the occurrence of beaded structures, the substrates were not used for further experiments. The increase in nanofiber thickness correlated with increasing PCL polymer concentration in solution and was less dependent on applied voltage. In addition, the nanofiber thickness was more homogeneous when less concentrated/viscous solutions were used.

The observed tendency is in line with the previously reported data for the optimization of electrospinning from a PCL solution in acetic acid using the Starter Kit 40 KV Web set-up, utilizing the needle electrospinning technology [36]. The advantage of needleless technology developed by Elmarco (used here) is a higher production rate and a bigger size of the produced samples. Regarding the optimized conditions, the polymer substrates electrospun from solutions with a PCL concentration of 9 wt.% and applied voltage of 50–55 kV were selected for plasma chemical depositions. From now on, the paper refers to this type of nanofiber simply as PCL-ref. The morphology of the nanofibers was not affected by the deposition of $Ar/CO_2/C_2H_4$ plasma layers and no damage was detected (Figure 3). The immobilization of PRP also had no effect on the morphology of the PCL nanofibers. No swelling or degradation of fibers can be observed in the SEM micrographs in Figure 3.

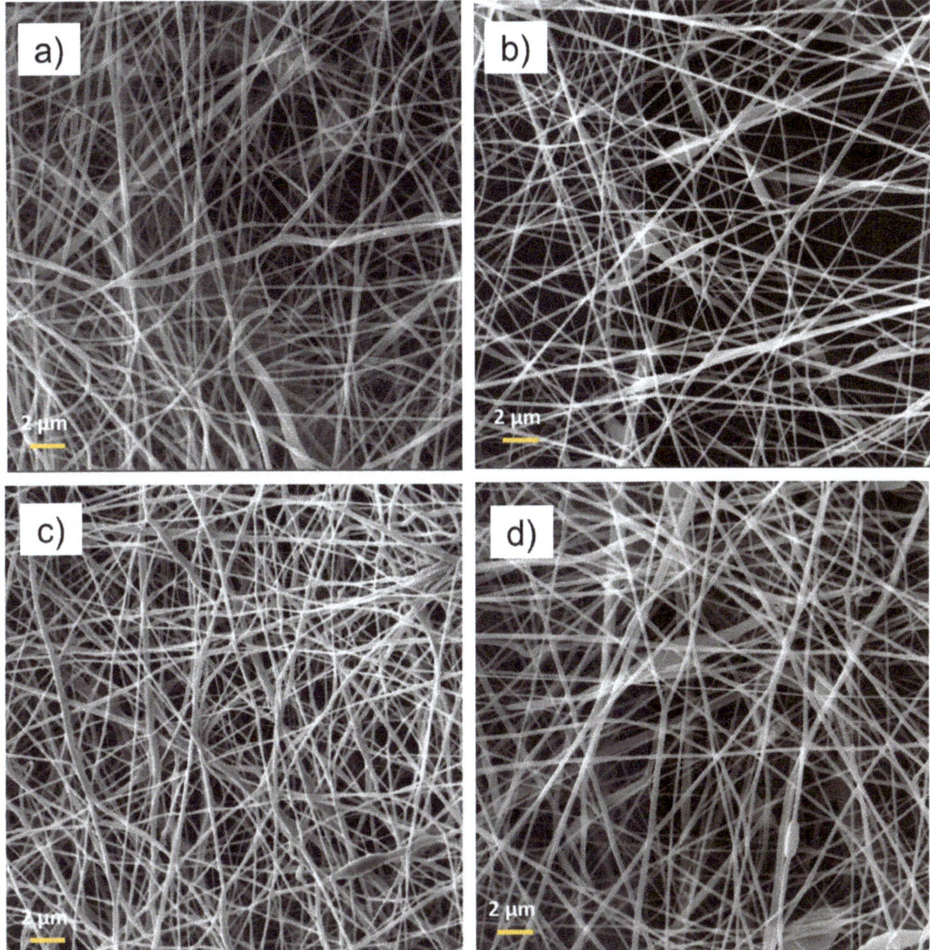

Figure 3. SEM micrographs of PCL-ref (**a**), PCL/platelet-rich plasma (PCL-PRP) (**b**), PCL-COOH (**c**), and PCL-COOH-PRP2 (**d**) The size of the bar corresponds to 1 μm.

3.2. Surface Characterization of the Samples

The effect of plasma coating was evident from the WCA measurements. PCL-ref exhibited a WCA of 121 ± 3°. The deposition of the $Ar/CO_2/C_2H_4$ plasma layer led to a decrease in WCA to 67 ± 7°. The increase in wettability of PCL nanofibers is related to the grafting of polar groups and it was also shown before for COOH modification of PCL nanofibers using the atmospheric pressure plasma deposition technique [5]. The chemical composition of the PCL nanofibers (PCL-ref) measured by XPS is reported in Table 1. The atomic percentages of the elements were quantified from the high-resolution spectra of each element. The survey spectra of the samples are reported in Figure S1 (Supplementary Materials). The XPS C *1s* spectrum of PCL-ref can be fitted with a sum of three components, namely hydrocarbons $\underline{C}H_x$ (BE = 285 eV), ether group $\underline{C}$–O (BE = 286.4 eV), and ester group $\underline{C}$(O)O (BE = 289.0 eV) (Figure 4a). The full width at the half maximum (FWHM) of $\underline{C}$–O was set to 1.35 eV, while $\underline{C}H_x$ and $\underline{C}$(O)O were fitted with the FWHM of 1.1 and 0.95 eV, respectively. The XPS O *1s* spectrum was fitted with a sum of two components: C=$\underline{O}$ (BE = 532.1 eV) and C–$\underline{O}$ (BE = 533.5 eV).

The relative concentrations of C=$\underline{O}$ and C–$\underline{O}$ components were 42% and 58%, respectively (see Figure S2, Supplementary Materials).

Table 1. Atomic composition (at.%) of samples derived from X-ray photoelectron spectroscopy (XPS) analysis. The traces of sodium, chlorine, and phosphorus were detected but were not taken into account. PCL—polycaprolactone; PRP—platelet-rich plasma.

Sample Name	C	O	N	S
PCL-ref	73.9	26.1	0.0	0.0
PCL-PRP	71.6	20.6	7.6	0.2
PCL-COOH	72.4	26.7	0.9	0.0
PCL-COOH-PRP1	71.3	18.5	10.0	0.2
PCL-COOH-PRP2	75.8	18.1	5.9	0.2

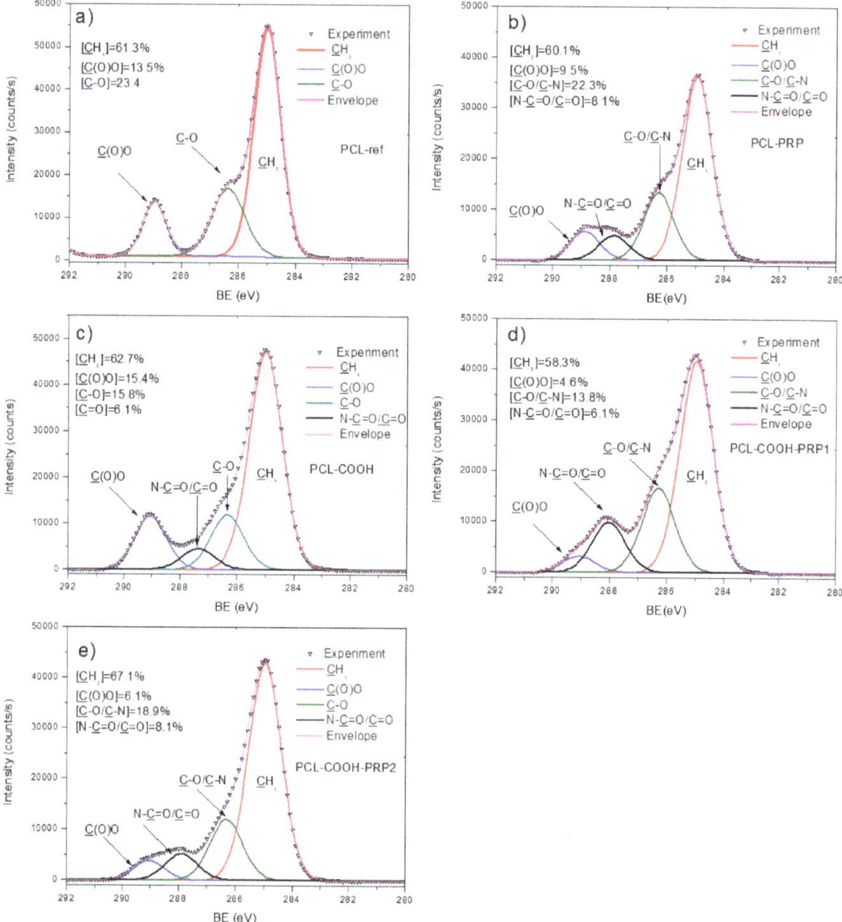

Figure 4. X-ray photoelectron spectroscopy (XPS) C *1s* curve fitting of PCL-ref (**a**), PCL-PRP (**b**), PCL-COOH (**c**), PCL-COOH-PRP1 (**d**), and PCL-COOH-PRP2 (**e**).

The deposition of Ar/CO$_2$/C$_2$H$_4$ plasma layers led to a slight increase in the O/C ratio. Interestingly, the observed changes in the XPS C *1s* spectrum were rather significant (Figure 4c). The XPS C *1s* spectrum of the Ar/CO$_2$/C$_2$H$_4$ plasma-coated nanofibers (PCL-COOH) was fitted with four carbon components: $\underline{C}H_x$ (BE = 285 eV), $\underline{C}$–O (BE = 286.4 eV), $\underline{C}$=O (BE = 287.4 eV), and $\underline{C}$(O)O (BE = 289.1 eV). The FWHM value for all components was set to 1.35 ± 0.05 eV. The concentrations of different carbon environments are summarized in Figure 4c. XPS analysis also revealed a small amount of nitrogen present on the PCL-COOH sample. The XPS N *1s* spectrum of PCL-COOH (Figure 5c) was fitted with two components: amide group $\underline{N}$–C=O (BE = 400.0 eV) and $\underline{N}$–O_x (BE = 402.0 eV). Most likely, the origins of these groups are related to the reaction with residual nitrogen in plasma or to the contact of PCL-COOH with the atmosphere. The O *1s* spectrum after the deposition of plasma layers was fitted with the same two components C–$\underline{O}$ and C=$\underline{O}$. However, the FWHM of these peaks became broader and the concentration of the $\underline{C}$=O component was increased by ~20% at the expense of the C–O component (see Figure S2, Supplementary Materials).

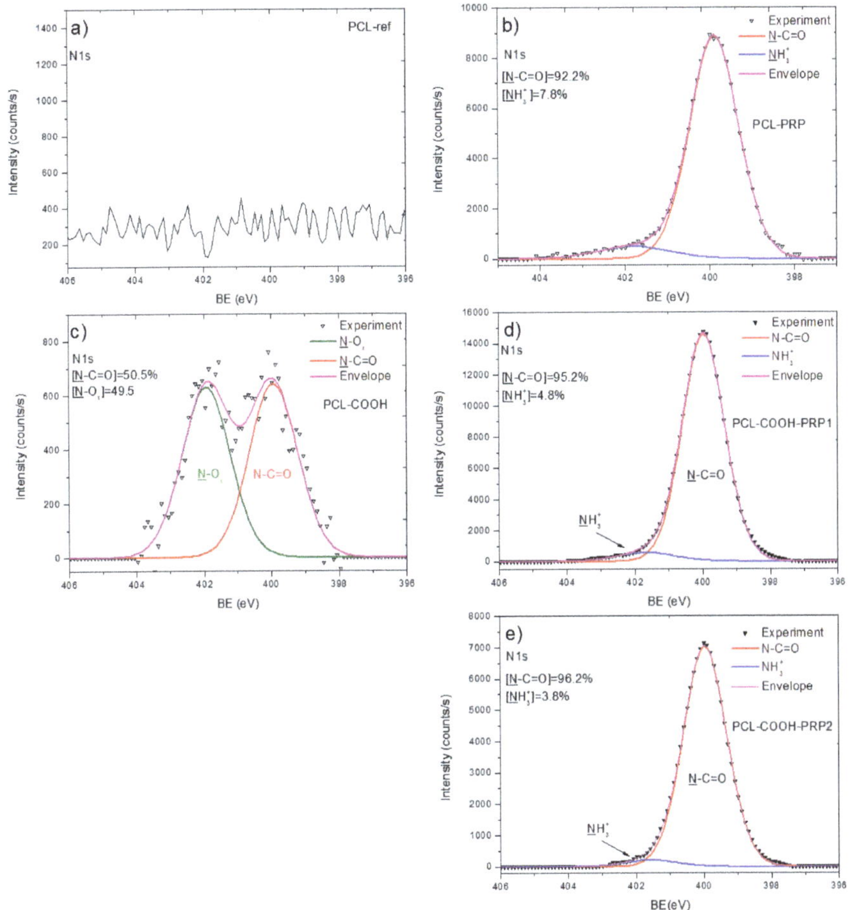

Figure 5. XPS N *1s* curve fitting of PCL-ref (**a**), PCL-PRP (**b**), PCL-COOH (**c**), PCL-COOH-PRP1 (**d**), and PCL-COOH-PRP2 (**e**).

Concerning the changes in the surface chemistry after the immobilization, according to the XPS results in Table 1, it is evident that the immobilization of PRP happens even without the plasma coating of the nanofibers. It is indicated by the relatively high concentrations of nitrogen found on the PCL-PRP, PCL-COOH-PRP1, and PCL-COOH-PRP2 samples. It is worth noting that the traces of sodium (<1 at.%), chlorine (<1 at.%), and phosphorus (<0.5 at.%) were also detected on survey scans (see

Figure S1, Supplementary Materials). These traces are the components of the PBS (NaCl and Na_2HPO_4) that we used for washing of the sample surface and, therefore, these elements were not taken into account in Table 1.

In addition, the ATR-FTIR analyses (not reported here) showed that there were no changes in the spectrum of PCL-PRP in comparison to the one of PCL-ref. The XPS N *1s* spectrum of PCL-PRP is presented in Figure 5b. The PCL-ref did not provide any N *1s* signal, whereas PCL-PRP, PCL-COOH-PRP1, and PCL-COOH- PRP yielded very similar spectra that were all fitted with a sum of two components, namely an amide group $\underline{N}$–C=O (BE = 400.0 eV) and protonated amines $\underline{N}H_3^+$ (BE = 401.7 eV). The FWHM for all components was set to 1.7 ± 0.1 eV. As shown in Figure 5b,d,e, for these samples, the main XPS N *1s* fit component is associated with amide groups, the building block of all proteins, thus confirming the immobilization of PRP. Interestingly, the highest concentration of nitrogen was detected for PCL-COOH-PRP1; this effect is most probably related to the steric hindrances induced by EDC activation as discussed in Section 4. It is worth noting that no unreacted EDC was detected on the surface, as no $\underline{N}$=C=$\underline{N}$ component at 398.5 eV was found in the N *1s* spectrum. The XPS O *1s* spectra of PCL-PRP, PCL-COOH-PRP1, and PCL-COOH-PRP2 were fitted with a sum of C–$\underline{O}$ and C=$\underline{O}$ components (Figure S2, Supplementary Materials). Interestingly the non-covalent bonding of PRP led to the shift of the C=$\underline{O}$ component from 532.0 to 531.5 eV (in PCL-PRP and PCL-COOH-PRP1), whereas the covalent immobilization of PRP did not affect the BE position of this component. The shift is related to the different neighboring atom adjacent to the C=$\underline{O}$ component. The BE of 531.5 eV most probably corresponds to the C=O group close to amine group (NH), whereas the BE of 532.0 eV corresponds to C=$\underline{O}$ close to C–$\underline{O}$ (e.g., O–C=$\underline{O}$).

3.3. Adhesion and Spreading of Fibroblasts Seeded on PCL-ref, PCL-COOH, and PCL-COOH-PRP

After 20 min of seeding cells on the studied scaffolds, the number of adhered cells showed no statistically significant differences. However, the largest cell size was found on PCL-COOH-PRP2 (up to 41 µm), while, on PCL-ref and PCL-COOH, the cell sizes reached a maximum of 35 µm (Figure 6).

After 2 h of cultivation, fibroblasts on all studied substrates formed filopodia. Nevertheless, as it can be seen from Figure 6, the number of cells on PCL-ref and on PCL-COOH was significantly lower compared to PCL-COOH-PRP2. Cells seeded on PCL-COOH-PRP2 had the biggest spreading area, multiple focal contacts, and filopodia, and a network of actin fibrils was formed (Figure 6 and Figure S3, Supplementary Materials).

After 24 h of incubation, the actin cytoskeleton of fibroblasts on PCL-ref was significantly pronounced at the edges of small rounded cells, whereas a network of actin fibers and stress fibrils was not observed. The fibroblasts adhered on the hydrophobic surface of PCL-ref, forming groups of cells.

The fibroblasts adhered onto PCL-COOH demonstrated a spread polygonal shape with a network of actin microfilaments. A part of the cells exhibited an elongated shape and stress fibrils (Figure 6). The fibroblasts were evenly distributed on the surface.

There was no significant difference in the number of cells after 24 h of incubation on PCL-ref and PCL-COOH. The binding of PRP to the substrate affected the adhesion of fibroblasts already at the initiation stage and, after one day, the cells had a well-spread polygonal shape with a pronounced cytoskeleton and lamellipodia (Figure 6).

Within three days, the number of adhered cells on PCL-ref did not increase significantly. The cells were of small size and exhibited multiple focal contacts. On the third day of incubation, the cells on PCL-COOH exhibited a relatively pronounced cytoskeleton and a flattened polygonal shape. However, the cytoskeleton was less pronounced than on PCL-COOH-PRP2 and there were no clear stress-fibrils and focal contacts. At this incubation time, the cells had a well-formed cytoskeleton, organized stress fibrils, focal contacts, and lamellipodia on PCL-COOH-PRP2.

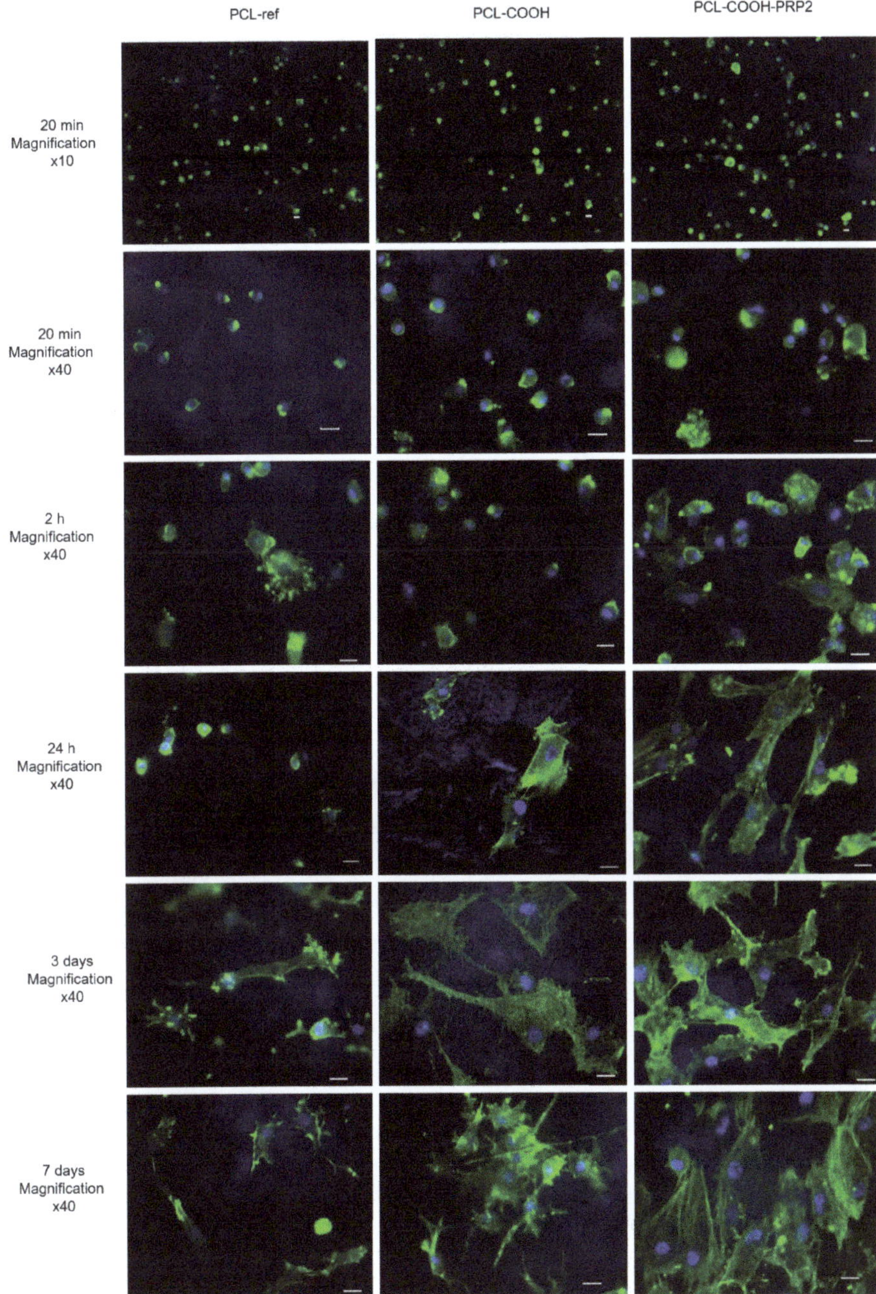

Figure 6. Fibroblast cell adhesion and spreading on the surface of PCL-ref, PCL-COOH, and PCL-COOH-PRP2 after 20 min, 24 h, and three and seven days of cultivation. The cytoskeleton actin filaments were stained by phalloidin (green), the cell nuclei were stained by Hoechst 33342 (blue). The size of the bar corresponds to 20 μm.

On PCL-ref, the number of cells after seven days of incubation did not differ statistically from that after three days. The cells revealed cytoplasmic micro-growths directed anisotropically. On PCL-COOH, despite the statistically important increase in the cell number after seven days of incubation, the cells also exhibited long, anisotropically directed cytoplasmic micro-growths. (Figure 6). The increased adhesion on the PCL-COOH-PRP2 substrates during early and late spreading stimulated cell proliferation, as, after seven days of incubation, a monolayer of fibroblasts was observed (Figure 6 and Figure S3, Supplementary Materials). There was a contact inhibition of the isotropic movement and, thus, the cells were co-directed (Figure 6).

3.4. The Modification of PCL Nanofibers with COOH Groups and Its Effect on Human Fibroblast Proliferation and the Cell Apoptosis

After 24 h of incubation, the number of adhered cells on PCL-ref and PCL-COOH did not differ significantly; however, on PCL-COOH, the cells were distributed more evenly. The number of cells on the third day of incubation on PCL-ref and PCL-COOH scaffolds made up 15 ± 7.4 and 18 ± 4.2 cells per field of view, respectively (Figure 7b). The level of cell proliferation on PCL-ref and PCL-COOH was not statistically different and was equal to $7.9 \pm 1.9\%$ and $8.2 \pm 1.6\%$, respectively (Figure 7a). The level of apoptosis on PCL-ref was high ($31 \pm 3.4\%$), and that on PCL-COOH was two times lower ($14 \pm 2.3\%$) (Figure 7a).

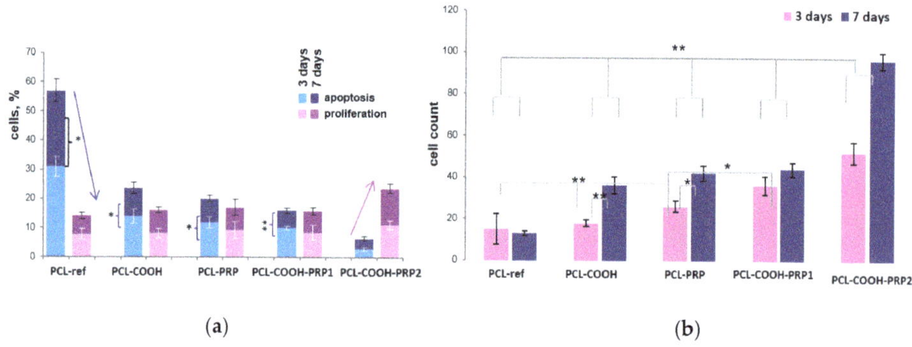

(a)   (b)

Figure 7. The influence of scaffold surface on (**a**) cell proliferation and apoptosis, and (**b**) cell count for three and seven days of cultivation. The level of proliferation rate was estimated by determining the ratio of total cells (Hoechst-positive) to the EdU-positive cells. The percentage of apoptotic cells was calculated as the ratio of the nuclei with chromatin condensation and nuclear fragmentation (kariorhexis) to nuclei with homogeneous coloration. The arrows indicate the relationship of the level of cell apoptosis and cell proliferation. Data are expressed as means $\pm$ standard errors; ** $p < 0.01$, * $p < 0.05$.

Due to the uneven density of cells on PCL-ref, the fields of view were divided into two regions with a high and low cell content. The level of proliferation was highly dependent on the cell number on the PCL-ref substrates. Thus, 10% EdU-positive cells were in the regions with a high cell density, while a low cell density demonstrated 4.5% proliferating cells.

At the same time, on PCL-COOH, the proliferation level was practically independent of the number of cells per field.

On the seventh day of incubation, the average number of cells on PCL-ref and PCL-COOH made up 13 ± 1.1 and 36 ± 4.2 cells per field of view, respectively (Figure 7b). The proliferation level on PCL-ref and PCL-COOH substrates did not change statistically. At the same time, the number of apoptotic cells seeded on PCL-ref and PCL-COOH decreased to $26 \pm 3.9\%$ and $9.5 \pm 1.9\%$, respectively (Figure 7a). The reduced apoptosis was probably due to the production of the extracellular matrix by viable cells.

Therefore, the COOH modification of the PCL nanofibers promotes a more uniform distribution of adhered cells and a better cell viability associated with a decrease in the level of apoptotic cell death, suggesting a better biocompatibility of this surface as compared to unmodified PCL nanofibers. The fibroblasts are capable of adhering and surviving on the hydrophobic surface of PCL-ref, maintaining a certain level of proliferation in the islands of high cell density.

3.5. The Effect of PRP Immobilization on Human Fibroblast Proliferation

We analyzed the proliferation of human fibroblasts on the PRP-modified PCL nanofibers with different forms of PRP bonding: non-covalent bonding of PRP to unmodified PCL-ref based on hydrophobic–hydrophobic interaction of the proteins with PCL nanofibers, ionic bonding of PRP to COOH group plasma-coated PCL scaffolds, and covalent bonding of PRP to the PCL-COOH using EDC.

Observation over three days showed that the addition of PRP to PCL significantly increased the number of cells. It was noted that the number of cells on PCL-PRP and PCL-COOH-PRP1 was lower compared to PCL-COOH-PRP2 (26 ± 2.6 and 32 ± 4.4 against 52 ± 5, respectively). The level of proliferative activity was also higher for PCL-COOH-PRP2, at 11.2 ± 1.5%, compared with 8.6 ± 1.32% and 9.4 ± 2.9% for PCL-COOH-PRP1 and PCL-PRP, respectively (Figure 7).

It should be noted that the level of apoptosis on PCL-COOH-PRP2 was three times lower compared to PCL-COOH-PRP1 and PCL-PRP (3 ± 0.7% versus 10 ± 0.6% and 12 ± 1.9%, respectively) (Figure 8). Thus, already on the third day, it can be seen that the surface with covalently bound PRP was most effective for maintaining the functional viability of fibroblasts.

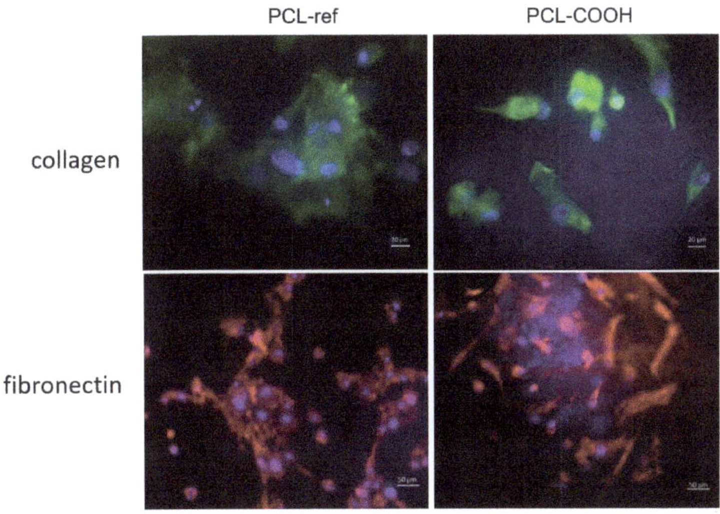

Figure 8. Fibronectin (FN) and collagen secretion by the fibroblasts seeded on PCL-ref and PCL-COOH. Cells were stained by antibody Alexa Fluor 594 (orange) to collagen (magnification 40×) and Alexa Fluor 488 to FN (magnification 20×).

After seven days on PCL-COOH-PRP1, the level of cell apoptosis decreased to 6 ± 0.9%, while maintaining cell proliferative activity (7.2 ± 1.3%). The percentage of apoptotic cell death on PCL-COOH-PRP2 remained at the same low level (3.5 ± 0.7%), while maintaining the level of proliferative activity of 12.4 ± 1.6%.

Thus, it is shown that the addition of PRP to PCL-ref and PCL-COOH nanofibers further increased the number of adhered cells and the level of cell proliferation, and decreased the level of apoptotic death. However, hydrophobic non-covalent and ionic bonding of PRP to PCL-COOH and PCL-ref led to a weak linkage between proteins and PCL and, as a result, only a short-term enhancement of the cell proliferation (Figure 7). The dynamics of the cell number growth and percentage of proliferating cells reduced after three days of cultivation. The covalent bonding of PRP to PCL-COOH ensured a constantly enhanced level of proliferation and a low level of fibroblast apoptosis even after seven days of cultivation.

3.6. The Secretion of Fibronectin and Type IV Collagen by Fibroblasts Seeded on PCL-ref

As shown earlier, the uneven cell distribution on PC-ref was associated with a high single-cell death and a good level of proliferation in the islands of high cell density. The cells on PCL-ref had a normal phenotype and a high expression level of fibronectin (FN) and collagen after three days of incubation in the areas of high cell density. On PCL-COOH, the fibroblasts were located evenly and secreted FN and collagen (Figure 8).

This pattern of ECM component secretion by fibroblasts also suggests that PCL-COOH is more suitable for cell viability.

4. Discussion

For the purpose of tissue engineering, as well as for regenerative medicine, it is necessary to create scaffolds suitable for the viability and functional activity of the cells [39]. The aim of this study was to create a nanofibrous scaffold suitable for the efficient cell adhesion, spreading and functional activity of fibroblasts. The normal cell adhesion and spreading on the substrate surface affect the later stages of the cell functional activity [14–17]. Therefore, the study of cell spreading is vital in the scaffold's screening for the purpose of regenerative medicine. The initial stage of cell adhesion is determined by the physico-mechanical interactions of the physical body (cell) and the substrate, and it does not depend on the molecular integrin–ECM interactions [15]. It is characterized by the blabbing of rounded cells, formation of a smooth contact area, and passive attachment to the substrate, and it lasts for 5–20 min after the cell seeding. We do not observe differences at this stage neither in the number nor in the morphological features of cells on the studied PCL scaffolds 20 min after the cell seeding (Figure 6). However, the bigger size of some of the cells on PCL-COOH-PRP2 was a result of better adhesion and an earlier spreading of fibroblasts.

The longer phase of early spreading lasted for approximately 120 min and was defined by the active cell attachment with the organization of a peripheral zone by actin fibrils and the formation of lamellipodia. At this phase, the polymerization of actin and the contraction of myosin are necessary for the polarization and movement of cells.

This phase was significantly extended until 24 h for PCL-COOH and until three days for PCL-ref. The electrophilicity of the PCL-COOH surface, together with the negative charge of the cells, led to a delay in their interaction. The presence of divalent cations in the cell culture medium of the PCL-COOH surface allowed the cells to adhere faster as compared to the hydrophobic surface of PCL-ref. The delay in the initial stage of adhesion and spreading on PCL-ref and PCL-COOH affected the proliferation rate and cell viability in the future and reduced the efficiency of PCL nanofiber colonization by fibroblasts (Figure 6 and Figure S3, Supplementary Materials).

It was previously shown that mesenchymal stem cells (MSCs), secreting mainly trophic factors, had good spread area on PCL-ref hydrophobic surfaces during the first hours after seeding but they did not survive on the PCL-ref surfaces during further cultivation [40]. Even though MSCs have highly promising regenerative potential, numerous MSC transplantation experiments showed that cells have a transient paracrine effect and do not survive in the pathologic area, losing their regenerative potential [41–43]. Fibroblasts have a morphology similar to MSCs and express a number of mesenchymal markers (cluster of differentiation 90 (CD90), CD105, CD44, and CD73). However, secretion of ECM components is the main function of these cells. In this work, the fibroblasts secrete ECM components, enriching PCL nanofibers with FN and collagen, thereby increasing the biocompatibility of the PCL-ref hydrophobic surface. On the third day of incubation, the fibroblasts cultivated on PCL-ref were found out to actively secrete FN and collagen in the cell islands. This promoted cell survival and proliferation, while the single cells died. Throughout the entire experiment, a high level of the fibroblast anoikis on PCL-ref was observed due to the surface adhesion deterioration [44] (see Figure 6 and Figure S3, Supplementary Materials). However, the observation of cells after seven days of cultivation revealed that the cells detached from the PCL nanofibers, despite the increase in cell density in the cell "islands" and the active secretion of fibronectin (data not shown). This can be explained by the fact that poor initial cell attachment to the substrate leads to the fibronectin turnover shift to the intercellular contacts instead of the cell–matrix contacts [29,45]. The increased cell-to-cell contact creates an area of an increased tension and leads to the cell detachment from the substrate [46].

The simple soaking of PCL-ref in PRP solution (non-covalent bonding) enabled immediate cell attachment to the substrate, thereby increasing the possibility of normal functioning of fibroblasts in the future. The grafting of PRP onto the uncoated PCL nanofibers can be explained by the interaction of the hydrophobic shell of proteins with the hydrophobic nanofibers. Such a process can potentially happen only on the very top surface of the fibers and it would lead to a very small immobilization of PRP. This was confirmed by the shape of the XPS C *1s* spectrum exhibiting many similarities to the spectrum of uncoated nanofibers (see Figure 4a,b). However, leaching of the PRP components from the surface of the nanofibers during cultivation resulted in a bare PCL-ref hydrophobic surface, leading to an increase in the anoikis of the cells.

In order to improve the biocompatibility of PCL nanofibers, an additional modification of their surface by the binding of freeze–thawed platelet-rich plasma was used.

The XPS data revealed that, after the immobilization of PRP, there was 0.2 at.% sulfur in the layer and the N *1s* signal consisted of $\underline{N}$–C=O and $\underline{N}H_3^+$. When compared to literature data for FN XPS analysis, the BE of the S *2p* peak, as well as the shape of the N *1s* peak, was very similar to the reported ones [47].

In order to understand how bioactive molecules attach to the PCL nanofibers, it can be useful to look at the immobilization process using the example of FN immobilization. When considering the immobilization of FN on the PCL nanofibers, one needs to remember that FN has a fibrillary structure with NH_2 and COOH groups at the "ends" of the long fibrils. In the absence of COOH groups on PCL nanofibers, the only possible binding of FN to PCL-ref would be the non-covalent binding via the interaction of the hydrophobic parts of FN with the hydrophobic PCL nanofibers. Only one monolayer of proteins can be immobilized on the top surface of PCL-ref and, certainly, the bonding will be very weak. The immobilization of FN directly on PCL-COOH (without EDC activator) would lead to the electrostatic coupling of FN to PCL-COOH, as shown in Figure 9. The absence of covalent bonding, a relatively long aliphatic chain between the "body" of protein and the amine group, and the excess of proteins in the PRP solution allow all available sites on the PCL-COOH-PRP1 to become "freely" occupied. The activation of the COOH groups by EDC leads to the grafting of carbodiimide bearing bulky fragments (see Figure 9). Most probably, these groups occupy more space on the surface during interaction in comparison to the amine group in the long aliphatic chain at the FN end. As a result, the concentration of nitrogen on PCL-COOH-PRP1 is higher by a factor of two than on PCL-COOH-PRP2. Hence, the density of the immobilized FN is higher for PCL-COOH-PRP1, but a weaker electrostatic bonding leads to the fast release of bioactive molecules, leaving the PCL-ref surface bare and with a hydrophobic nature. The formation of an area on the surface free of PRP molecules leads to a decrease in the survival of cells, thus increasing the apoptosis on the PCL-COOH-PRP1 and PCL-PRP samples (Figure 6).

Regardless of the type of bonding with PCL, the immobilization of PRP to PCL-ref and PCL-COOH leads to the normal cell adhesion and spreading due to the high biological activity of PRP components. The immobilization has a positive effect on the proliferative activity of fibroblasts and reduces the cell death by anoikis. However, the covalent bonding of PRP to PCL-COOH-PRP2 not only promotes the early stages of cell adhesion and spreading, but also significantly reduces the degree of cell apoptosis as compared to PCL-PRP and PCL-COOH-PRP1.

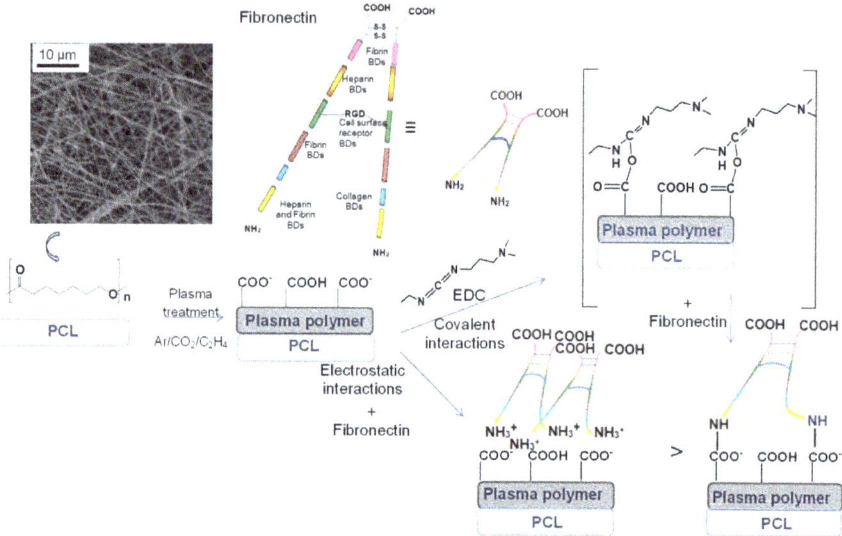

Figure 9. Scheme of "electrostatic" and covalent immobilization of FN on PCL-COOH.

Most likely, the covalent bonding of PRP occurs over the entire volume of the porous plasma polymer coating and provides a stable fixation of biologically active substances of PRP, causing the consistently high cell proliferation rate throughout the study. Additionally, the confocal layer-by-layer imaging of fibroblasts seeded on PCL showed that the cells fill the entire thickness of the nanofibers, which, in this case, became "friendly for the cell". The use of such matrices with covalently immobilized PRP for wound healing prolongs the biological activity of the immobilized molecules and protects them from the aggressive media of the wound. The proteins with a weak, non-covalent bonding to the PCL surface will be quickly washed out from the surface by exudation of the wound, leaving the PCL-ref surface bare and with a hydrophobic nature. Such a hydrophobic surface is not "cell-friendly" as discussed above.

Considering that the modification with COOH groups allows any type of proteins to be covalently attached, such a technique considerably widens the potential of the PCL nanofiber application in tissue engineering.

5. Conclusions

The reported modification of PCL nanofibers with COOH groups and the subsequent covalent binding of protein molecules is a rather simple and technologically accessible procedure. It was demonstrated that covalent binding of PRP occurs without the loss of its biochemical activity, allowing a significantly reduced fibroblast apoptosis and an increased cell proliferation on PCL-COOH-PRP2 compared with PCL-ref, PCL-COOH and PCL-COOH-PRP1 to be obtained.

The reported research findings reveal the potential of PCL matrices for application in tissue engineering, while the plasma modification with COOH groups and their subsequent covalent binding with NH_2 groups of proteins expand this potential.

Supplementary Materials: The following are available online at http://www.mdpi.com/2079-4991/9/4/637/s1: Figure S1: XPS survey spectra of PCL-ref, PCL-PRP, PCL-COOH, PCL-COOH-PRP1 and PCL-COOH-PRP-2; Figure S2: XPS O 1s curve fitting of PCL-ref, PCL-PRP, PCL-COOH, PCL-COOH-PRP1 and PCL-COOH-PRP-2; Figure S3: The influence of dynamic disturbance of cell adhesion and spreading on cell proliferation. Representative images of cell adhesion and spreading stage on PCL-ref and PCL-COOH-PRP after 20 min and 2 hours. The second stage of adhesion (2 hours) is significantly slower on PCL-ref. The cells have the same morphology as on the early stage of adhesion (20 min). The cells seeded on PCL-COOH-PRP have a well-spread polygonal shape with a pronounced cytoskeleton and lamellipodia. As a consequence, the level of cell proliferation and the number of cells are the highest for PCL-COOH-PRP.

Author Contributions: A.M., A.S. and D.V.S. conceived and designed the experiments; A.S., V.T. and S.M. performed the biological experiments; E.P. ATR-FTIR and immobilization studies, S.E. analyzed XPS data, P.K.-K. performed the plasma deposition experiments, A.M. and A.S. analyzed the data, L.Z. and E.D. performed the SEM analyses.

Funding: This research was funded by the Russian Science Foundation (grant No. 18-75-10057).

Acknowledgments: The authors gratefully acknowledge the financial support of the Russian Science Foundation (grant No. 18-75-10057). L.Z. and E.K. acknowledge gratefully acknowledge the project CEITEC 2020 (LQ1601) with financial support from the Ministry of Education, Youth, and Sports of the Czech Republic (MEYS CR) under the National Sustainability Program II for supporting scanning electron microscopy measurements. We thank the Microscopic Center of the Siberian Branch of the Russian Academy of Sciences for granting access to microscopic equipment.

Conflicts of Interest: The authors declare no conflicts of interest.

References

1. Lev-Tov, H.; Li, C.-S.; Dahle, S.; Isseroff, R.R. Cellular versus acellular matrix devices in treatment of diabetic foot ulcers: Study protocol for a comparative efficacy randomized controlled trial. *Trials* **2013**, *14*, 8. [CrossRef] [PubMed]

2. Watt, S.M.; Pleat, J.M. Stem cells, niches and scaffolds: Applications to burns and wound care. *Adv. Drug Deliv. Rev.* **2018**, *123*, 82–106. [CrossRef] [PubMed]

3. Dickinson, L.E.; Gerecht, S. Engineered Biopolymeric Scaffolds for Chronic Wound Healing. *Front. Physiol.* **2016**, *7*, 623. [CrossRef] [PubMed]

4. Chaudhari, A.A.; Vig, K.; Baganizi, D.R.; Sahu, R.; Dixit, S.; Dennis, V.; Singh, S.R.; Pillai, S.R.; Hardy, J.G. Future Prospects for Scaffolding Methods and Biomaterials in Skin Tissue Engineering: A Review. *Int. J. Mol. Sci.* **2016**, *17*, 1974. [CrossRef] [PubMed]

5. Manakhov, A.; Kedroňová, E.; Medalová, J.; Černochová, P.; Obrusník, A.; Michlíček, M.; Shtansky, D.V.; Zajíčková, L. Carboxyl-anhydride and amine plasma coating of PCL nanofibers to improve their bioactivity. *Mater. Des.* **2017**, *132*, 257–265. [CrossRef]

6. Al-Enizi, A.M.; Zagho, M.M.; Elzatahry, A.A. Polymer-Based Electrospun Nanofibers for Biomedical Applications. *Nanomaterials* **2018**, *8*, 259. [CrossRef] [PubMed]

7. Phan, L.T.; Yoon, S.M.; Moon, M.-W. Plasma-Based Nanostructuring of Polymers: A Review. *Polymers* **2017**, *9*, 417. [CrossRef]

8. Ivanova, A.A.; Syromotina, D.S.; Shkarina, S.N.; Shkarin, R.; Cecilia, A.; Weinhardt, V.; Baumbach, T.; Saveleva, M.S.; Gorin, D.A.; Douglas, T.E.L.; et al. Effect of low-temperature plasma treatment of electrospun polycaprolactone fibrous scaffolds on calcium carbonate mineralisation. *RSC Adv.* **2018**, *8*, 39106–39114. [CrossRef]

9. Permyakova, E.S.; Polčák, J.; Slukin, P.V.; Ignatov, S.G.; Gloushankova, N.A.; Zajíčková, L.; Shtansky, D.V.; Manakhov, A. Antibacterial biocompatible PCL nanofibers modified by COOH-anhydride plasma polymers and gentamicin immobilization. *Mater. Des.* **2018**, *153*, 60–70. [CrossRef]

10. Savelyeva, M.S.; Abalymov, A.A.; Lyubun, G.P.; Vidyasheva, I.V.; Yashchenok, A.M.; Douglas, T.E.L.; Gorin, D.A.; Parakhonskiy, B.V. Vaterite coatings on electrospun polymeric fibers for biomedical applications. *J. Biomed. Mater. Res. Part A* **2017**, *105*, 94–103. [CrossRef]

11. Saveleva, M.; Ivanov, A.; Kurtukova, M.; Atkin, V.; Ivanova, A.; Lyubun, G.; Martyukova, A.; Cherevko, E.; Sargsyan, A.; Fedonnikov, A.; et al. Hybrid PCL/CaCO3 scaffolds with capabilities of carrying biologically active molecules: Synthesis, loading and in vivo applications. *Mater. Sci. Eng. C* **2018**, *85*, 57–67. [CrossRef] [PubMed]

12. Chen, C.-H.; Chen, S.-H.; Kuo, C.-Y.; Li, M.-L.; Chen, J.-P. Response of Dermal Fibroblasts to Biochemical and Physical Cues in Aligned Polycaprolactone/Silk Fibroin Nanofiber Scaffolds for Application in Tendon Tissue Engineering. *Nanomaterials* **2017**, *7*, 219. [CrossRef] [PubMed]

13. Kiran, A.S.K.; Kumar, T.S.; Sanghavi, R.; Doble, M.; Ramakrishna, S. Antibacterial and Bioactive Surface Modifications of Titanium Implants by PCL/TiO$_2$ Nanocomposite Coatings. *Nanomaterials* **2018**, *8*, 860. [CrossRef] [PubMed]

14. Bereiter-hahn, J.; Luck, M.; Mdebach, T.; Stelzer, H.K.; Voth, M. Spreading of trypsinized cells: Cytoskeletal dynamics and energy requirements. *J. Cell Sci.* **1990**, *96*, 171–188. [PubMed]

15. Cuvelier, D.; Théry, M.; Chu, Y.-S.; Dufour, S.; Thiery, J.-P.; Bornens, M.; Nassoy, P.; Mahadevan, L. The Universal Dynamics of Cell Spreading. *Curr. Boil.* **2007**, *17*, 694–699. [CrossRef] [PubMed]

16. Tvorogova, A.; Saidova, A.; Smirnova, T.; Vorobjev, I. Dynamic microtubules drive fibroblast spreading. *Boil. Open* **2018**, *7*, bio038968. [CrossRef] [PubMed]

17. Pollard, T.D.; Borisy, G.G. Cellular Motility Driven by Assembly and Disassembly of Actin Filaments. *Cell* **2003**, *113*, 549. [CrossRef]

18. McGrath, J.L. Cell Spreading: The Power to Simplify. *Curr. Boil.* **2007**, *17*, R357–R358. [CrossRef] [PubMed]

19. Olsen, B.R. *Matrix Molecules and Their Ligands*, 4th ed.; Elsevier: Amsterdam, The Netherlands, 2014; ISBN 9780123983589.

20. Bershadsky, A.D.; Balaban, N.Q.; Geiger, B. Adhesion-Dependent Cell Mechanosensitivity. *Annu. Cell Dev. Boil.* **2003**, *19*, 677–695. [CrossRef]

21. Pelham, R.J., Jr.; Wang, Y. Cell locomotion and focal adhesions are regulated by substrate flexibility. *Proc. Natl. Acad. Sci. USA* **1997**, *94*, 13661–13665.

22. Zhou, G.; Liu, S.; Ma, Y.; Xu, W.; Meng, W.; Lin, X.; Wang, W.; Wang, S.; Zhang, J. Innovative biodegradable poly(L-lactide)/collagen/hydroxyapatite composite fibrous scaffolds promote osteoblastic proliferation and differentiation. *Int. J. Nanomed.* **2017**, *12*, 7577–7588. [CrossRef] [PubMed]

23. Kiran, S.; Nune, K.C.; Misra, R.D.K. The significance of grafting collagen on polycaprolactone composite scaffolds: Processing-structure-functional property relationship. *J. Biomed. Mater. A* **2015**, *103*, 2919–2931. [CrossRef] [PubMed]

24. Daelemans, L.; Steyaert, I.; Schoolaert, E.; Goudenhooft, C.; Rahier, H.; De Clerck, K. Nanostructured Hydrogels by Blend Electrospinning of Polycaprolactone/Gelatin Nanofibers. *Nanomaterials* **2018**, *8*, 551. [CrossRef] [PubMed]

25. Lin, S.J.; Jee, S.H.; Hsaio, W.C.; Lee, S.J.; Young, T.H. Formation of melanocyte spheroids on the chitosan-coated surface. *Biomaterials* **2005**, *26*, 1413–1422. [CrossRef] [PubMed]

26. Salević, A.; Prieto, C.; Cabedo, L.; Nedović, V.; Lagaron, J.M. Physicochemical, Antioxidant and Antimicrobial Properties of Electrospun Poly(ε-caprolactone) Films Containing a Solid Dispersion of Sage (Salvia officinalis L.) Extract. *Nanomaterials* **2019**, *9*, 270. [CrossRef] [PubMed]

27. DeFrates, K.; Moore, R.; Lin, G.; Hu, X.; Beachley, V.; Borgesi, J.; Mulderig, T. Protein-Based Fiber Materials in Medicine: A Review. *Nanomaterials* **2018**, *8*, 457. [CrossRef] [PubMed]

28. Zardi, L.; Cecconi, C.; Barbieri, O.; Carnemolla, B.; Picca, M.; Santi, L. Concentration of Fibronectin in Plasma of Tumor-bearing Mice and Synthesis by Ehrlich Ascites Tumor Cells. *Cancer Res.* **1979**, *39*, 3774–3777. [PubMed]

29. Sottile, J.; Hocking, D.C.; Schwartz, M. Fibronectin Polymerization Regulates the Composition and Stability of Extracellular Matrix Fibrils and Cell-Matrix Adhesions. *Mol. Boil. Cell* **2002**, *13*, 3546–3559. [CrossRef] [PubMed]

30. Sottile, J.; Hocking, D.C.; Swiatek, P.J. Fibronectin matrix assembly enhances adhesion-dependent cell growth. *J. Cell Sci.* **1998**, *111*, 2933–2943. [PubMed]

31. Agrawal, A.A. Evolution, current status and advances in application of platelet concentrate in periodontics and implantology. *World J. Clin. Cases* **2017**, *5*, 159–171. [CrossRef]

32. Guszczyn, T.; Surażyński, A.; Zaręba, I.; Rysiak, E.; Popko, J.; Pałka, J. Differential effect of platelet-rich plasma fractions on β 1-integrin signaling, collagen biosynthesis, and prolidase activity in human skin fibroblasts. *Drug Des. Dev. Ther.* **2017**, *11*, 1849–1857. [CrossRef] [PubMed]

33. Chen, B.; Ding, J.; Zhang, W.; Zhou, G.; Cao, Y.; Liu, W. Tissue Engineering of Tendons: A Comparison of Muscle-Derived Cells, Tenocytes, and Dermal Fibroblasts as Cell Sources. *Plast. Reconstr. Surg.* **2016**, *137*, 536e–544e. [CrossRef] [PubMed]

34. Li, J.; Chen, M.; Wei, X.; Hao, Y.; Wang, J. Evaluation of 3D-Printed Polycaprolactone Scaffolds Coated with Freeze-Dried Platelet-Rich Plasma for Bone Regeneration. *Materials* **2017**, *10*, 831. [CrossRef] [PubMed]

35. Beamson, G.; Briggs, D. *High Resolution XPS of Organic Polymers*; John Wiley & Sons: Chichester, UK, 1992.

36. Liverani, L.; Boccaccini, A.R.; Armentano, I. Versatile Production of Poly(Epsilon-Caprolactone) Fibers by Electrospinning Using Benign Solvents. *Nanomaterials* **2016**, *6*, 75. [CrossRef] [PubMed]

37. Lee, K.; Kim, H.Y.; Bang, H.; Jung, Y.; Lee, S. The change of bead morphology formed on electrospun polystyrene fibers. *Polymer* **2003**, *44*, 4029–4034. [CrossRef]

38. Pillay, V.; Dott, C.; Choonara, Y.E.; Tyagi, C.; Tomar, L.; Kumar, P.; Du Toit, L.C.; Ndesendo, V.M.K. A Review of the Effect of Processing Variables on the Fabrication of Electrospun Nanofibers for Drug Delivery Applications. *J. Nanomater.* **2013**, *2013*, 1–22. [CrossRef]

39. Netti, M.V.; Netti, P.A. Engineering Cell Instructive Materials To Control Cell Fate and Functions through Material Cues and Surface Patterning. *Acs Appl. Mater. Interfaces* **2016**, *8*, 14896–14908.

40. Solovieva, A.; Miroshnichenko, S.; Kovalskii, A.; Permyakova, E.; Popov, Z.; Dvořáková, E.; Kiryukhantsev-Korneev, P.; Obrosov, A.; Polčak, J.; Zajíčková, L.; et al. Immobilization of Platelet-Rich Plasma onto COOH Plasma-Coated PCL Nanofibers Boost Viability and Proliferation of Human Mesenchymal Stem Cells. *Polymer* **2017**, *9*, 736. [CrossRef] [PubMed]

41. Eggenhofer, E.; Luk, F.; Dahlke, M.H.; Hoogduijn, M.J. The Life and Fate of Mesenchymal Stem Cells. *Front. Immunol.* **2014**, *5*, 5. [CrossRef]

42. Chen, L.; Tredget, E.E.; Wu, P.Y.G.; Wu, Y. Paracrine Factors of Mesenchymal Stem Cells Recruit Macrophages and Endothelial Lineage Cells and Enhance Wound Healing. *PLoS ONE* **2008**, *3*, e1886. [CrossRef]

43. Gharaibeh, B.; Lavasani, M.; Cummins, J.H.; Huard, J. Terminal differentiation is not a major determinant for the success of stem cell therapy—Cross-talk between muscle-derived stem cells and host cells. *Stem Cell Ther.* **2011**, *2*, 31. [CrossRef] [PubMed]

44. Chiarugi, P.; Giannoni, E. *Anoikis*: A necessary death program for anchorage-dependent cells. *Biochem. Pharm.* **2008**, *76*, 1352–1364. [CrossRef] [PubMed]

45. Lefort, C.T.; Wojciechowski, K.; Hocking, D.C. N-cadherin Cell-Cell Adhesion Complexes Are Regulated by Fibronectin Matrix Assembly. *J. Biol. Chem.* **2011**, *286*, 3149–3160. [CrossRef] [PubMed]

46. Brennan, J.R.; Hocking, D.C. Cooperative effects of fibronectin matrix assembly and initial cell–substrate adhesion strength in cellular self-assembly. *Acta Biomater.* **2016**, *32*, 198–209. [CrossRef]

47. Sousa, S.R.; Moradas-Ferreira, P.; Barbosa, M.A.; Sousa, S. TiO$_2$ type influences fibronectin adsorption. *J. Mater. Sci. Mater. Med.* **2005**, *16*, 1173–1178. [CrossRef] [PubMed]

 *nanomaterials*

Review

Electrospun Polyvinylidene Fluoride-Based Fibrous Scaffolds with Piezoelectric Characteristics for Bone and Neural Tissue Engineering

Yuchao Li [1,*], Chengzhu Liao [2] and Sie Chin Tjong [3,*]

1   Department of Materials Science and Engineering, Liaocheng University, Liaocheng 252000, China
2   Department of Materials Science and Engineering, Southern University of Science and Technology, Shenzhen 518055, China
3   Department of Physics, City University of Hong Kong, Tat Chee Avenue, Kowloon, Hong Kong, China
*   Correspondence: liyuchao@lcu.edu.cn (Y.L.); aptjong@gmail.com (S.C.T.)

Received: 16 May 2019; Accepted: 15 June 2019; Published: 30 June 2019

Abstract: Polyvinylidene fluoride (PVDF) and polyvinylidene fluoride-trifluoroethylene (P(VDF-TrFE) with excellent piezoelectricity and good biocompatibility are attractive materials for making functional scaffolds for bone and neural tissue engineering applications. Electrospun PVDF and P(VDF-TrFE) scaffolds can produce electrical charges during mechanical deformation, which can provide necessary stimulation for repairing bone defects and damaged nerve cells. As such, these fibrous mats promote the adhesion, proliferation and differentiation of bone and neural cells on their surfaces. Furthermore, aligned PVDF and P(VDF-TrFE) fibrous mats can enhance neurite growth along the fiber orientation direction. These beneficial effects derive from the formation of electroactive, polar β-phase having piezoelectric properties. Polar β-phase can be induced in the PVDF fibers as a result of the polymer jet stretching and electrical poling during electrospinning. Moreover, the incorporation of TrFE monomer into PVDF can stabilize the β-phase without mechanical stretching or electrical poling. The main drawbacks of electrospinning process for making piezoelectric PVDF-based scaffolds are their small pore sizes and the use of highly toxic organic solvents. The small pore sizes prevent the infiltration of bone and neuronal cells into the scaffolds, leading to the formation of a single cell layer on the scaffold surfaces. Accordingly, modified electrospinning methods such as melt-electrospinning and near-field electrospinning have been explored by the researchers to tackle this issue. This article reviews recent development strategies, achievements and major challenges of electrospun PVDF and P(VDF-TrFE) scaffolds for tissue engineering applications.

Keywords: piezoelectricity;   scaffold;   polyvinylidene   fluoride;   polyvinylidene fluoride-trifluoroethylene; tissue engineering; osteoblast; neuron; stem cell; electrospinning; aligned fiber

---

1. Introduction

The design of novel biomaterials for applications in hard tissue such as bone and soft tissue like nerves presents a big challenge for chemists, materials scientists and biomedical engineers. Advanced functional scaffolds for hard and soft tissue engineering applications should possess certain requirements including biocompatible, hydrophilic, porous and electro-mechanical characteristics. Piezoelectric materials can generate electrical charges in response to an applied stress or minute mechanical deformation, thus eliminating the need for external power sources for electrical stimulation. Typical piezoelectric materials include barium titanate, lead titanate and lead zirconate titanate ceramics as well electroactive polymers such as polyvinylidene fluoride (PVDF) [1–5]. However, the toxicity and environmental impacts of lead-based piezoceramics preclude their use for biomedical applications.

In this respect, piezoelectric polymers exhibit several distinct advantages over piezoceramics including nontoxic, excellent flexibility, light weight and ease of fabrication. Therefore, PVDF and polyvinylidene fluoride-trifluoroethylene (P(VDF-TrFE) copolymer with excellent piezoelectricity and good processability can be tailored to form smart scaffolds to stimulate cell growth for tissue engineering applications [2–4,6]. By applying mechanical stresses to piezoelectric scaffolds, electrical stimulation is activated, and is transmitted to the neighboring cells, thereby enhancing cell signaling pathways for tissue regeneration [7,8]. Consequently, the biological electric field in the host tissues plays important roles in cellular membrane function, tissue growth and regeneration [9]. Electrospun PVDF fibrous mats that mimic the structure and electrical response of natural extracellular matrix (ECM) are considered of clinical importance. This is because electrospun PVDF scaffolds can promote bone generation, osteogenic and neural stem cell differentiation [10–17]. Human bones generally exhibit piezoelectric properties [18,19]. Bones consist of collagen fibrils and nanohydroxyapatite (nHA) that control biomechanical motion of humans [20]. The properties of collagen are dictated by the peptide bond. The carbonyl oxygen of peptide has a negative charge, while the amide nitrogen with a positive charge, thus establishing a small electric dipole [21]. When collagen is mechanically deformed, electric charges are produced, and the electrical potential generated promotes bone growth and regeneration [22–26].

Recently, bone defects and diseases remain a major health issue globally because of the rise in injuries due to osteoporosis, bone cancer, sport activity and traffic accident [27–32]. This leads to an urgent need in the biomedical sector to develop biocompatible substitutes to replace and regenerate defective bone tissues [33]. Moreover, neurotrauma due to traumatic brain injury and neurodegenerative diseases, such as Alzheimer's, Parkinson's, and stroke, are becoming increasingly prevalent among elderly patients [34–36]. Degenerative diseases are caused by a progressive loss of function of nerve cells of human nervous system. Regeneration of the nervous system requires the repair or replacement of nerve cells that have been damaged by injury or disease. However, human central nervous system has limited capability to regenerate damaged nerve cells [37]. As such, stem cell therapy is a promising route to treat neurological disorders because stem cells can differentiate into multiple cell types including neurons, thus serving as a source for cell replacement of damaged cells [38]. In this respect, tissue engineering and regenerative medicine provide potential solutions for creating novel smart scaffolds for self-repairing, remodeling and regeneration of bone and nerve tissues [39–42].

As is widely recognized, electrical charge stimulation greatly influences cellular behavior by affecting ion channel function across the cell membrane, monitoring the membrane potential and regulating the intracellular signal transduction pathways. Therefore, the cells can respond directionally to an applied electric field in vitro and in vivo [43]. Conductive polymers such as polypyrrole, polyaniline, poly(3,4-ethylenedioxythiophene), polythiophene, etc., have been shown to enhance and direct neurite outgrowth on their surfaces [44]. However, the application of those polymers has been hampered in the clinical sector due to the requirement of wired electrical stimulation with external power sources, thus increasing the risk of infection and inflammation [45]. Accordingly, electroactive PVDF and its copolymers, which can generate electrical charges on its surface upon mechanical or electrical stimulation, are very attractive for tissue engineering applications [3,46]. The generated charges and electrical dipoles would stimulate bone remodeling and growth through the opening of voltage-gated calcium ion channels [47]. Thus, the calcium/calmodulin pathway of bone cells is activated, facilitating osteogenic differentiation and proliferation (Figure 1) [8]. Tissue engineering is a multidisciplinary field combining cell biology, biochemistry, materials science, medicine and engineering disciplines to restore, maintain and repair damaged tissues and organs.

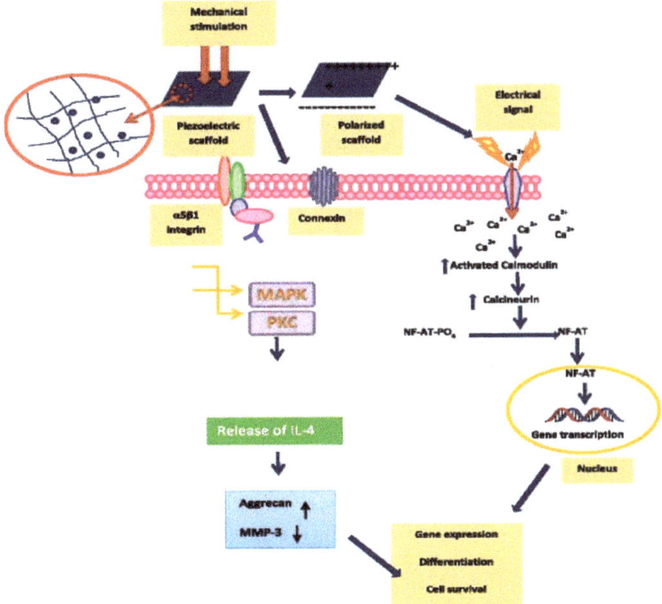

Figure 1. Schematic diagram showing the activation of Ca^{2+} signal transduction pathway and other miscellaneous pathways in response to the electrical and mechanical stimulations. Ca^{2+} is an important signal transducer.

PVDF exhibits excellent electroactive behaviors, good biocompatibility, excellent chemical resistance, and thermal stability, rendering it an attractive material for biomedical, electronic, environmental and energy harvesting applications [3,4,48–52]. For instance, PVDF with good biocompatibility and processability can be used to make monofilament suture material for cardiovascular surgery [52]. PVDF is a semicrystalline polymer having five crystalline polymorphs, including α-, β-, γ- δ- and ε-phases [2,53–55]. Among these, the β-PVDF-phase exhibits excellent piezoelectric and ferroelecric properties. The β-phase can be promoted in PVDF by either mechanical drawing, annealing, electrical poling or nanoparticle additions [54–60]. Aided by recent advances in nanotechnology, a wide range of nanomaterials can be synthesized for biomedical and industrial applications [61–72]. In particular, clay nanoplatelets, carbon nanotubes, graphene/graphene oxide and silica nanoparticles have been reported to be very effective to induce β-phase in PVDF [57,59,73–82]. As is generally recognized, electrospinning can fabricate micro- and nanofibers with interconnecting pores, resembling the natural ECM in tissues [83]. Thus, it is an effective technique for preparing smart PVDF fibrous scaffolds with piezoelectric characteristics. Sundaray et al. indicated that electrospinning could induce a change of crystalline structure of PVDF from non-polar α-phase to polar β-phase. This was ascribed to the intense stretching of the PVDF jet during the electrospinning process [84]. The β-phase content can be enhanced by monitoring electrospinning parameters [85]. This article provides the state-of-the art review on the development and piezoelectric properties of electrospun PVDF and P(VDF-TrFE) scaffolds for tissue engineering applications.

2. Structural Behavior

PVDF, with the chemical formula (CH$_2$–CF$_2$)n, possesses good piezoelectricity. Its molecular chain consists of highly electronegative fluorine atoms compared to the carbon and hydrogen atoms. This leads to the formation of polar C–F bonds, and each C–F bond possesses a significant dipole moment. PVDF exhibits five crystalline polymorphs including nonpolar α- and ε-phases, and polar β-, γ- and δ-phases depending on the crystallization and processing conditions [2,48–50]. Figure 2 shows a typical chain conformation for α, β and γ crystalline phases of PVDF [50]. The most common polymorph is α-PVDF that crystallized readily from the melt [51]. Nonpolar α-PVDF is monoclinic with its unit cell containing two chains in an alternating trans-gauche-trans-gauche' (TGTG') conformation. As a result, the net dipole moment cancels out due to its antiparallel molecular chain arrangement (Figure 3). The α-PVDF can be transformed into three other polymorphic forms under the application of mechanical stress (cold drawing), electrical field or annealing treatment. The δ-phase exhibits the same TGTG' conformation of macromolecular chains, but all dipoles are arranged parallel to each other, resulting in ferroelectric behavior. This phase can be produced by poling α-PVDF at an applied electric field of 100–150 MV/m. The hydrogen and fluorine atoms flip with respect to the carbon backbone. However, this process often leads to the breakdown of both the electrode and the polymer [86]. Moreover, δ-phase can transform to β-phase, with the hydrogen, fluorine, and carbon atoms all moving to produce the all-trans configuration under a high electric field (~500 MV/m) (Figure 3) [55]. The β-PVDF phase is orthorhombic with all-trans (TTTT) planar zigzag conformation, having all dipoles aligned in the same direction normal to the chain axis. So β-PVDF phase can generate the highest spontaneous polarization, showing strong ferroelectric and piezoelectric properties. It is unlikely to form β-phase from the melt, because of the high energy barrier of the all-trans conformation. The α-phase can transform to the β-phase under mechanical drawing, annealing treatment at high pressure and electrical poling (Figure 3). The γ-phase also has an orthorhombic unit cell, and is characterized by a sequence of trans and gauche conformation (T3GT3G'). This phase can be obtained by high-temperature drawing of ultrahigh molecular weight PVDF [87].

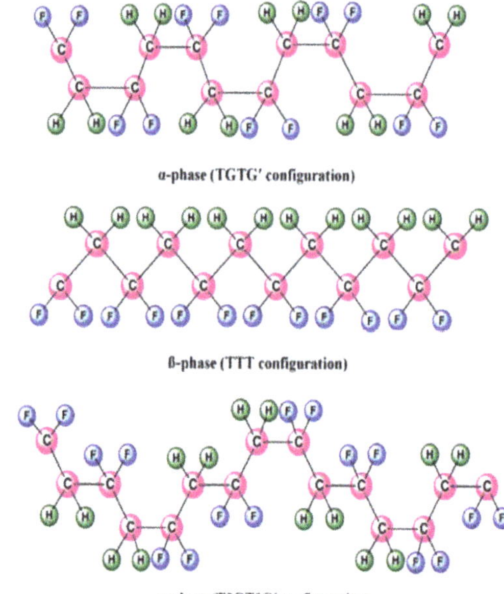

Figure 2. Primary polymorphic crystalline phases of PVDF. Reproduced with permission from [50], published by Wiley-VCH, 2019.

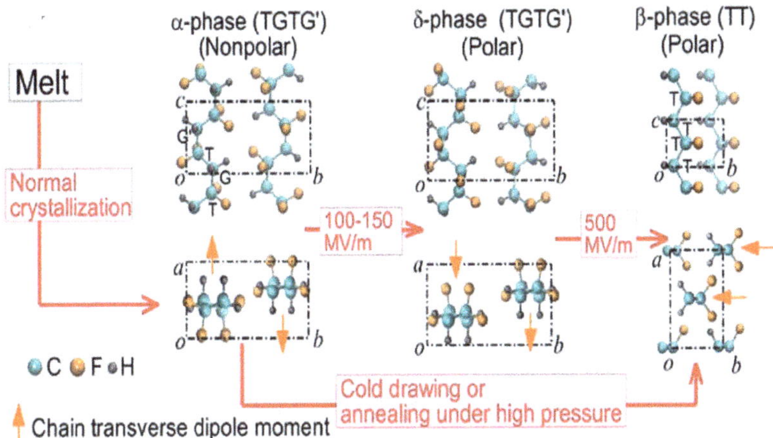

Figure 3. Electric field-induced phase transitions of PVDF. Electric poling aligns the dipoles along the electric field by applying a very high voltage. The transverse dipole moment of each polymer chain is shown using an orange arrow that points from the negatively charged fluorine atoms to the positively charged hydrogen atoms. T-trans; G-gauche. Reproduced with permission from [55], published by AIP Publishing, 2016.

PVDF can be copolymerized with TrFE, i.e.,–(CHF–CF$_2$)–, and tetrafluoroethylene (TeFE) [–(CF$_2$–CF$_2$)–], in a random sequence. P(VDF-TrFE) crystallizes readily from the melt and forms the β-phase through copolymerization without mechanical stretching or drawing. This is due to the extra fluorine atoms introduced into the molecular chain causing a steric hindrance effect, thus preventing the formation of α-phase. The co-monomer extends the interchain distance and reduces the activation energy for the α-phase to β-phase. Additional annealing, mechanical stretching or electrical poling treatments can lead to a further increase in the degree of crystallinity and alignment of the CF$_2$ dipoles, thereby producing higher piezo- and pyro-electric effects than PVDF [54]. P(VDF-TrFE) generally exhibits all trans conformation having TrFE content ranging 20–50 mol%. At TrFE content <20 mol%, P(VDF-TrFE) possesses mixed phases of α, β, and γ [88–90]. Thus, the structural, ferroelectric and piezoelectric properties of P(VDF-TrFE) depend greatly on the TrFE content [91].

As mentioned above, nanoparticle additions can promote formation of the β-phase in PVDF by acting as effective nucleating agents. Nanofillers with high surface-to-volume ratios can enhance their interactions with the polymer matrices. These nanomaterials include carbon nanofiber, carbon nanotube, graphene oxide (GO), and barium titanate [57–60,74–81]. Furthermore, metal (e.g., silver) and metal oxide nanoparticles like zinc oxide, and titania can also induce β-PVDF formation [82,92–95]. The extent of β-phase induced depends greatly on the properties of nanomaterials (e.g., type, shape, and size), surface chemistry and dispersion state. For instance, Ke et al. functionalized multiwalled carbon nanotubes (MWCNTs) with amino, carboxyl and hydroxyl groups, and then melt compounded with PVDF. They reported that amino group-functionalized MWCNTs (NH$_2$–MWCNTs) induced the highest β-phase content (17.4%), followed by those with hydroxyl groups (11.6%) and unmodified MWCNTs (9.4%). The nanocomposites with carboxyl-functionalized nanotubes exhibited the lowest β-phase content (4.7%) (Figure 4) [96]. The interfacial interactions between the amino or hydroxyl groups of MWCNTs with the CH$_2$/CF$_2$ dipoles of PVDF facilitates the formation of hydrogen bonding. Moreover, the combined effects of the MWCNTs dispersion and the nanotube–PVDF interaction were responsible for the formation of β-phase in PVDF (Figure 4). El Achaby et al. prepared PVDF/GO nanocomposites by means of solvent casting [77]. They reported that the strong interfacial interactions between carbonyl group of GO and fluorine group (CF$_2$) of PVDF led to the homogeneous dispersion of GO sheets in the PVDF matrix. Consequently, a small amount of GO (0.1 wt%) was needed to induce electroactive β-phase.

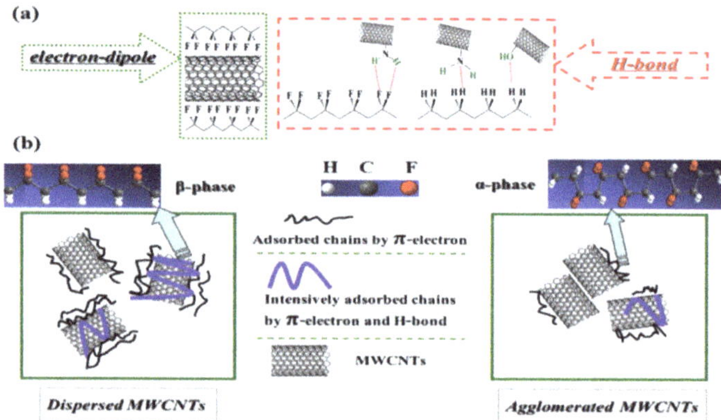

Figure 4. Schematic showing the effect of MWCNTs on the β-phase formation in PVDF: (**a**) Hydrogen bonding between functionalized MWCNTs and PVDF chains; (**b**) the adsorbed chains of PVDF on the surface of MWCNTs influenced by the dispersion of MWCNTs. Reproduced with permission from [96], published by Elsevier, 2014.

The polymorphs of PVDF and its copolymers are typically identified by means of X-ray diffraction (XRD), Fourier transform infrared spectroscopy (FTIR), and differential scanning calorimetry (DSC), as shown in Figure 5 [2]. The XRD pattern of α-PVDF shows the presence of diffraction peaks at 17.7°, 18.3°, 19.9° and 26.5°, corresponding to the (100), (020), (110), and (021) reflections. The XRD pattern of β-PVDF displays a characteristic peak at 20.26°, corresponding to the diffracting planes of (110) and (200). The γ-PVDF exhibits characteristic peaks at 18.5°, 19.2° and 20.04°, assigned to the (020), (002) and (110) reflections, respectively (Figure 5a) [2]. FTIR is particularly useful for identifying vibrational modes of the molecular chains of PVDF with different polymorphs, and for quantifying the amount of β-phase content. From Figure 5b, the vibrational bands of α-PVDF are located at 530 cm^{-1} (CF$_2$ bending), 615 cm^{-1} (CF$_2$ bending and skeletal bending), 766 cm^{-1} (CF$_2$ bending and skeletal bending), 795 cm^{-1} (CH$_2$ rocking), 855 cm^{-1} (CH out-of-plane deformation) and 976 cm^{-1} (CH out-of-plane deformation). The bands of β-PVDF appear at 510 cm^{-1} (CF$_2$ bending), 840 cm^{-1} (CH$_2$ rocking) and 1279 cm^{-1} (CF out-of-plane deformation). The characteristic bands of γ-PVDF are located at 776 cm^{-1} (CH$_2$ rocking), 812 cm^{-1} (CH$_2$ out-of-plane wag), 833 cm^{-1}, 840 cm^{-1} (CH$_2$ rocking), and 1234 cm^{-1} (CF out-of-plane deformation) [97]. Furthermore, the fraction of β-phase, F(β), can be quantitatively determined from the following equation [98],

$$F(\beta) = A_\beta/[(K_\beta/K_\alpha)A_\alpha + A_\beta] \tag{1}$$

where A$_\alpha$ and A$_\beta$ are the absorbance of α- and β-phases at 766 and 840 cm^{-1}, and K$_\alpha$ (6.1 × 10^4 cm^2/mol) and K$_\beta$ (7.7 × 10^4 cm^2/mol) are the absorption coefficients at the respective wavenumbers [2]. DSC is a powerful tool to characterize the melting and crystallization behaviors of PVDF [99]. The melting temperature (Figure 5c), crystallization temperature and the degree of crystallinity can be obtained from the DSC scans. All these parameters depend greatly on the structural conformation and molecular weight of PVDF.

Very recently, Lanceros-Méndez and coworkers prepared PVDF/silica (17 nm) nanocomposites at different processing temperatures. They also fabricated porous nanocomposite mats using electrospinning [100]. The β-phase contents of electrospun nanocomposite mats with oriented (O-17P) and random (R-17P) fibers determined from Equation (1) were about 80% and 79.5%, respectively

(Figure 6). A similarly high β-phase content was observed in porous nanocomposite film processed at room temperature (FTrt-17P). The high β-phase contents in porous electrospun nanocomposite mats derived from the solvent evaporation at room temperature and macromolecular chain stretching during the fiber formation. However, nonporous composite film prepared at 210 °C (F210-17NP) through a melting and recrystallization had a very low β-phase content; this film crystallized mainly into the α-phase. It appeared that silica nanoparticles had little influence on inducing the β-phase. Thus, processing temperature and electrospinning conditions are the main factors affecting the β-phase content in those nanocomposites [100].

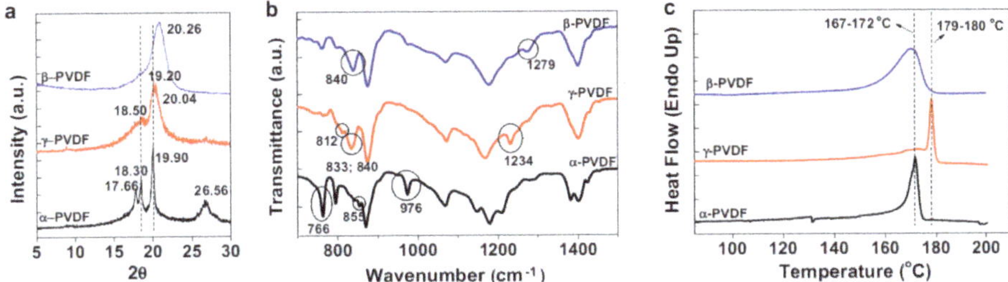

Figure 5. (**a**) XRD patterns, (**b**) FTIR spectra, and (**c**) DSC curves of the polymorphs of PVDF. Reproduced with permission from [2], published by Elsevier, 2014.

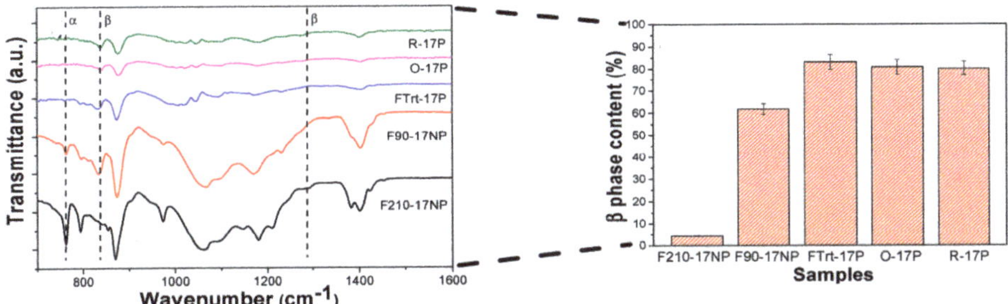

Figure 6. FTIR spectra of nonporous PVDF/silica (17 nm) nanocomposite films processed at 90 °C (F90-17NP) and at 210 °C (F210-17NP), porous nanocomposite film processed at room temperature (FTrt-17P), electrospun nanocomposite mats with oriented (O-17P) and random (R-17P) fibers. Right panel displays the β-phase content of these nanocomposites.

3. Scaffold Fabrication

The scaffolds for tissue engineering applications should be bioactive and biocompatible with sufficient porosity, pore size and highly interconnected pore network for the repair and regeneration of tissues. They should also possess high mechanical strength for supporting cell adhesion, growth and proliferation, and for transporting nutrient and metabolic waste [40,101–103]. PVDF with good processability, flexibility and low-cost can be readily fabricated into porous scaffolds for tissue engineering applications. Functional scaffolds based on PVDF and its copolymers can be prepared using conventional processing techniques, including solvent-casting/particulate leaching, solvent-casting and 3D polymer template, non-solvent induced phase separation (NIPS), and thermally induced phase separation (TIPS) [3,4,104–108]. The solvent casting/particulate leaching is a relatively simple process for forming scaffolds with a high porosity (up to 93%) and pore sizes of up to 500 μm [109]. In this process, a polymer is first dissolved in an organic solvent, mixed with porogens such as salt or sugar, and then cast into a mold. Subsequently, the mold is dipped in a water bath to dissolve the porogens. The pore volume, pore size and pore shape are governed by the amount, size and shape of porogens added. The main drawbacks of this technique include a wide variation of the pore sizes, poor pore interconnectivity, and irregular pore geometry [110]. Accordingly, several approaches have been developed to address some of these issues. For instance, combination of solvent casting and polymer templates (e.g., polyvinyl alcohol or nylon) can yield PVDF scaffolds with interconnected and well-distributed pores [104,105]. Alternatively, a phase inversion method can be used to fabricate microporous scaffolds with interconnected pores. Phase inversion occurs when a change takes place in the stability of a polymer solution as a result of the temperature variation, solvent evaporation, or mass exchange with nonsolvent [111,112]. In the NIPS or immersion precipitation approach, the polymer/solvent solution is immersed in a coagulation bath containing nonsolvent (e.g., water). This leads to a mass transfer between the solvent and nonsolvent, i.e., the solvent diffuses to the nonsolvent, while nonsolvent penetrates into the polymer solution (Figure 7). Consequently, the polymer solution becomes thermodynamically unstable, resulting in the phase separation, either by liquid-liquid or solid liquid demixing. This causes precipitation of a polymer-rich phase and a polymer lean-phase [113]. Upon the removal of the solvent, polymer-rich phase develops into a continuous matrix of the scaffold, while the polymer lean-phase forms the porous tunnels within the matrix, resulting in an interconnected porous network.

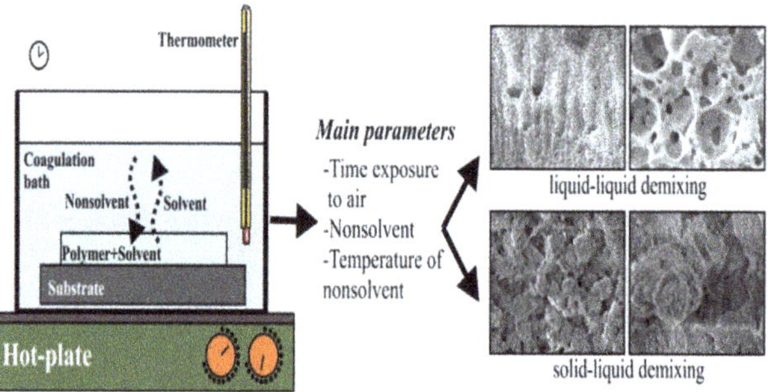

Figure 7. Porous scaffolds produced by non-solvent induced phase separation (NIPS). Reproduced with permission from [112], published by Elsevier, 2015.

Very recently, Abzan et al. adopted NIPS to fabricate three-dimensional piezoelectric PVDF and PVDF-GO scaffolds for nerve tissue engineering applications [81,106]. As recognized, GO possesses a range of oxygenated functional groups such as epoxy, hydroxyl, carbonyl and carboxyl. Carbonyl and carboxyl groups are mainly attached at the edge of basal plane of graphene sheet and hydroxyl groups on the plane. GO is produced by chemical oxidation of graphite flakes in strong oxidizing solution containing sulfuric acid, sodium nitrate, and potassium permanganate [114,115]. So hydrophilic oxygen functional groups of GO can render hydrophobic PVDF with improved hydrophilicity and enhanced GO-PVDF interfacial interactions. Accordingly, the incorporation of 0.5–3 wt% GO into PVDF reduces its water contact angle, especially with the 3 wt% GO addition. The water contact angle of hydrophobic PVDF is 117.2° ± 5.4°, but reduces markedly to 71.3° ± 6.4° by adding 3 wt% GO. As a result, PVDF/3 wt% GO scaffold is hydrophilic, thus favoring the attachment and growth of PC12 nerve cells. Moreover, GO additions also facilitate the formation of β-PVDF phase due to the interactions between carbonyl group (C=O) of GO and fluorine group ($>CF_2$) of PVDF. The β-phase content in neat PVDF, PVDF/0.5 wt% GO (P-0.5GO), PVDF/1 wt% GO (P-1GO) and PVDF/3 wt% GO (P-3GO) scaffolds determined from Equation (1) is 64.5, 77.4, 64.6 and 70.15%, respectively. Apparently, GO additions can induce a high amount of β-phase in PVDF, leading to enhanced piezoelectric effect. The piezoelectric response can be determined by applying a mechanical force through a human finger imparting on the PVDF-based scaffold sandwiched between the copper electrodes (Figure 8a). The generation of average electrical voltage on neat PVDF, P-0.5GO, P-1GO, P-3GO and P-5GO scaffolds is 1.36 V, 1.78 V, 1.40 V, 1.49 V and 1.58 V, respectively (Figure 8b).

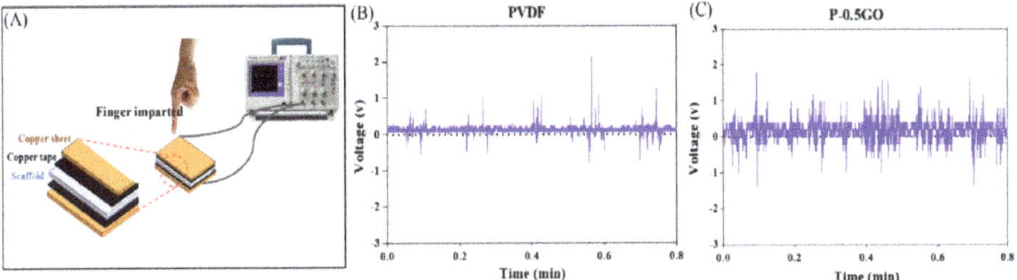

Figure 8. (**A**) Schematic of piezoelectricity measurement. Output voltage generation from representative (**B**) PVDF and (**C**) PVDF/0.5 wt% GO scaffolds. Reproduced with permission from [81], published by Elsevier, 2019.

3.1. Electrospinning

Compared to conventional processed scaffolds, electrospun nanofibrous mats with a large surface to volume ratio are attractive for tissue engineering applications because they mimic the fibrillar structure of natural ECM secreted by the cells (Figure 9) [116]. Fibers from nano- to micrometer scale can be prepared by electrospinning. The fabrication process involves the application of a high electric field to the polymer/solvent solution. Above a critical voltage, electrostatic repulsion overcomes the surface tension of polymer droplet developed at a needle tip attached to the syringe pump. Therefore, a charged polymer jet is ejected from the needle tip towards a grounded collector, leading to the formation of random or aligned fibers. The electrically charged jet experiences bending instability and whipping motion resulting in a randomly oriented fiber deposition on the collector. Moreover, the jet also undergoes stretching due to electric field and solvent evaporation, thus causing jet thinning. The fiber diameter and porosity of the fibrous scaffolds depend on the processing parameters such as applied voltage, solution flow rate, type of solvent, polymer concentration in the solution, solution conductivity, and the distance between the needle and collector [117–119]. Because of the chaotic trajectory of the polymer jet, the fibers collected on a grounded collector generally exhibit random orientation, so depositing as a non-woven scaffold. Electrospun fibers can be aligned by controlling the rotation speed of the mandrel, magnetic field or electric field at the gap between a pair of electrodes (Figure 10) [118,120,121].

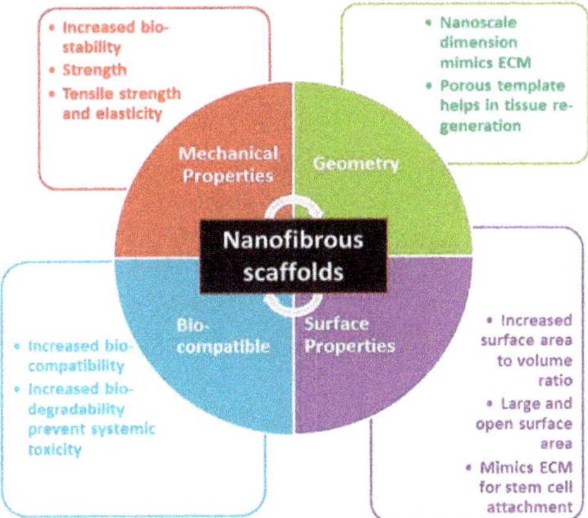

Figure 9. Schematic representation of the required properties of nanofibrous scaffolds including geometry, mechanical competence, biocompatibility and surface behavior.

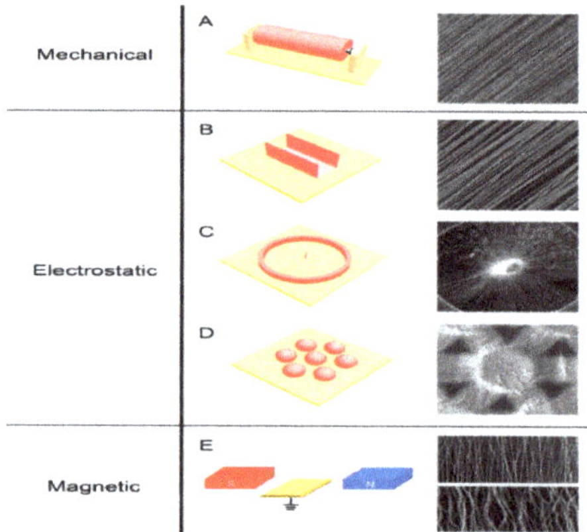

Figure 10. Schematics of electrospinning methods to direct fiber orientation by means of mechanical rotation of mandrel (**A**), electrostatic forces through the use of a metallic staple (**B**), a metallic ring (**C**), and an array of metallic beads (**D**), as well as magnetic forces through the use of a pair of permanent magnets (**E**). The yellow plates are grounded conductive electrodes. Scanning electron micrographs in the right panel show the morphologies of aligned nanofibers depositing at the target via different methods. Reproduced with permission from [120], published by Wiley-VCH, 2012. Reproduced with permission from [121], published by American Chemical Society, 2010.

3.1.1. Electrospun PVDF Scaffolds

As mentioned above, piezoelectric properties of β-PVDF facilitate the generation of electrical potential on its surface due to mechanical deformation. The stretching of the polymer jet induces the β-PVDF phase, i.e., transforming nonpolar α-phase into polar β-phase [17,122–125]. So piezoelectric β-PVDF fibrous scaffolds that generate electrical stimulation are particularly attractive for bone and neural tissue engineering applications [15,17,46,125,126]. From the literature, existing studies are mainly focused on the effect of electrospinning parameters on the fiber morphology, fiber diameter and the β-phase formation. These parameters include PVDF concentration, solvent type, applied voltage, spinning distance, stationary or rotating collector, etc. [17,122–128]. In general, the fiber diameter increases with increasing PVDF concentration due to the higher solution viscosity and stronger intermolecular interactions. The selection of the appropriate solvent is crucial for forming smooth continuous fibers without beads. Volatile solvents with low boiling point and fast evaporation rate are generally preferred because they facilitate dehydration and solidification of the polymer jet. However, highly volatile solvents with very low boiling points can lead to the clogging and obstructing the flow-rate of polymer solution [128]. Apparently, smooth continuous PVDF fibers can be achieved by selecting a particular solvent combination, i.e., *N,N*-dimethylformamide (DMF; boiling point: 152 °C) and acetone (boiling point: 56 °C), or *N,N*-dimethylacetamide (DMAC; boiling point: 165 °C) and acetone. The slower evaporation rate of DMF or DMAC enables polymer jets to stretch and facilitate the transformation of α-phase to β-phase, whereas acetone can prevent the bead formation [17]. By adding more acetone (low DMF/acetone ratio), the evaporation rate of the polymer/solvent solution tends to increase, resulting in the formation of more α-phase in PVDF mats. In contrast, higher DMF/acetone ratios favor the formation of β-PVDF fibers with fined diameters. Figure 11 summarizes the 3D plots showing the β-phase content induced in electrospun fibrous mats as a function of PVDF solution concentration and DMF/acetone ratio. Typical SEM images of PVDF fibers prepared at different DMF/acetone ratios under a PVDF concentration of 25% are shown in Figure 12a,b.

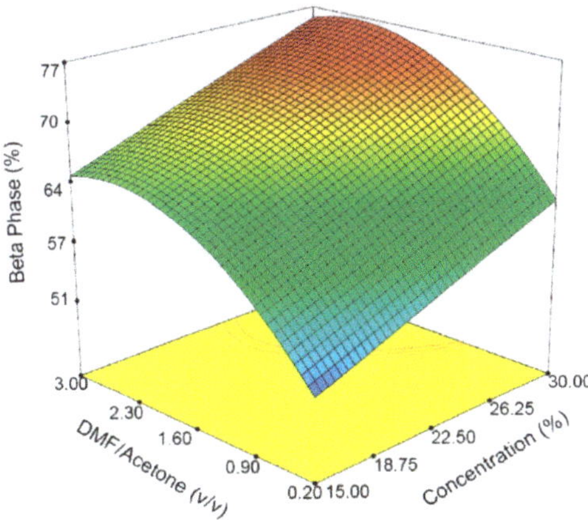

Figure 11. Three-dimensional plots of beta phase content vs. PVDF concentration and DMF/acetone ratio. Reproduced with permission from [124], published by Springer, 2018.

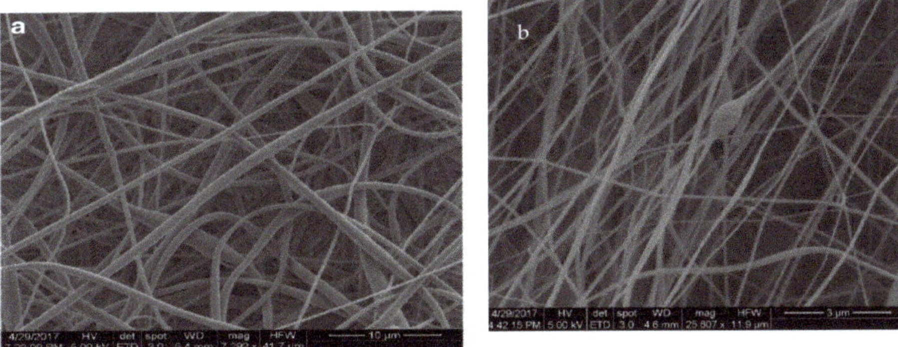

Figure 12. Scanning electron micrographs of electrospun nanofibers with a PVDF concentration of 25% and (**a**) DMF/acetone ratio of 1, and (**b**) DMF/acetone ratio of 3. Reproduced with permission from [124], published by Springer, 2018.

Recently, Shao et al. studied the processing-structural relationship of PVDF nanofibers prepared by electrospinning [129]. Increasing the applied voltage from 9 to 15 kV at a 20% PVDF solution leads to a higher charge density or electrostatic force for stretching the polymer jet, thereby decreasing the fiber diameter. Above 15 kV, the increase in fiber diameter is caused by the intensive bending instability (Figure 13a). The β-phase increases from 76.7% to 85.9% by increasing the applied voltage from 9 kV to 15 kV. Above 15 kV, the β-phase content decreases slowly with increasing applied voltage. Lanceros-Méndez and coworkers also demonstrated that the applied voltage can produce a higher stretching of the polymer jet, favoring the formation of β-phase [123]. In addition, the strong electric field employed also plays the role of fiber poling. A strong dependence of the fraction of β-phase in PVDF fibers on the applied voltage has also been reported by Sengupta et al. [130]. From Figure 13a, electrical (voltage/current) outputs of electrospun mats follow the increasing/decreasing trend of the β-phase content with the applied voltage. The fiber size and β-phase content of PVDF fibrous scaffolds also depend on the spinning distance (needle tip to collector distance) (Figure 13b). A long spinning distance offers a large space for jet stretching and more time for solvent evaporation, thus giving rise to fibers with fined diameters. For a 20% PVDF solution under an applied voltage of 15 kV, the fiber diameter decreases from about 550 nm to 458 nm as the spinning distance increases from 9 cm to

13 cm. So coarse fibers are produced at such short distances. At a spinning distance of 15 cm, the fiber size decreases significantly to 284 nm. This is a critical distance range to form nanofibers with fine diameters. Accordingly, the highest β-phase content of 85.9% is achieved at this critical distance. The voltage/current outputs of PVDF fibrous follow a similar changing trend of the β-phase content with the spinning distance.

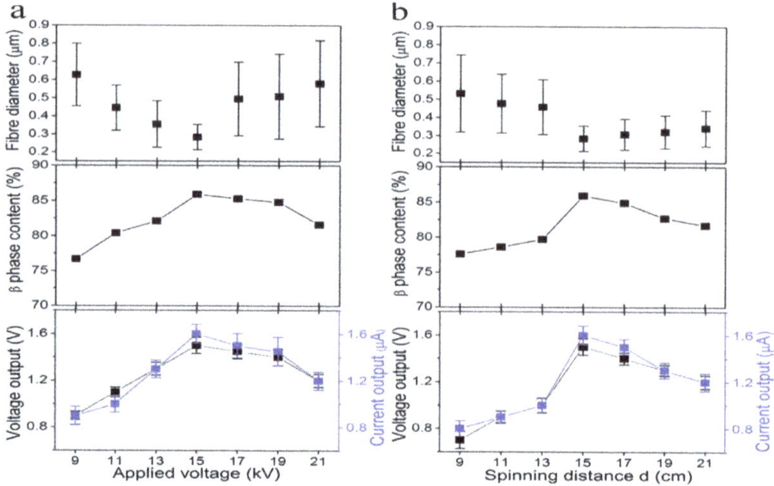

Figure 13. Effects of (**a**) applied voltage and (**b**) spinning distance on the fiber diameters, β-phase contents and electrical outputs of PVDF nanofiber mats (PVDF concentration 20%; nanofiber mat thickness 100 µm). Reproduced with permission from [129], published by Royal Society of Chemistry, 2015.

Uniaxially Aligned Nanofibers

Electrospun nanofibers can be aligned into uniaxial arrays through mechanical rotation of the collector (e.g., drum, disk), the gap method and magnetic electrospinning [118,120,121,131]. From the literature, several research groups have prepared aligned PVDF fibers by rotating a collecting drum at various speeds due to its simplicity [15,132–134]. For instance, Lins et al. classified the alignment of PVDF fibers obtained at different drum speeds as non-aligned (50 rpm), low-aligned (1000 rpm), medium-aligned (2000 rpm) and high-aligned (3000 rpm) on the basis of SEM observations [15]. Apparently, a drum speed of 50 rpm is relatively low to initiate the fiber alignment. The fiber orientation of electrospun scaffolds increases with increasing drum speed from 1000 to 3000 rpm. Increasing rotating speed also reduces the fiber diameter and porosity of the scaffolds. The average diameter of PVDF fibers under rotating speeds of 50, 1000, 2000, and 3000 rpm is 1.98 ± 0.55, 1.95 ± 0.60, 1.56 ± 0.60, and 1.51 ± 0.68 µm, respectively. The porosity level of scaffolds fabricated under these speeds is 86, 79, 54 and 46%, respectively. Thus, fiber alignment results in a decrease in the porosity, but an increase in the pore size. The pore size of highly aligned (300 rpm) scaffold is 8.5 ± 3.2 µm, while that of non-aligned (50 rpm) scaffold is 4.2 ± 0.4 µm. The aligned PVDF nanofibers also contain the β-phase as expected. Wu et al. demonstrated that the β-phase content of aligned PVDF fibers reaches 88% by rotating a drum collector, being much higher than that of randomly oriented PDVF fibers deposited on a stationary collector (79%) [134].

In general, the rotating collector method is a time-consuming process, and requires proper manipulation of the speed to ensure the as-spun nanofibers with desired properties. In contrast, the gap method is an effective route to fabricate uniaxially aligned arrays by using two parallel conductive silicon strips or bars separated by a gap acting as a collector. In this context, the fibers can be aligned in the gap between two conducting electrodes (Figure 14) [135]. In the process, the electric field lines in the vicinity of the collector are split into two directions toward the opposite edges of the gap. Thus, the charged nanofibers are pulled toward both edges of the conductors, and stretched across the gap to form a parallel, uniaxial array mat. Very recently, Shebata et al. prepared aligned PVDF and PVDF/MWCNT fibrous mats using the gap method. They reported that the two-bar collector system provides a superior alignment for the PVDF nanofibers [136]. Figure 15a,b shows typical SEM images of randomly oriented and aligned PVDF nanofibers. Randomly oriented fibers are deposited on a grounded metal plate covered with aluminum foil.

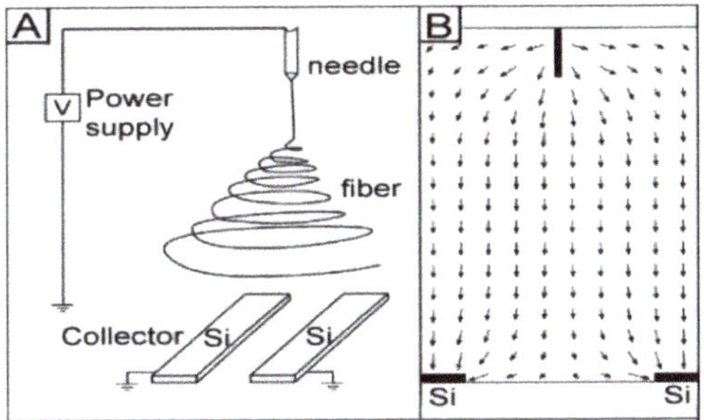

Figure 14. Schematic diagram showing electrospinning setup for forming uniaxially aligned nanofibers. (**A**) A collector with two pieces of conductive silicon stripes separated by a gap. (**B**) Electric field strength around the needle and the collector. The arrows denote the direction of the electrostatic field lines. Reproduced with permission from [135], published by American Chemical Society, 2003.

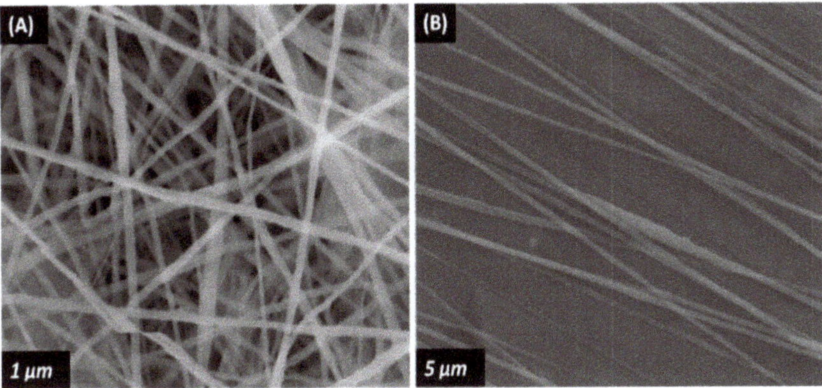

Figure 15. (**A**) Randomly oriented PVDF nanofibers deposited on a stationary metal plate covered with aluminum foil. (**B**) Aligned PVDF nanofibers deposited at the gap between two metallic bars of a collector.

PVDF Nanocomposite Scaffolds

The incorporation of functional nanoparticles into PVDF has a large influence on its structural properties, especially conducting nanomaterials such as MWCNTs, graphene sheets and silver nanoparticles (AgNPs). It is well established that the conductivity of polymer solutions increases dramatically by adding an appropriate salt such as sodium chloride. The salt increases the number of ions in the polymer solution, thereby enhancing surface charge density of the polymer solution and the electrostatic force generated by the applied electric field [127,128,137]. This leads to spun fibers with fine diameters. As recognized, nanoparticle additions to the polymer solutions can increase their viscosity and fiber diameter. For conducting AgNPs and MWCNTs, there exists a balance between the increase in solution conductivity and solution viscosity. In recent years, graphene derivatives such as graphene oxides and reduced graphene oxide (rGO) have found potential biomedical applications for making biosensors, tissue engineering scaffolds and orthopedic implants [138,139]. Graphene oxide is an electrical insulator due to the presence of oxygen functional groups. Its electrical conductivity can be resumed by either treating with reducing agents such as hydrazine and sodium borohydride to form reduced graphene oxide (rGO) [140], or via rapid heating in a furnace at 1050 °C to generate thermally reduced graphene oxide (TRG) [141]. Existing literature reports reveal that AgNPs, rGO or MWCNT nanomaterials are very effective for inducing the β-phase in electrospun PVDF fibrous mats by serving as the nucleating sites [92,142,143]. Therefore, both the conducting nanofillers and electrostatic field stretching during electrospinning contribute to the formation of β-PVDF.

As is widely recognized, piezoelectricity is reversible, because mechanical energy can be converted to electrical energy and vice versa. The strain S_j induced in a piezoelectric material by an applied electric field E can be written as [144],

$$S_j = d_{ij} \cdot E_i \tag{2}$$

where d_{ij} is the piezoelectric coefficient; the subscript indices 1–3 describe components along the x, y, and z axis of a rectangular coordinate system, and indices 4–6 represent shear components of the strain. Accordingly, the longitudinal piezoelectric coefficient (d_{33}) can be determined from the following Equation:

$$S = d_{33} \cdot E \tag{3}$$

So d_{33} describes the strain induced by an applied electric field (E_3; V/m), having a unit of pm/V. An alternate way to express d_{33} is as an induced polarization in the direction 3 (C/m^2) per unit applied stress (N/m^2), yielding an equivalent unit of pC/N [145]. PVDF and P(VDF-TrFE) generally show an unusual negative longitudinal piezoelectric effect, i.e., a contraction in their lattice constants under the application of an electric field. Therefore, a negative sign is often used to describe such a parameter [146].

Bose and coworkers fabricated electrospun PVDF, PVDF/1 wt% MWCNT and PVDF/1 wt% (MWCNT-AgNP) fibrous mats with randomly oriented fibers [147]. Carboxyl functionalized MWCNT and AgNP decorated MWCNT were added to PVDF for enhancing its piezoelectric effects. The incorporation of both types of MWCNTs into electrospun PVDF fibers led to an increase in the β-phase content. As a result, the d_{33} value of the PVDF/1 wt% MWCNT and PVDF/1 wt% (MWCNT-AgNP) fibrous mats increased markedly, especially the latter, with MWCNT-AgNP nanofillers (Table 1). In a recent study, Bose and coworkers employed carboxylated GO (CGO) and fluorinated GO (FGO) nanofillers for further enhancing d_{33} value of PVDF fibrous mats. The highest β-phase content was achieved in the PVDF/1 wt% FGO mat due to the presence of highly electronegative fluorine [148]. Consequently, PVDF/1 wt% FGO fibrous mat had a high d_{33} value of 63 pm/V. So the addition of very low loading level of GO-based nanofillers into electrospun PVDF scaffolds led to enhanced piezoelectricity. Combining GO addition with electrospinning offers the opportunity to fabricate PVDF fibrous mats with tunable piezoelectric characteristics. Very recently, Wu et al. deposited aligned PVDF and PVDF/MWCNT fibers on a rotating drum, and randomly oriented PDVF fibers on a stationary collector, respectively. They then determined d_{33} values of electrospun PVDF fibers with different orientations under mechanical force deformations [134]. Compared to random fibers, aligned PVDF/MWCNT fibers exhibited the largest d_{33} value of 31.3 ± 2.1 pC/N due to the presence of the highest amount of the β-phase content (Table 1).

Table 1. Effect of β-phase content on piezoelectric coefficient of electrospun PVDF and its nanocomposite mats.

Materials	β-Phase Content	Fiber Diameter, nm	d_{33} Value, pm/V	d_{33} Value, pC/N	Ref.
PVDF random fibers	NA	1410	—	24.90	[39]
PVDF random fibers	79 ± 3	155 ± 17	—	16.8 ± 1.4	[134]
PVDF aligned fibers	88 ± 1	118 ± 23	—	27.4 ± 1.5	[134]
PVDF/MWCNT aligned fibers	89 ± 2	116 ± 21	31.3 ± 2.1	31.3 ± 2.1	[134]
PVDF random fibers	78	1500	30	—	[147]
PVDF/1% MWCNT random fibers	84	300	35	—	[147]
PVDF/1% (MWCNT-AgNP) random fibers	89	800	54	—	[147]
PVDF/1% GO random fibers	70	623	40	—	[148]
PVDF/1% CGO random fibers	79	622	46	—	[148]
PVDF/1% FGO random fibers	89	619	63	—	[148]
Human bones	—	—	—	7–8	[26]

NA: Not available.

3.1.2. Electrospun P(VDF-TrFE) Scaffolds

As mentioned, the presence of extra fluorine atoms in the TrFE monomer means that the β-phase can crystallize easily from P(VDF-TrFE) melt without mechanical stretching. Additional annealing, mechanical stretching or electrical poling can lead to a further enhancement of the β-phase content. Generally, annealing is particularly useful for enhancing the degree of crystallinity, reducing the porosity and eliminating the residual solvent of the P(VDF-TrFE) films [149]. From an economic point of view, P(VDF-TrFE) is more expensive than PVDF, owing to the production risk of explosion of TrFE monomer during co-polymerization [150]. Nevertheless, P(VDF-TrFE) has found wide applications for energy harvesters [151,152], and bone tissue engineering [8].

Jiang et al. prepared aligned P(VDF-TrEF) fibers (400–550 nm) with a rotating drum under a rotation speed of 2500 rpm followed by annealing at 130 °C and 140 °C [153]. Furthermore, spin-coated P(VDF-TrEF) film was also fabricated for the purpose of comparison. The XRD pattern of the as-spun fibers revealed the presence of a strong characteristic peak at about 20°, corresponding to the (110) and (200) planes of the β-phase. This characteristic peak became more intense in the spun fibers by annealing at 130 °C and 140 °C (Figure 16a,b). FTIR spectra also showed that the intensity of the bands due to the β-phase (846, 1285 and 1431 cm^{-1}) of the annealed sample was stronger than that of the same bands of the unannealed specimen, especially at 846 and 1285 cm^{-1} (Figure 16c). Therefore, the β-phase content was further increased by the annealing treatment.

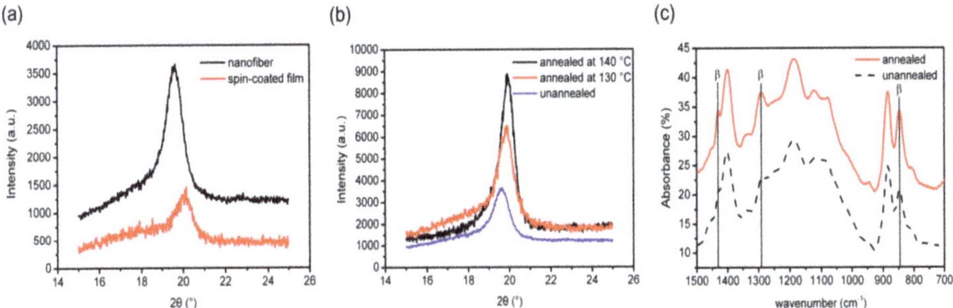

Figure 16. (**a**) X-ray diffraction (XRD) patterns of spin-coated P(VDF-TrFE) film and electrospun P(VDF-TrFE) nanofibers. (**b**) XRD patterns of the P(VDF-TrFE) nanofibers before and after annealing at 130 °C and 140 °C for 2 h. (**c**) FTIR spectra of the P(VDF-TrFE) nanofibers before and after annealing at 140 °C for 2 h.

Jiang et al. also employed piezoresponse force microscopy (PFM) to examine piezoelectric response of a single P(VDF-TrEF) fiber [153]. PFM is a contact-mode atomic force microscopy (AFM) in which the cantilever tip is used to detect piezoelectric response of a ferroelectric material. By applying an AC voltage to the cantilever tip, the material experiences expansion or contraction. So the piezoresponse of a sample can be determined through the deflection of the cantilever associated with the strain of mechanical deformation. The piezoresponse is detected and quantified in terms of its amplitude and phase by a lock-in amplifier. The PFM amplitude and phase signals characterize the magnitude of the piezoelectricity and the polarization direction, respectively [154]. Figure 17a,b shows the respective PFM amplitude and phase images of a single P(VDF-TrFE) nanofiber. The amplitude image reveals a strong piezoelectric contrast because of the deflection caused by the applied AC field. By superimposing a DC bias voltages of ±50 V, a butterfly-like amplitude loop and a well-defined hysteresis phase loop are generated, as shown in Figure 17c,d. The butterfly loop is a typical characteristic of piezoelectric materials [154]. So P(VDF-TrFE) nanofiber exhibits excellent piezoelectricity due to the presence of electroactive β-PVDF phase.

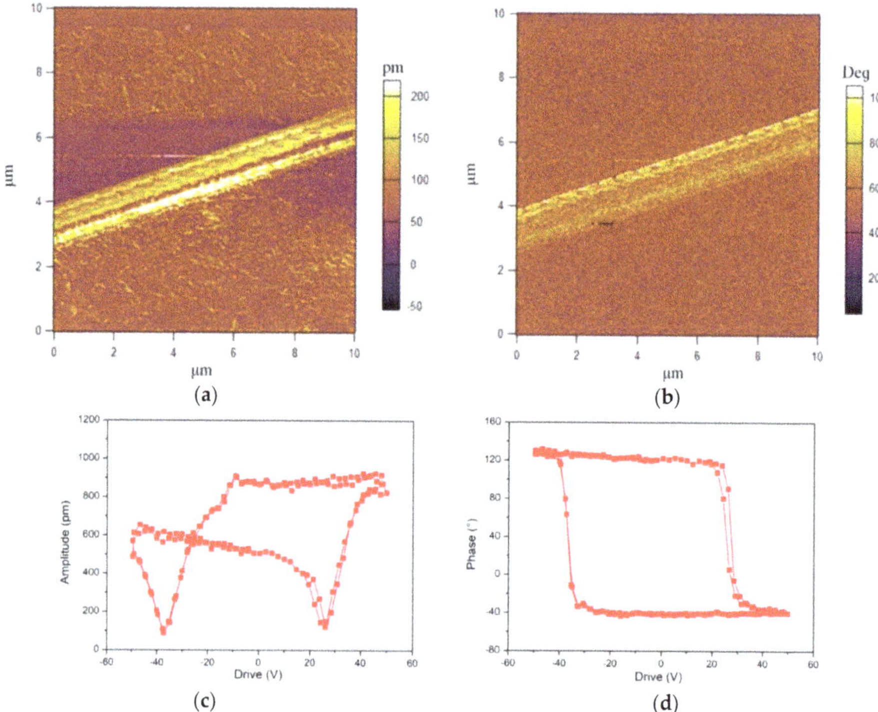

Figure 17. Piezoresponse force microscopy: (**a**) amplitude and (**b**) phase images of a single P(VDF-TrFE) nanofiber. (**c**) PFM amplitude and (**d**) PFM phase of the P(VDF-TrFE) nanofiber as functions of DC bias for two cycles, displaying good repeatability for forward and reverse scans.

Ico et al. prepared electrospun P(VDF-TrFE) mats with fiber diameters ranging from 1 μm to nanometer scale [152]. Such a fiber size reduction led to an increase of electroactive phase content and the degree of crystallinity. Furthermore, they also employed PFM to determine the d_{33} value of a single P(VDF-TrFE) fiber. The d_{33} value of electrospun P(VDF-TrFE) fiber was found to be size dependent, and reached 55 pm/V for the fiber with a few nanometer size. Thus, a reduction of fiber diameters of P(VDF-TrFE) from micro- to nanoscale dimensions resulted in higher piezoelectricity (Figure 18a). The fiber size dependency of piezoelectric coefficient correlated well with the amount electroactive β-PVDF phase induced in the fibrous mat during electrospinning. The amount of β-phase induced was also dependent on the fiber diameter. The β-phase content determined from the FTIR spectroscopy increased considerably as the fiber diameter reduced to <100 nm (Figure 18b).

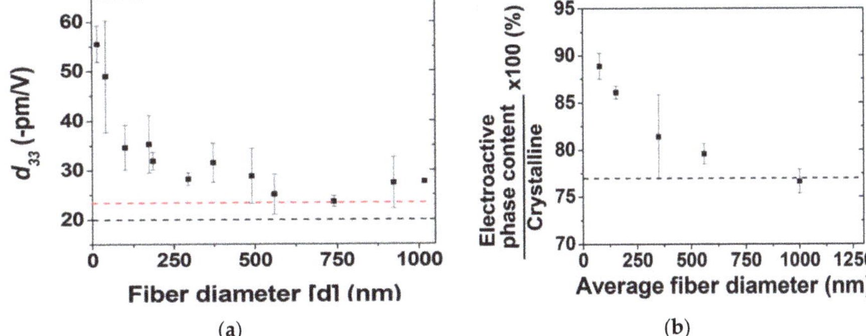

Figure 18. (**a**) Measured d_{33} as a function of fiber diameter from PFM. The red dashed line corresponds to the measured d_{33} of a 80 μm thick film, and the black dashed line is the d_{33} of bulk P(VDF-TrFE). Thick P(VDF-TrFE) film (80 μm) was made by drop-casting and employed as a reference. (**b**) Electroactive phase content determined by FTIR as a function of mean fiber diameter. The black dashed line represents the measured electroactive active content of a thick film prepared by drop-casting. Reproduced with permission from [152], published by Royal Society of Chemistry, 2016.

3.1.3. Melt Electrospinning

Electrospun fibrous mats can be prepared either using polymer solution or melt. Solution electrospinning has a main drawback involving the use of organic solvents to dissolve polymers. Most organic solvents are harmful to biological cells and may cause detrimental effects in tissue engineering applications. In this respect, electrospinning of polymer melts offers an advantage for fabricating fibrous structures without using organic solvents. Melt-electrospinning has been studied to a significantly lesser extent than solution electrospinning, because this process results in larger fiber diameters of several to hundreds of micrometers [155]. In the process, polymer melts with high viscosity and low conductivity suppress whipping motion, allowing the jet travels in a straight path from the needle to the collector. So the melt viscosity plays a crucial role in determining the spinnability and fiber diameter of a polymer. In general, low melt viscosity would lead to the formation of beads. At a very high melt viscosity, the electric field cannot overcome the viscosity resistance for spinning [156]. The melt viscosity of polymers is controlled by their molecular weight and heating temperature. The electrical heating system is the most commonly used for melting polymers. However, the electrical heating source can interfere with the high-voltage spinning system. Accordingly, laser heating has some advantages, such as fast melting of polymers due to intense laser beams, and noninterference with the high-voltage spinning system. The melt-electrospun studies in the literature are focused mainly on the fabrication of polycaprolactone (PCL) fibers [157–159].

Very recently, Asai et al. prepared electrospun PVDF fibers by means of laser-melt electrospinning (MES) [160]. In the MES approach, PVDF was melted by a CO_2 laser beam, and the molten jet was accelerated at different applied voltages towards a rotating collector of various speeds. Figure 19 shows the SEM images of PVDF fibers formed at different applied voltages and collector rotation speeds. Aligned PVDF fibers formed only at a low voltage of −10 kV for different rotating speeds. The average diameter of fibers spun at −10 kV and 1000 rpm was 4.2 ± 3.4 μm. By increasing the voltage to −25 kV and collector speeds to 500–1000 rpm, ribbon-like flat fibers were produced. The XRD results revealed that non-polar α-phase was a dominant structure in the PVDF fibers. As such, melt-electrospun PVDF fibers had a very low d_{33} value of 0.03 pC/N.

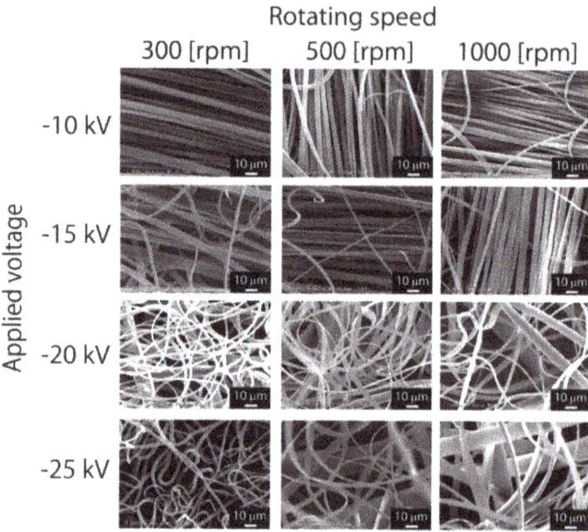

Figure 19. SEM images of PVDF fibrous mats prepared by melt-electrospinning at various applied voltages and collector rotating speeds. The laser output power and the polymer feed rate were fixed at 53 W and 1 mm min^{-1}, respectively. Reproduced with permission from [160], published by Royal Society of Chemistry, 2017.

4. In Vitro and In Vivo Models

4.1. Bone Tissue Engineering

Human mesenchymal stem cells (hMSCs) can differentiate into osteoblasts, adipocytes and chondrocytes. Osteoblasts produce bone matrix proteins and mineralize the matrix into bones. Preosteoblast differentiation proceeds through 3 stages including proliferation, extracellular matrix maturation, and mineralization [161]. In the first stage, osteoblasts proliferate by secreting bone matrix proteins, such as collagen type 1 alpha 1 (Col1α1), osteopontin and fibronectin. In stage 2, they begin to differentiate and mature the ECM with alkaline phosphatase (ALP) and collagen. The final bone matrix mineralization occurs by expressing osteocalcin, leading to the deposition of calcium phosphate [162].

Human bones have the ability of self-remodeling through an electromechanical mechanism due to the piezoelectric effect [163]. Thus, mechanical stimulation of bones induces their growth and regeneration as a result of the generation of electrical potential [3,23]. The application of electrical stimulation has been found to be effective in enhancing rat bone marrow mesenchymal stem cells (rBMSCs) and rat adipose-derived mesenchymal stem cells (AT-MSCs) migration, proliferation, differentiation, and stimulates high levels of osteogenic expressions. Electrically stimulated rBMSCs and AT-MSCs exhibit an increased osteogenic differentiation, as indicated by high expression levels of osteogenic markers, including collagen 1, osteopontin, Runx 2 and calmodulin [164,165]. Furthermore, electrical stimulation of AT-MSCs also promotes bone repair and bone regeneration in vivo as manifested by implanted scaffolds with AT-MSCs in the rat femur defects [166]. Thus, human MSCs capable of differentiating into osteoblasts show great potential for healing of bone defects [167].

Electroactive PVDF can be used for modifying the surface of titanium in order to improve its bioactivity. Titanium (Ti) and its alloys are widely used as load-bearing implant materials in orthopedics. Polarization of the PVDF film coated on Ti has been reported to promote osteoblastic cell adhesion and proliferation [2]. More recently, Zhou et al. deposited PVDF film on Ti through corona discharge at 100 °C. They demonstrated that polarized PVDF-Ti sample promotes osteogenic differentiation of rBMSCs [161]. Figure 20 shows the real-time polymerase chain reaction (RT-PCR) analysis, revealing the expression of osteogenic differentiation-related genes of rBMSCs on polarized PVDF-Ti (PPTi) and nonpolarized PVDF-Ti (NPTi) samples. Apparently, the cells on the PPTi sample exhibit significantly higher gene expression than those on the NPTi after 14 days of incubation. This is due to the presence of surface charges on PPTi as a result of the polarization.

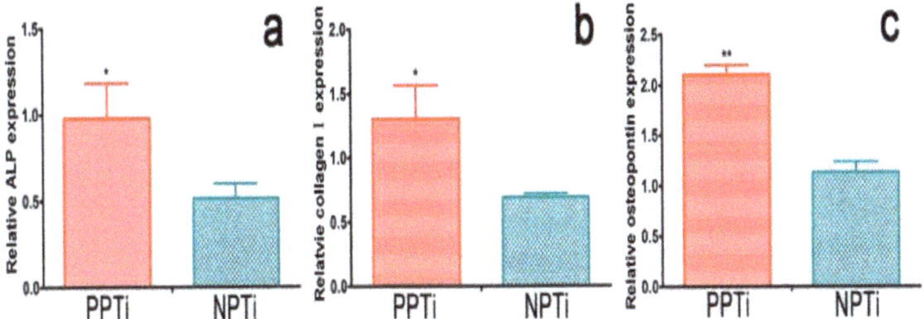

Figure 20. (**a**) Relative ALP, (**b**) collagen I, and (**c**) osteopontin gene expressions of rBMSCs cultured on PPTi and NPTi samples for 14 days ($n = 3$). * $p < 0.05$, ** $p < 0.01$ compared with NPTi.

4.1.1. In Vitro Cell Cultivation

Lanceros-Méndez and coworkers carried out a series of studies on the interactions between the PVDF and cells [13,48,105,112,168,169]. In those studies, solvent-cast dense films, solvent-cast particulate-leaching films, and NIPS porous membranes were fabricated. Particular attention was paid to the effect of the surface charge of PVDF induced by electrical poling on the fibronectin adsorption, osteoblastic cell attachment and proliferation [13,168]. Solvent-cast PVDF films (α-PVDF) were mechanically drawn to induce β-PVDF. The β-PVDF films were further poled by corona discharge to induce negative and positive electrical surface charge on the cell culture side, respectively [168]. The results showed that electrical poling decreased the water contact angle of β-PVDF films. The dipoles of β-PVDF would interact with water molecules, thereby enhancing their wettability. The positively and negatively poled β-PVDF films exhibited a water contact angle of 31.8° and 51.1°, respectively, thus showing hydrophilic behavior. Therefore, poled β-PVDF films favored fibronectin protein adsorption, thereby facilitating MC3T3-E1 osteoblastic cell adhesion and proliferation. It is generally known that the surface hydrophilicity and topography modulate protein adhesion, thereby affecting cellular response accordingly [170]. Apparently, the combination of surface wettability and piezoelectricity was effective for promoting osteoblastic cell attachment and proliferation. In another study, it was also found that the differentiation of human adipose-derived stem cells (hASCs) into osteogenic lineage was affected by substrate polarization of β-PVDF. Negatively poled β-PVDF promoted higher osteogenic differentiation, as evidenced by higher ALP activity (Figure 21) [13,171].

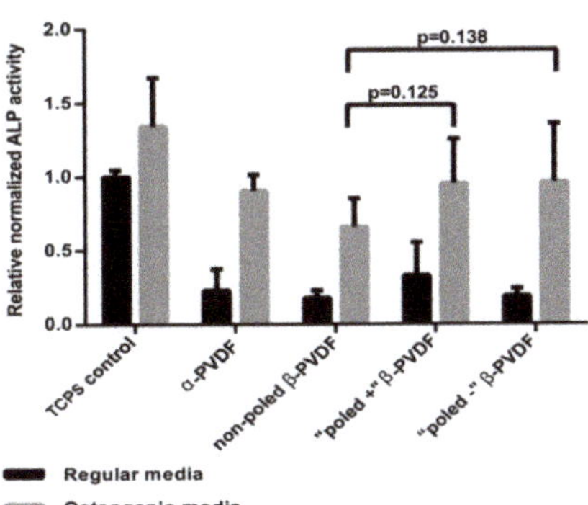

Figure 21. Relative alkaline phosphatase activity of hASCs on different PVDF films and tissue culture polystyrene (TCPS) control. The ALP activity was normalized against the DNA content of the cells. Reproduced with permission from [171], published by Wiley, 2015.

Electrospun Fibrous Scaffolds

Arinzeh and coworkers studied the beneficial effect of piezoelectric scaffolds for tissue engineering applications [172,173]. They prepared nonwoven PVDF fibrous mats by using electrospinning under applied voltages of 12–30 kV [17]. The fibrous mat formed at 25 kV had the highest 72% β-phase, while the mat fabricated at 12 kV contained 68% β-phase. Therefore, hMSCs cultivated on the PVDF-25 kV scaffold had higher levels of ALP activity and biomineralization when compared to the PVDF-12 kV mat. They also studied osteogenic differentiation of hMSCs on nonwoven P(VDF-TrFE) fibrous mats under dynamic compression at 1 Hz with 10% deformation to mimic physiological strain conditions [14]. Two power sources were employed to fabricate P(VDF-TrFE) fibrous mats with a large thickness of 3 mm, porosity of 90% and 64% β-phase. These fibrous mats were further annealed at 135 °C to increase the degree of crystallinity and β-phase content (i.e., 75%). Osteogenic markers of ALP activity, mineralization, and osteocalcin of all mats were examined accordingly. They reported that the ALP activity and matrix mineralization of the as-spun P(VDF-TrFE) mat were considerably lower than those of annealed P(VDF-TrFE) and PCL control at day 28. More recently, Kitsara et al. treated electrospun PVDF nanofibrous scaffolds with oxygen plasma for improving their hydrophilicity [174]. As a result, osteoblasts cultivated on hydrophilic PVDF fibrous scaffolds had better cell spreading over the non-treated counterparts as expected.

Wang et al. fabricated electrospun P(VDF-TrFE) mats with aligned fibers followed by annealing and electrical poling. The mean diameter of electrospun nanofibers was 590 nm ± 26 nm [175]. The β-phase content of the as-spun, annealed and poled P(VDF-TrFE) mats was 43.1%, 46.6% and 69.2%, respectively. They studied the effect of dynamic electrical stimulation on the adhesion and proliferation of mouse osteoblastic cells (MC3T3-E1) on annealed P(VDF-TrFE) and electrically poled P(VDF-TrFE) mats. Annealed P(VDF-TrFE) was designated as A-NFM, and P(VDF-TrFE) poled with the electric field of 80 MV/m and 100 MV/m were labeled as P80-NG and P100-NG, respectively. The set-up of flexible-bottomed culture plate was given in Figure 22. A speaker attached to the bottom of the cell culture plate would generate mechanical vibration at a frequency of 2 Hz for mimicking low-frequency biomechanics. The CCK-8 assay was used to examine the proliferation of MC3T3-E1 osteoblasts. Fluorescence microscopy images of MC3T3 osteoblasts on A-NFM, P80-and P100-NG revealed that the cells were elongated and oriented along the direction of nanofibers (Figure 23a). Moreover, P100-NG and P80-NG exhibited a higher cell proliferation rate than A-NFM. So poled fibrous mats with piezoelectricity increased osteoblastic cell proliferation considerably (Figure 23b).

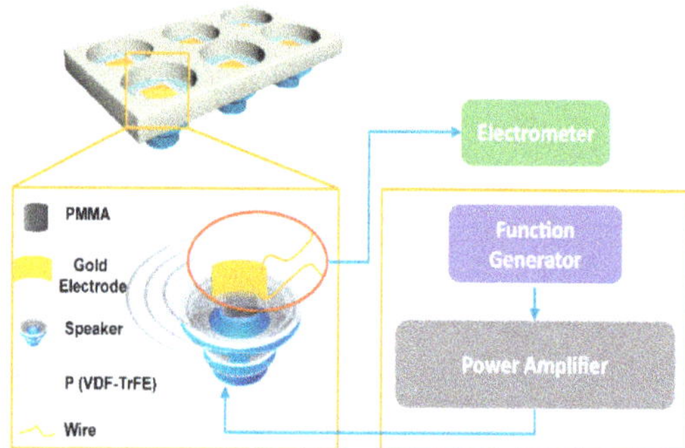

Figure 22. A simple set-up of flexible-bottomed culture plate together with an attached speaker for inducing mechanical vibration.

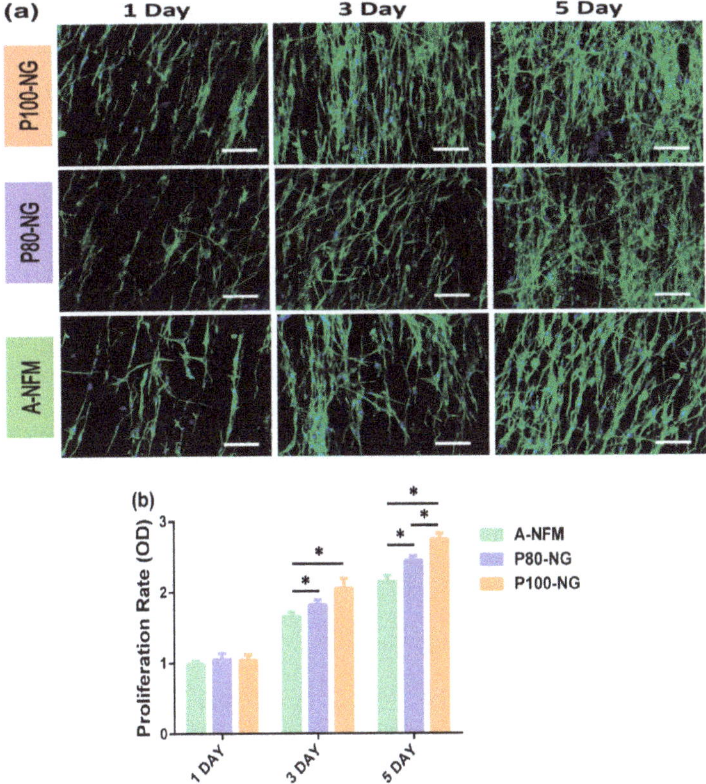

Figure 23. (**a**) Fluorescence micrographs of MC3T3 cells cultured on fibrous A-NFM, P80-NG, and P100-NG mats for 1, 3, and 5 days. The scale bar is 100 μm. (**b**) Proliferation of MC3T3 cells on P80-NG, P100-NG, and A-NFM mats. All data represent the mean standard deviation (n = 3, * $p < 0.05$).

Szewczyk et al. prepared PVDF(+) and PVDF(−) fibrous scaffolds by applying a constant voltage of 15 kV with positive and negative polarities to the stainless needle during electrospinning. Human osteoblast-like cell line (MG63) was then cultivated on those scaffolds [10]. The surface potential value of the PVDF(+) and PVDF(−) fibers was −173 mV and −65 mV, respectively as determined by the Kelvin probe force microscopy. Increased cell viability/proliferation was found in the PVDF(−) samples at different time points on the basis of Alamar Blue results (Figure 24a). Furthermore, PVDF(−) fibers also exhibited a much higher number of cells in comparison with the PVDF(+) fibers (Figure 24b). They attributed this to the surface potential of PVDF(−) fibers (i.e., −65 mV) was very close to the membrane potential of MG63 cell having a value of −60 mV. Collagen fibrils with an average diameter of 0.15 μm were readily seen on the osteoblasts cultured on the PVDF(+) and PVDF(−) fibers (Figure 25a,b) for three days. At day 7, small round nodules were observed on the surfaces of osteoblasts cultured on both scaffolds. However, the PVDF(−) fibers had a higher density of round nodules (Figure 25c,d). Those nodules were calcium phosphate as evidenced by the presence of Ca and P signals in the X-ray energy dispersive spectrum, thus showing collagen mineralization. Accordingly, the formation of mineralized collagen fibrils on the PVDF fibers can be tailored by monitoring the surface potential of electrospun scaffolds.

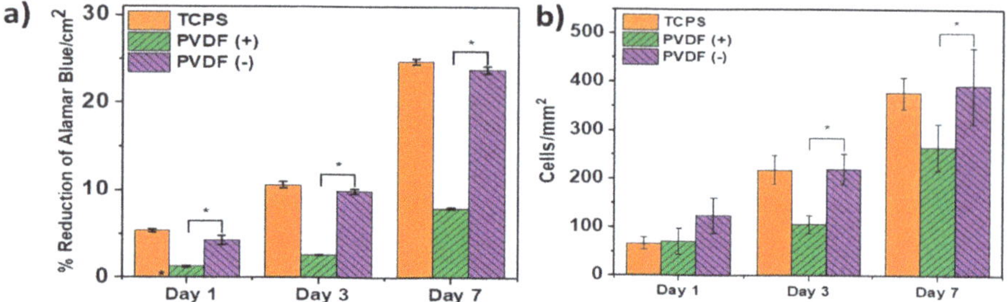

Figure 24. (a) Cell proliferation based on Alamar Blue assay of MG63 osteoblasts cultured on PVDF(+), and PVDF(−) scaffolds, as well as tissue culture polystyrene (TCPC) control for 1, 3 and 7 days. (b) Number of cells/mm^2 growing on TCPS control, PVDF(+) and PVDF(−) fibers after cultivation of MG63 cells for 1, 3 and 7 days. * Significant difference between PVDF(+) and PVDF(−) samples determined with Tukey test ($p < 0.05$). Reproduced with permission from [10], published by American Chemical Society, 2019.

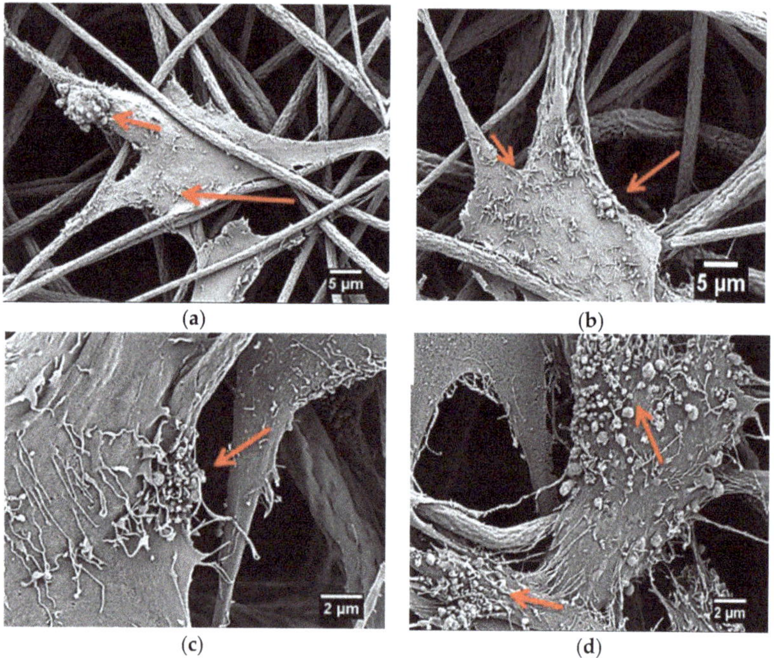

Figure 25. SEM images showing the formation of collagen fibrils on MG63 osteoblasts cultured on electrospun (a) PVDF(+) and (b) PVDF(−) scaffolds for three days. Accumulation of collagen fibrils and calcium phosphate nodules on MG63 osteoblasts cultured on (c) PVDF(+) and (d) PVDF(−) scaffolds for seven days. Red arrows indicate collagen fibrils present on cell surfaces. Reproduced with permission from [10], Copyright American Chemical Society, 2019.

It is noteworthy that the incorporation of nanoparticles (e.g., ZnO, GO and barium titanate) into electrospun PVDF-based scaffolds enhances the adhesion, proliferation and differentiation of hMSCs [176–178]. This is because those nanoparticles serve as effective nucleating sites for forming electroactive β-phase [77]. Augustine et al. reported that ZnO nanoparticles have a positive influence on the cellular behavior of electrospun P(VDF-TrFE) scaffolds. Both the hMSCs and human umbilical vein endothelial cells (HUVECs) cultivated on P(VDF-TrFE)/ZnO nanocomposite scaffolds showed higher cell adhesion and proliferation compared to the same cells cultured on pure P(VDF-TrFE) scaffolds [177]. Saburi et al. demonstrated that GO nanofillers of electrospun PVDF-GO nanofibers enhance osteogenic differentiation of human induced pluripotent stem cells (iPSCs). The ALP, Runx2, osteocalcin and osteonectin gene expression levels of the iPSCs cultivated on the PVDF-GO fibrous scaffold were significantly higher than those cultured on neat PVDF fibrous mat [178].

4.1.2. In Vivo Models for Bone Tissue Engineering

In vivo animal models can be used to assess biocompatibility of polymeric materials that directly contact with different living tissue types including nerve, bone and blood. Compared to in vitro cell culture studies, limited information is available on the animal bone tissue responses to piezoelectric PVDF-based scaffolds in vivo [39,177,179,180]. Guo et al. performed subcutaneous implantation of PVDF/polyurethane (1:1, *v/v*) scaffolds into Sprague-Dawley (SD) rats. They reported that such scaffolds showed higher fibrosis level due to the piezoelectric stimulation as a result of random rat movements followed by mechanical deformation of the scaffolds [39]. Lanceros-Méndez and coworkers implanted non-poled β-PVDF films, poled β-PVDF films (d_{33} = −24 pC/N) and randomly oriented β-PVDF fibrous mats into bone defects created in each femur of Wistar rats [179]. After 4 weeks, the femurs were removed from the sacrificed rats and subjected to histological examinations. Bone defects treated with randomly oriented β-PVDF fibers displayed obvious bone regeneration. This was revealed by the presence of organized fibers and trabecular formation. Moreover, poled β-PVDF films also showed the formation of bone marrow and trabecular bone. Very recently, Wang et al. investigated piezoelectric responses of polarized, aligned P(VDF-TrFE) nanofibrous scaffolds into the subcutaneous thigh region of Sprague Dawley (SD) rats (Figure 26a). To simulate the movement of SD rats, a linear motor system was employed to gently pull a leg of the SD rat under 0.5 N with 1 Hz frequency [180]. By pulling the leg of a rat gently, piezoelectric current and voltage were generated in vivo (Figure 26b,c).

Augustine et al. electrospun P(VDF-TrFE) and P(VDF-TrFE)/ZnO fibrous mats; the fiber diameter of neat copolymer was 1 µm, while that of P(VDF-TrFE)/1 wt% ZnO and P(VDF-TrFE)/2 wt% ZnO nanocomposite mats was 1.05 and 1.19 µm, respectively [177]. Those fibrous mats with or without hMSCs were subcutaneously implanted into abdominal region of Wistar rats (Figure 27a). Extensive collagen fiber networks were observed in all fibrous scaffolds after implantation for 7 days. Moreover, newly formed blood vessels were readily seen in the nanocomposite scaffolds with 1 wt% and 2 wt% ZnO nanoparticles. This was further increased by pre-seeding with hMSCs in the nanocomposite scaffolds prior to implantation (Figure 27b). Thus, ZnO nanoparticles of the nanocomposite scaffolds promoted angiogenesis and favored integration of the scaffolds into the surrounding tissue [177]. In this respect, reduced neovascularization of conventional nanofibrous scaffolds can be tackled by using piezoelectric P(VDF-TrFE)/ZnO fibrous scaffolds.

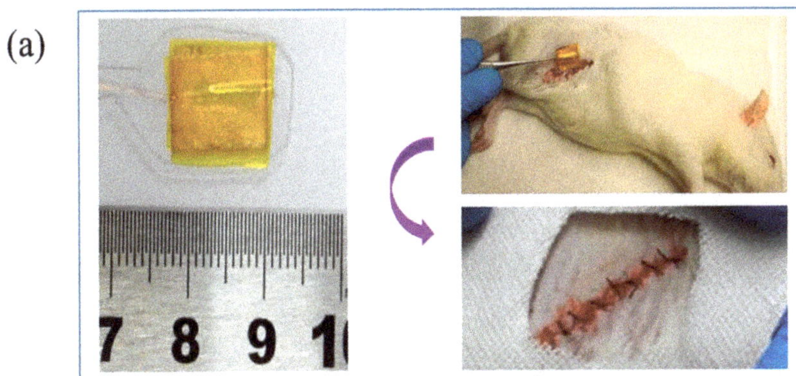

Figure 26. *Cont.*

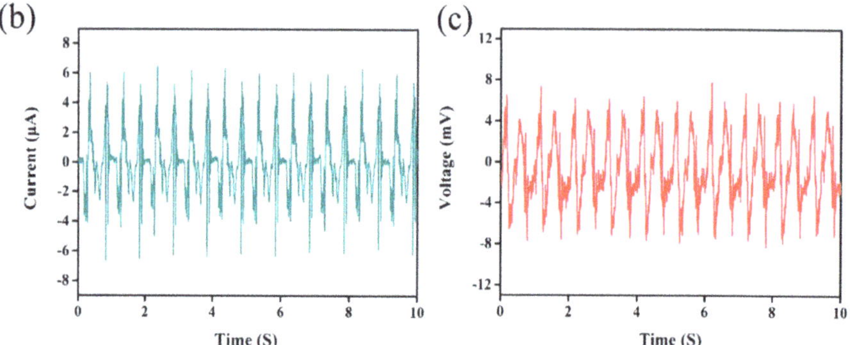

Figure 26. (a) Photographs showing the dimension of a poled P(VDF-TrFE) scaffold before implantation (left), the process of implanting piezoelectric scaffold into subcutaneous thigh region of a SD rat (upper right), and the implanting site after suturing (lower right). (b) Current and (c) voltage outputs of electrospun P(VDF-TrFE) nanofibrous scaffold after implantation under pulling. Reproduced with permission from [180], published by Elsevier, 2018.

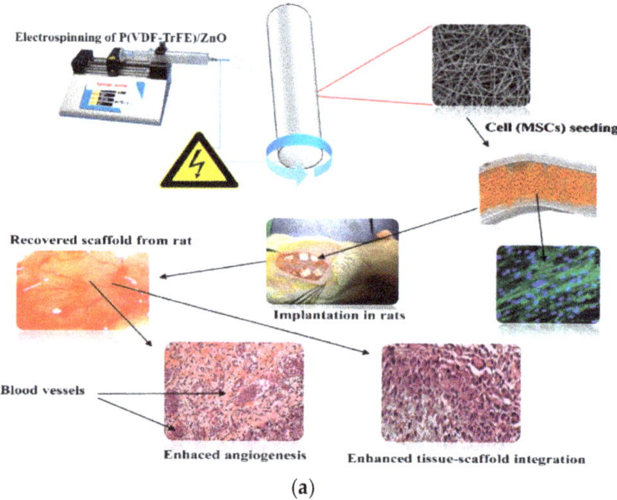

Figure 27. *Cont.*

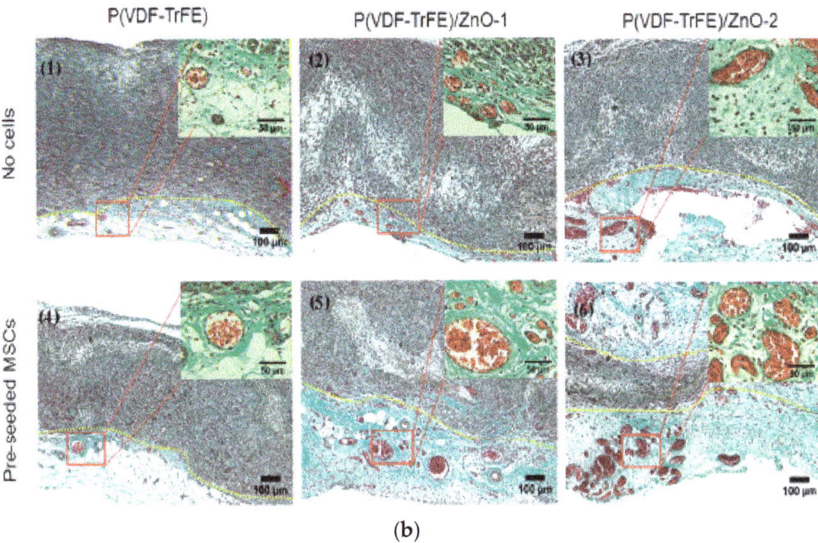

Figure 27. (**a**) Schematic showing the fabrication of electrospun P(VDF-TrFE) and P(VDF-TrFE)/ZnO scaffolds, hMSC seeding, and subsequent implantation into Wistar rats. (**b**) Histological examinations of fibrous scaffolds with or without pre-seeded hMSCs after implantation in rats for 7 days, and stained with Masson's trichrome. Blood vessels developed in connective tissue adjacent to scaffolds as distinguished by yellow dashed lines, and collagen was found in all scaffolds (green). P(VDF-TrFE)/ZnO-1 and P(VDF-TrFE)/ZnO-2 contained 1 wt% and 2 wt% ZnO, respectively. Reproduced with permission from [177], published by Springer, 2017.

4.2. Neural Tissue Engineering

As is widely recognized, different cell types can respond in different ways to a biomaterial surface. So different tissues need different microenvironments for sufficient cell–cell interaction, cell migration, proliferation, differentiation and regeneration [181]. In particular, piezoelectric PVDF-based materials are effective sites for the attachment, growth and differentiation of neurons with the involvement of electrical activity. In the context of neural tissue engineering, aligned electrospun fibrous mats offer a distinct advantage over the scaffolds with randomly oriented fibers. This is because highly aligned fibers provide spatial guidance for neurite outgrowth and axonal elongation.

4.2.1. In Vitro Model

Genchi et al. added 60 wt% barium titanate nanoparticles (BTNPs) to solvent-cast P(VDF-TrFE) film for improving its piezoelectric properties. The BTNPs addition further enhanced the β-phase content of P(VDF-TrFE) from 30% to 50%. As a result, P(VDF-TrFE)/BTNP film promoted the viability and differentiation of human neuroblastoma cells (SH-SY5Y). Furthermore, ultrasound stimulation (US) was used to promote the adhesion and differentiation of SH-SY5Y cells cultivated on the P(VDF-TrFE) and P(VDF-TrFE)/BTNP specimens (Figure 28a,b) [182]. From Figure 28b, β3-tubulin was used as a marker for the differentiation of SH-SY5Y cells. The percentage of β3-tubulin positive cells increased markedly after US treatment, especially for the P(VDF-TrFE)/BTNP film. Ultrasound stimulation also increased neurite length of SH-SY5Y cells cultured on both P(VDF-TrFE) and P(VDF-TrFE)/BTNP films as a result of the activation of calcium channels (Figure 28c). Similarly, Hoop et al. demonstrated that ultrasonic stimulation of PC12 neuronal cells cultured on poled β-PVDF membrane activates their calcium channels, thus increasing neurite length greatly [46].

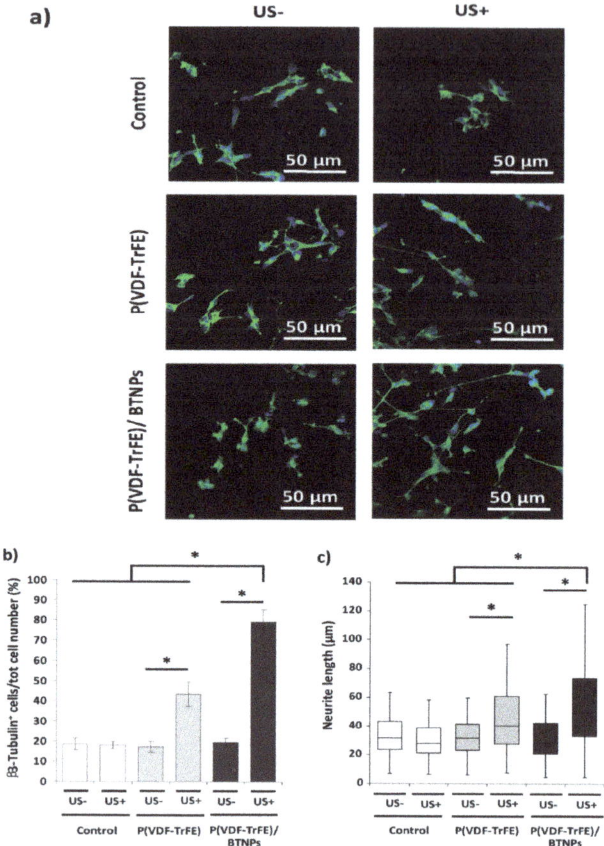

Figure 28. (**a**) Confocal fluorescence microscopic images of SH-SY5Y cells on solvent-cast P(VDF-TrFE) and P(VDF-TrFE)/BTNP films, as well as Ibidi (control) at the end of differentiation period of 6 days. SH-SY5Y cells were treated with or without ultrasound stimulation (US+ or US−). β3-tubulin was stained in green, nuclei in blue. (**b**) Percentages of β3-tubulin positive cells. (**c**) Neurite lengths are expressed as median values ± confidence interval at 95%. * $p < 0.05$. Reproduced with permission from [182], published by Wiley-VCH, 2016.

Fibrous Scaffolds for Neural Tissue Engineering

As mentioned, highly aligned electrospun fibers provide spatial guidance for neurite outgrowth and axonal elongation. Accordingly, neurite migration, attachment, proliferation and differentiation on aligned nanofibers are directed along the nanofiber orientation [183,184]. Moreover, electrical stimulation has been found to be particularly useful for neurite growth. Corey et al. demonstrated that neurons from the primary rat dorsal root ganglia cultured on electrospun poly ʟ-lactide (PLLA) fibrous mats with highly aligned fibers were 20% longer than the neurites on random fibers [185]. Koppes et al. investigated neurite outgrowth on laminin-coated PLLA films and electrospun microfibers with or without electrical stimulation. The electrical stimulation increased neurite outgrowth by 32% on the films or fibers when compared to unstimulated films. In addition, neurite extension increased by 74% on the aligned fibers compared to the control film specimen [186]. PLLA also exhibits piezoelectric effect with a smaller piezoelectric constant; the shear piezoelectric constant (d_{14}) of uniaxially stretched PLLA film is about 6–10 pC/N [187].

Lins et al. fabricated PVDF fibrous mats by rotating the drum collector for achieving non-aligned (50 rpm), low-aligned (1000 rpm), medium-aligned (2000 rpm) and high-aligned (3000 rpm) fibers. The effect of PVDF fiber alignment on the cellular responses of monkey neural stem cells (NSCs) was studied [15]. NSCs are multipotent stem cells capable of differentiating into all types of neural lineage such as neuronal and glial cells. For immunofluorescent microscopy examinations, cells were stained with β3-tubulin and glial fibrillary acidic protein (GFAP), i.e., an intermediate protein expressed in glial cell (Figure 29a). For low-aligned (1000 rpm) scaffold, the highest number of cells (143.5 cells per field) was observed in which nearly 64% of the cells expressed β3-tubulin and 8% expressed GFAP (Figure 29b). The average number of cells expressing β3-tubulin was smaller in other PVDF scaffolds. Thus, low-aligned PVDF scaffold favored the differentiation of NSCs. Figure 30 showed the SEM images of stem cell, neuronal cell, and glial cell cultured on the PVDF mats with different fiber orientations. Stem and glial cells showed irregular features, and neuronal cells had elongated morphology. The neurons were oriented parallel to the fiber direction of medium- and high-aligned mats, while they oriented randomly on low-aligned and nonaligned fibrous scaffolds.

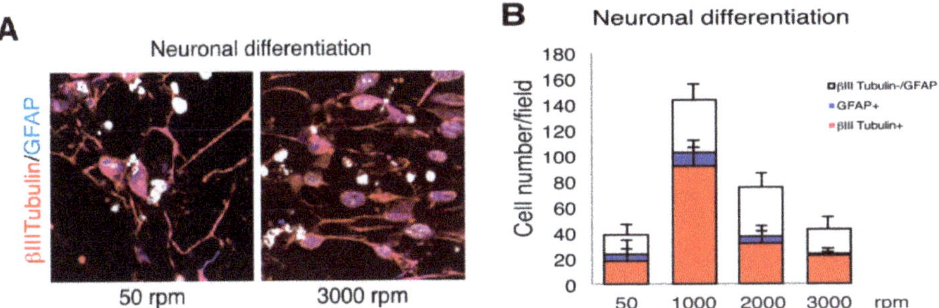

Figure 29. (**A**) Immunofluorescent staining for β3-tubulin (red) and GFAP (blue), and (**B**) mean percentage of cells expressing β3-tubulin and GFAP upon NSCs differentiation. n = 3–5. Reproduced with permission from [15], published by Wiley, 2017.

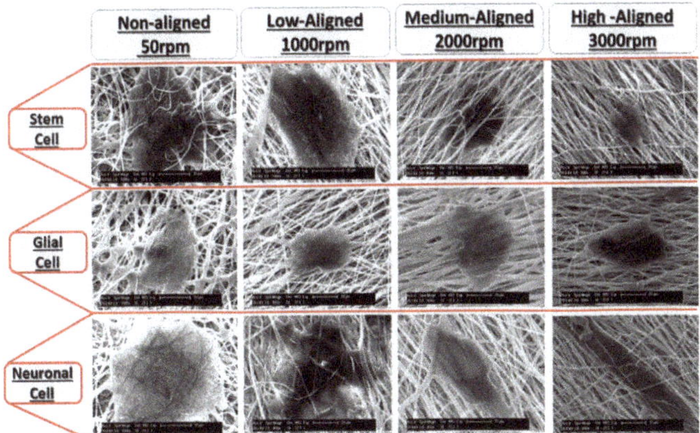

Figure 30. Scanning electron micrographs showing the morphologies of stem cell, neuronal cell, and glial cell cultivated on electrospun PVDF scaffolds with different fiber orientations. Reproduced with permission from [15], published by Wiley, 2017.

Arinreh and coworkers electrospun PVDF and P(VDF-TrFE) fibrous mats having both random and aligned fibers with mean diameters of micron (3.32 μm) and nanometer (750 nm) dimensions [6,188]. The random fibers were collected on a grounded metal plate, and aligned fibers were deposited on a rotating drum. The as-spun scaffolds were annealed at 135 °C for 96 h and quenched with ice water. The porosity and pore size of the micron-sized, as-spun mats were 58.08% and 4.8 µm, while those of annealed counterparts were 79.63% and 8.0 µm, respectively. The porosity and pore size of the nano-sized, as-spun mats were 43.92% and 1.7 µm, while those of the annealed counterparts were 67.55% and 1.5 µm, respectively [6]. For a given fiber size, annealed and aligned mats had higher elastic modulus and degree of crystallinity than the random and as-spun counterparts, leading to annealed mats with higher tensile strength [188]. In general, elastic modulus and tensile strength of polymers increased with increasing the degree of crystallinity [189–191]. Rat dorsal root ganglia (DRG) neurons were cultured on all fibrous mats for 4 days. Annealed and aligned P(VDF-TrFE) mats with micron-sized fibers showed the largest neurite extension in comparison with random mats (Figure 31a–c). This was ascribed to annealing treatment increased the β-phase content and piezoelectricity of P(VDF-TrFE) mats. Furthermore, neurite length on micron-sized, aligned and annealed mats was slightly longer than that of nano-sized counterparts. From Figure 31a, neurons with sizes of about 2000 µm including the cell body and dendrites favored micron-sized, annealed and aligned fibers for their attachment and outgrowth. Such fibrous mats exhibited the highest porosity and pore size values of 79.63% and 8.0 µm. In another study, human neural stem/progenitor cells cultured on annealed and aligned P(VDF-TrFE) mats with micron-sized fibers differentiated into neuron-liked cells as revealed by positive β3-thenubulin staining [188]. So those fibrous mats show potential applications for nerve regeneration, because in vivo piezoelectric activity in these fibers can be activated by the bulk deformations associated with cerebrospinal fluid circulation. In this respect, piezoelectric P(VDF-TrFE) mats with enhanced human stem cell differentiation are effective biomaterials for nerve tissue repair or replacement of damaged nerve cells due to the injury or disease.

From the literature, aligned electrospun PCL fibers can enhance Schwann cell maturation [192]. Those fibers can mimic the natural ECM of the spinal cord, providing an ideal microenvironment at the injury site to facilitate neural repair. Schwann cells (SCs) are the glial cells of the peripheral nervous system (PNS), and cover the surface of axons of motor/sensory neurons to produce a myelin sheath and enhance nerve repair/regeneration. During injury, SCs remove damaged axons/myelin debris, promote axonal regrowth, and remyelinate axons [193]. Apparently, aligned PVDF-based scaffolds with piezoelectricity are effective biomaterials to promote Schwan cell maturation and nerve regeneration in the PNS. Very recently, Arinzeh and coworkers electrospun aligned P(VDF-TrFE) mats and then cultured SCs and DRG explants on those scaffolds [194,195]. They reported that aligned P(VDF-TrFE) mats enhance SCs growth and neurite extension, especially for Matrigel coated scaffolds (Figure 32a,b). Matrigel contains multiple extracellular matrix proteins and bioactive components extracted from mouse tumor cells [195]. Growth factors of different types are frequently tested for its influence on peripheral nerve regeneration. Therefore, aligned P(VDF-TrFE) scaffolds loaded with Matrigel are beneficial to treat peripheral nerve injury.

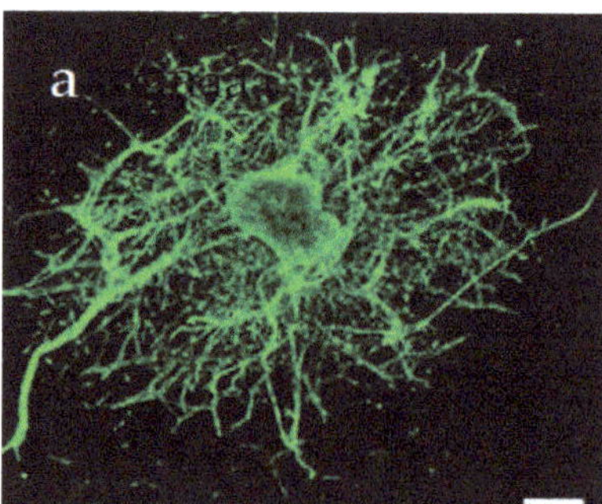
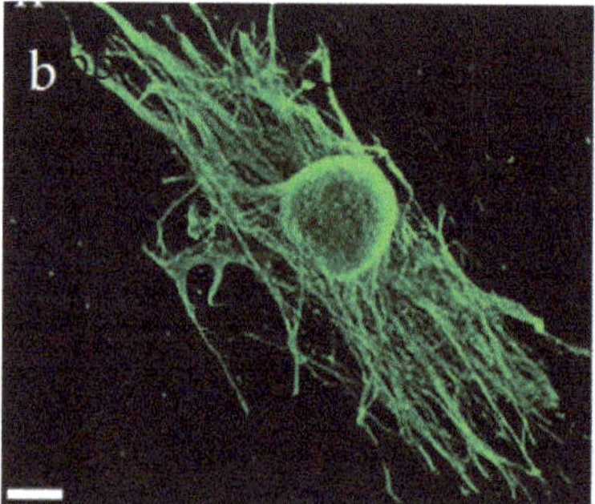

Figure 31. *Cont.*

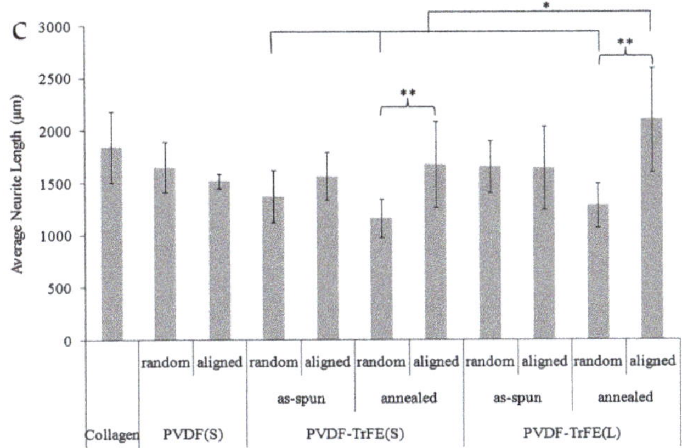

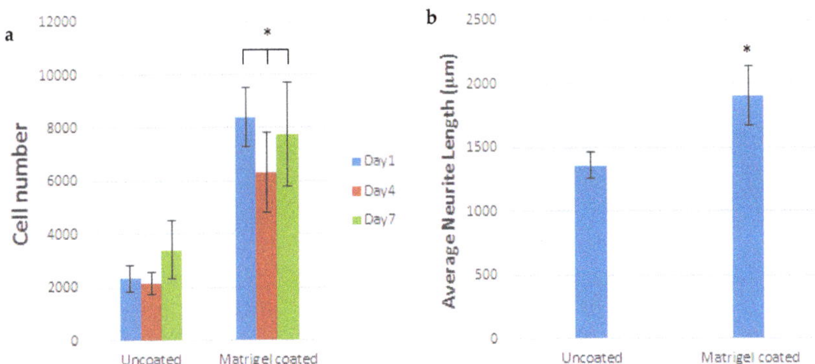

Figure 31. Confocal fluorescent images of DRG stained with phalloidin (actin) on micron-sized, annealed (**a**) random and (**b**) aligned P(VDF-TrFE) mats. Scale bar: 300 μm. (**c**) Average neurite length of DRG neurons cultured on micron-sized (L) and nano-sized (S) P(VDF-TrFE) mats with random and aligned fibers under as-spun and annealed conditions. Mean neurite length cultured on nano-sized (S) PVDF mats with random and aligned fibers, and collagen control are also shown for comparison. * and ** denote statistically significant difference between the sample groups; $p < 0.05$. Reproduced with permission from [6], published by Elsevier, 2011.

Figure 32. (**a**) SC number on aligned PVDF-TrFE fibrous scaffolds with or without Matrigel coating. Cell number on Matrigel coated scaffolds was significantly higher than on uncoated scaffolds at all time points (* $p < 0.05$). (**b**) Average neurite extension of DRGs cultured on uncoated and Matrigel coated PVDF-TrFE scaffolds. Neurite extension on Matrigel coated scaffolds was significantly higher than uncoated scaffolds (* $p < 0.05$). Reproduced from [194], published by Frontiers, 2018.

4.2.2. In Vivo Models for Neural Tissue Engineering

Peripheral nerve injury due to trauma remains a big challenge for the researchers nowadays. Autologous nerve grafts are typically used by medical doctors to treat the nerve damage. The use of autologous nerve grafts has some drawbacks, including donor site morbidity, limited availability, mismatch of nerves and complicated surgeries. Accordingly, synthetic nerve guide conduits have been developed to facilitate axonal growth and nerve regeneration [196]. The inclusion of electrospun PCL fibers into nerve guide conduits generally supports axonal regeneration in vivo [197].

Very scarce in vivo models have been conducted on the neural tissue responses to the PVDF-based scaffolds. Aebischer et al. fabricated piezoelectric PVDF nerve guidance channels. Nerves regenerated in poled PVDF channels had a higher number of myelinated axons than those regenerated in unpoled channels [198]. Very recently, Arinzeh and coworkers assessed the potential use of PVDF-TrFE conduits with SCs for spinal cord repair in vivo. In their study, the spinal cords of adult rats were completely transected. The conduits, with random or aligned fibrous inner walls, were transplanted into transected rat spinal cords for 3 weeks. The conduits with aligned fibers promoted greater axon regeneration over random fibers [199].

5. Major Challenges

Remarkable achievements have been made in recent years in the development of electrospun fibrous PVDF-based scaffolds with piezoelectric effects for bone and nerve tissue engineering. Those nanofibrous mats have a very large volume to surface ratio, thus favoring adhesion, proliferation and differentiation of osteoblasts, hMSCs and neuronal cells. However, electrospun fibrous scaffolds also have some drawbacks, including small pore size, limited mat thickness, reduced neovascularization, etc. For instance, the as-spun P(VDF-TrFE) mats have a very small pore size of 1.7 μm [6]. As the pore sizes of fibrous mats are much smaller than the dimensions of individual cells, thus bone or nerve cells can only attach and proliferate on the mat surfaces. This leads to the formation of a single cell layer on the scaffold surfaces. In this respect, researchers have spent much effort developing electrospun scaffolds with larger pore sizes and porosities for cell infiltration and ingrowth. These include cryogenic electrospinning, electrospinning with salt leaching, and electrospinng with gas foaming [83,200,201].

In cryogenic electrospinning, the collector is immersed in liquid nitrogen such that ice crystals are formed in the polymer fibers. The ice particles increase the distance between the fibers acting as void spacers during the fiber deposition. They are subsequently removed by sublimation, leaving void spaces behind. This technique is based on thermally induced phase separation (TIPS) approach for inducing polymer-rich and polymer-poor regions in the fibers. The resulting porosities can reach up to 99.5% [202,203]. The salt leaching strategy involves the introduction of salt particles to the Taylor cone during electrospinning, and the particles are removed by leaching in a water bath after fiber deposition [204]. Despite the increase in the pore sizes of fibrous mats by these techniques, conventional electrospinning process can only generate two-dimensional (2D) fibrous mats with irregular pore size network due to the whipping instability. As is widely recognized, 3D electrospun fibrous scaffolds, which can mimic the natural tissue structures, are considered to be of clinical importance. This motivates the researchers to develop a near-field electrospinning (NFES) to deposit well-aligned fibers in a short distance between the needle and collector (e.g., few millimeters) to suppress bending instability [205,206]. Thus, NFES can produce complex 3D fibrous scaffolds with desirable patterns and geometries because it allows a precise control of the fiber deposition in a direct writing (DW) mode. Designed patterns can be made by monitoring the translation of the collector, and the resulting fibers are stacked to form 3D scaffold [207]. Limited studies have been conducted on the PVDF-based mats prepared by NFES. Liu et al. employed NFES to prepare PVDF/MWCNT mats with enhanced piezoelectricity [208]. Very recently, Lee et al. employed NFES to form well-aligned, 3D-PVDF fibrous mats for sensor applications [209]. Till to present, there exists no literature reports relating in vitro and in vivo behaviors of 3D fibrous scaffolds prepared by NFES.

The main drawback of the solution-based electrospinning for biomedical applications is the use of organic solvents for dissolving polymers and for evaporating polymer jets to form fibers. In particular, DMF and DMAc solvents used for fabricating PVDF-based fibers are highly toxic and harmful to human cells [210]. Thus, melt-electrospinning can be used to fabricate fibrous mats without using toxic solvents. However, coarse fibers in the tens to hundreds of micrometers are produced in melt-electrospinning due to the high melt viscosity and low melt conductivity [155]. The low surface area to volume ratio of coarse fibers can reduce cell attachment on their surfaces. In general, the melt fluid flow with a straight flight path can be tailored to create 3D scaffolds with controlled porosity through a direct writing process, terming as the 'melt electrospinning writing' (MEW) [211]. MEW has been used only to create 3D PCL scaffolds due to the low melting temperature of PCL (i.e., 60 °C) [157,211,212]. By optimizing the processing parameters of PCL, direct-write PCL scaffolds with fine filaments of 817 ± 165 nm can be produced. As mentioned, non-polar α-phase is the dominant crystalline structure found in melt-electrospun PVDF fibers [160]. Therefore, great effort is required to solve the technical issues for forming melt-electrospun 3D PVDF fibrous mats containing electroactive β-phase for tissue engineering applications.

Most literature studies have been confined to the use of 2D PVDF-based fibrous mats for cell cultivation in vitro. A systematic investigation on the cellular response to electrospun PVDF-based scaffolds in vivo is lacking, especially for the bone and neural defects. At present, only a few studies have been conducted on in vivo animal models of electrospun PVDF-based scaffolds and conduits [39,177,179,199]. For bone tissue engineering, an in-depth study and comprehensive understanding of animal models treated with the PVDF-based fibrous scaffolds is also lacking. The literature relating neuronal cell responses to the PVDF-based fibrous scaffolds in vivo is scarce.

6. Future Direction

For the successful implementation of electrospun PVDF-based scaffolds and conduits for bone and neural tissue engineering, bacterial infection of those scaffolds will become an issue and cannot be ignored. Surgical procedures involving scaffolds and implants are complicated by bacterial infection. Bacterial infection often leads to the morbidity and mortality of patients globally. Device-related infection is resulted from the bacterial adhesion, colonization, and biofilm formation. This is mostly caused by gram-positive *Staphylococcus aureus* (*S. aureus*). Healing of bone defects with fibrous scaffolds can be complicated by the presence of *S. aureus*, especially methicillin resistant *Staphylococcus aureus* (MRSA) [213]. Electrospun polymer scaffolds with a large surface/volume ratio can load antibiotics such as vancomycin and gentamicin in preventing infection of the bone/joint tissue and implant biofilm formation [214,215]. To prevent bacteria from developing drug resistance, we can load silver nanoparticles into electrospun PVDF-based scaffolds having no bactericidal activity to form antibacterial nanocomposites. Silver nanoparticles (AgNPs) are known to resist a wide variety of bacteria strains including gram-positive S. aureus, MRSA, and gram-negative Pseudomonas aeruginosa and Escherichia coli [51,216]. However, AgNPs can induce cytotoxicity on human neural cells and fibroblasts in a dose-dependent manner [217,218]. So AgNPs behave like a double-edged sword, having bactericidal activity, but also cytotoxicity on some human cells. Therefore, it is necessary to study the effects of AgNPs additions on the bactericidal activity, biocompatibility and electroactive β-phase formation on electrospun PVDF and P(VDF-TrFE) fibrous mats in detail.

As mentioned above, very scarce information is available in the literature on in vivo animal studies of piezoelectric PVDF-based scaffolds for bone and neural tissue engineering. Those studies were typically carried out on mouse models for assessing bone regeneration and spinal cord repair using PVDF-based scaffolds [179,199]. However, mouse models have some drawbacks, because they are not helpful for long-term investigations in which multiple biopsies or blood samples are required. These arise from short life expectancy, and relatively small tissue and blood sample volumes of mice when compared with larger animal models such as goats and pigs [219]. In this context, more in vivo animal studies using both the mouse and large animal models are needed in the near future to investigate the biocompatibility of electrospun PVDF-based scaffolds, and an inhibiting effect of bacterial biofilm on wound healing using PVDF-based scaffolds containing AgNPs nanofillers.

7. Conclusions

Electrospinning merges wet chemical solution processing, electric field poling and stretching into a single step procedure for forming piezoelectric β-PVDF fibers. The incorporation of TrFE monomer into PVDF can stabilize the β-phase without mechanical stretching or electrical poling. However, mechanical stretching or electrical poling can further enhance β-phase in the P(VDF-TrFE). Piezoelectric

PVDF and P(VDF-TrFE) fibrous mats find attractive applications for bone and neural tissue engineering. Bone is a natural nanocomposite consisting of collagen fibrils and nHA. Minute mechanical deformation of collagen fibrils in human bones can induce electrical potential that is crucial for bone healing and regeneration. Accordingly, piezoelectric PVDF and P(VDF-TrFE) fibrous mats promote osteoblastic adhesion, proliferation and differentiation on their surfaces in vitro. The hMSCs cultivated on PVDF scaffolds under electrical stimulation exhibit high ALP activity and biomineralization. Furthermore, aligned PVDF and P(VDF-TrFE) fibrous mats can direct neurite growth, promote neural stem cell differentiation and support the growth of Schwann cells. Therefore, those fibrous mats can be used as potential biomaterials for making nerve guidance conduits for treating peripheral nerve damage.

The main drawbacks of the electrospinning process for making piezoelectric PVDF-based scaffolds for tissue engineering applications are the formation of very small pores and the use of organic solvents. Conventional two-dimensional PVDF-based scaffolds with very small pores prevent the infiltration of bone and neuronal cells into the scaffolds. Consequently, those cells can only attach and proliferate on the scaffold surfaces, forming a single cell layer with limited cell infiltration. Near-field electrospinning designed for sensor applications can create 3D PVDF fibrous mats for tissue engineering [209]. However, there are no literature articles reporting in vitro and in vivo behaviors of 3D fibrous PVDF-based scaffolds prepared by NFES. The main limitation of NFES for tissue engineering is the use of organic solvents. The solvents typical used for fabricating PVDF fibers are DMF and DMAc, which are toxic to human cells. Melt electrospinning with a direct writing process can create 3D scaffolds with controlled porosity without using organic solvents. However, MEW process is currently confined to make PCL fibrous mats due to the ease of processing as PCL has a low melting temperature. Much efforts are needed by the researchers to develop solvent free, electrospun 3D PVDF-based scaffolds with interconnecting pore networks for bone and neural tissue engineering applications.

Author Contributions: S.C.T. conceived the idea and planned the text content. Y.L., C.L. and S.C.T. wrote the review article.

Funding: The authors would like to thank Natural Science Foundation of Shandong Province, China (Project No. ZR2019MB053) for supporting this research.

Conflicts of Interest: The authors declare no conflict of interest.

References

1. Acosta, M.; Novak, N.; Patel, S.; Vaish, R.; Koruza, J.; Rossetti, R.A.; Rodel, J. BaTiO$_3$-based piezoelectrics: Fundamentals, current status, and perspectives. *Appl. Phys. Rev.* **2017**, *4*, 041305. [CrossRef]
2. Martins, P.; Lopes, A.C.; Lanceros-Méndez, S. Electroactive phases of poly(vinylidene fluoride): Determination, processing and applications. *Prog. Polym. Sci.* **2014**, *39*, 683–706. [CrossRef]
3. Ribeiro, C.; Sencadas, V.; Correia, D.M.; Lanceros-Méndez, S. Piezoelectric polymers as biomaterials for tissue engineering applications. *Colloids Surf. B Biointerfaces* **2015**, *136*, 46–55. [CrossRef] [PubMed]
4. Cardoso, V.F.; Correia, D.M.; Ribeiro, C.; Fernandes, M.M.; Lanceros-Méndez, S. Fluorinated polymers as smart materials for advanced biomedical applications. *Polymers* **2018**, *10*, 161. [CrossRef] [PubMed]
5. Soulestin, T.; Ladmiral, V.; Dos Santos, F.D.; Ameduri, B. Vinylidene fluoride- and trifluoroethylene-containing fluorinated electroactive copolymers. How does chemistry impact properties? *Prog. Polym. Sci.* **2017**, *72*, 16–60. [CrossRef]
6. Lee, Y.S.; Collins, G.; Arinzeh, T.L. Neurite extension of primary neurons on electrospun piezoelectric scaffolds. *Acta Biomater.* **2011**, *7*, 3877–3886. [CrossRef] [PubMed]
7. Hitscherich, P.; Wu, S.; Gordan, R.; Xieh, L.H.; Arinzeh, T.; Lee, E.J. The effect of PVDF-TrFE scaffolds on stem cell derived cardiovascular cells. *Biotechnol. Bioeng.* **2016**, *113*, 1577–1585. [CrossRef] [PubMed]
8. Jacob, J.; More, N.; Kalia, K.; Kapusetti, G. Piezoelectric smart biomaterials for bone and cartilage tissue engineering. *Inflamm. Regen.* **2018**, *38*, 2. [CrossRef] [PubMed]
9. Meng, S.; Rouabhia, M.; Zhang, Z. Electrical Stimulation in Tissue Regeneration. In *Applied Biomedical Engineering*; Gargiulo, G., McEwan, A., Eds.; IntechOpen: London, UK, 2011; Chapter 3; pp. 38–62.

10. Szewczyk, P.K.; Metwally, S.; Karbowniczek, J.E.; Marzec, M.M.; Stodolak-Zych, E.; Gruszczyński, A.; Bernasik, A.; Stachewic, U. Surface-potential-controlled cell proliferation and collagen mineralization on electrospun polyvinylidene fluoride (PVDF) fiber scaffolds for bone regeneration. *ACS Biomater. Sci. Eng.* **2019**, *5*, 582–593. [CrossRef]

11. Przekora, A. Current trends in fabrication of biomaterials for bone and cartilage regeneration: Materials modifications and biophysical stimulations. *Int. J. Mol. Sci.* **2019**, *20*, 435. [CrossRef] [PubMed]

12. More, N.; Kapusetti, G. Piezoelectric material—A promising approach for bone and cartilage regeneration. *Med. Hypotheses* **2017**, *108*, 10–16. [CrossRef] [PubMed]

13. Ribeiro, C.; Parssinen, J.; Sencadas, V.V.; Correia, V.V.; Miettinen, S.; Hytonen, V.P.; Lanceros-Mendez, S.; Pärssinen, J.; Sencadas, V.V.; Correia, V.V.; et al. Dynamic piezoelectric stimulation enhances osteogenic differentiation of human adipose stem cells. *J. Biomed. Mater. Res. Part A* **2015**, *103*, 2172–2175. [CrossRef] [PubMed]

14. Damaraju, S.M.; Shen, Y.; Elele, E.; Khusid, B.; Eshghinejad, A.; Li, J.; Jaffe, M.; Arinzeh, T.L. Three-dimensional piezoelectric fibrous scaffolds selectively promote mesenchymal stem cell differentiation. *Biomaterials* **2017**, *149*, 51–62. [CrossRef] [PubMed]

15. Lins, L.C.; Wianny, F.; Livi, S.; Dehay, C.; Duchet-Rumeau, J.; Gérard, J.F. Effect of polyvinylidene fluoride electrospun fiber orientation on neural stem cell differentiation. *J. Biomed. Mater. Res. B Appl. Biomater.* **2017**, *105*, 2376–2393. [CrossRef] [PubMed]

16. Schellenberg, A.; Ross, R.; Abagnale, G.; Joussen, S.; Schuster, P.; Arshi, A.; Pallua, N.; Jockenhoevel, S.; Gries, T.; Wagner, W. 3D Non-woven polyvinylidene fluoride scaffolds: Fibre cross section and texturizing patterns have impact on growth of mesenchymal stromal cells. *PLoS ONE* **2014**, *9*, e94353. [CrossRef] [PubMed]

17. Damaraju, S.M.; Wu, S.; Jaffe, M.; Arinzeh, T. Structural changes in PVDF fibers due to electrospinning and its effect on biological function. *Biomed. Mater.* **2013**, *8*, 045007. [CrossRef]

18. Anderson, J.; Eriksson, C. Piezoelectric properties of dry and wet bone. *Nature* **1970**, *227*, 491–492. [CrossRef]

19. Halperin, C.; Mutchnik, S.; Agronin, A.; Molotskii, M.; Urenski, P.; Salai, M.; Rosenman, G. Piezoelectric effect in human bones studied in nanometer scale. *Nano Lett.* **2004**, *4*, 1253–1256. [CrossRef]

20. Reis, J.; Silva, F.C.; Queiroga, C.; Lucena, S.; Potes, J. Bone mechanotranduction: A review. *J. Biomed. Bioeng.* **2011**, *2*, 37–44.

21. Jonsson, A.L.; Roberts, M.A.; Kiappes, J.L.; Scott, K.A. Essential chemistry for biochemists. *Essays Biochem.* **2017**, *61*, 401–427. [CrossRef]

22. Rodriquez, R.; Rangel, D.; Fonseca, G.; Gonzalez, M.; Vargas, S. Piezoelectric properties of synthetic hydroxyapatite-based organic-inorganic hydrated materials. *Results Phys.* **2016**, 925–932. [CrossRef]

23. Kang, H.; Hou, Z.; Qin, Q.H. Experimental study of time response of bending deformation of bone cantilevers in an electric field. *J. Mech. Behav. Biomed. Mater.* **2018**, *77*, 192–198. [CrossRef] [PubMed]

24. Ahn, A.C.; Grodzinsky, A.J. Relevance of collagen piezoelectricity to "Wolff's law": A critical review. *Med. Eng. Phys.* **2009**, *31*, 733–741. [CrossRef] [PubMed]

25. Fernández, J.R.; García-Aznar, J.M.; Martínez, R. Piezoelectricity could predict sites of formation/resorption in bone remodelling and modelling. *J. Theor. Biol.* **2012**, *292*, 86–92. [CrossRef] [PubMed]

26. Bassett, C.A.; Becker, R.O. Generation of electrical potentials by bone in response to mechanical stress. *Science* **1962**, *137*, 1063–1064. [CrossRef]

27. Ma, L.; Li, Y.; Wang, J.; Zhu, H.; Yang, W.; Cao, R.; Qian, Y.; Feng, M. Quality of life is related to social support in elderly osteoporosis patients in a Chinese population. *PLoS ONE* **2015**, *10*, e0127849. [CrossRef]

28. Albergaria, B.; Chalem, M.; Clark, P.; Messina, O.D.; Pereira, R.M.; Vidal, L.F. Consensus statement: Osteoporosis prevention and treatment in Latin America—Current structure and future directions. *Arch. Osteporos.* **2018**, *13*, 90. [CrossRef]

29. Coleman, R.; Body, J.J.; Aapro, M.; Hadji, P. Bone health in cancer patients: ESMO clinical practice guidelines. *Ann. Oncol.* **2014**, *25*, iii124–iii137. [CrossRef]

30. Poidvin, A.; Carel, J.C.; Ecosse, E.; Levy, D.; Michon, J.; Coste, J. Increased risk of bone tumors after growth hormone treatment in childhood: A population-based cohort study in France. *Cancer Med.* **2018**, *7*, 3465–3473. [CrossRef]

31. Ekegren, C.L.; Beck, B.; Simpson, P.M.; Gabbe, B.J. Ten-year incidence of sport and recreation injuries resulting in major trauma or death in Victoria, Australia, 2005–2015. *Orthop J. Sports Med.* **2018**, *6*. [CrossRef]

32. Singh, A.; Tetreault, L.; Kalsi-Ryan, S.; Nouri, A.; Fehlings, M. Global prevalence and incidence of traumatic spinal cord injury. *Clin. Epidemiol.* **2014**, *6*, 309–331. [CrossRef] [PubMed]

33. Winkler, T.; Sass, F.A.; Duda, G.N.; Schmidt-Bleek, K. A review of biomaterials in bone defect healing, remaining shortcomings and future opportunities for bone tissue engineering. *Bone Jt. Res.* **2018**, *7*, 232–243. [CrossRef] [PubMed]

34. Griesbach, G.S.; Masel, B.E.; Helvie, R.E.; Ashley, M.J. The Impact of traumatic brain injury on later life: Effects on normal aging and neurodegenerative diseases. *J. Neurotrauma* **2018**, *35*, 17–24. [CrossRef] [PubMed]

35. Esopenko, C.; Levine, B. Aging, neurodegenerative disease, and traumatic brain injury: The role of neuroimaging. *J. Neurotrauma* **2015**, *32*, 209–220. [CrossRef] [PubMed]

36. McKee, A.C.; daneshvar, D.H. The neuropathology of traumatic brain injury. *Handb. Clin. Neurol.* **2015**, *127*, 45–66. [CrossRef]

37. Gumera, C.; Rauck, B.; Wang, Y. Materials for central nervous system regeneration: Bioactive cues. *J. Mater. Chem.* **2011**, *21*, 7033–7051. [CrossRef]

38. Song, C.G.; Zhang, Y.Z.; Wu, H.N.; Cao, X.L.; Guo, C.J.; Li, Y.Q.; Zheng, M.H.; Han, H. Stem cells: A promising candidate to treat neurological disorders. *Neural Regen. Res.* **2018**, *13*, 1294–1304. [PubMed]

39. Guo, H.F.; Li, Z.S.; Dong, S.W.; Chen, W.J.; Deng, L.; Wang, Y.F.; Ying, D.J. Piezoelectric PU/PVDF electrospun scaffolds for wound healing applications. *Colloids Surf. B Biointerfaces* **2012**, *96*, 29–36. [CrossRef]

40. Chocholata, P.; Kulda, V.; Babuska, V. Fabrication of scaffolds for bone-tissue regeneration. *Materials* **2019**, *12*, 568. [CrossRef]

41. Dzobo, K.; Thomford, N.E.; Senthebane, D.A.; Shipanga, H.; Rowe, A.; Dandara, C.; Pillay, M.; Motaung, K.S. Advances in regenerative medicine and tissue engineering: Innovation and transformation of medicine. *Stem Cells Int.* **2018**, *2018*, 2495848. [CrossRef]

42. Perez, J.R.; Kouroupis, D.; Li, D.J.; Best, T.M.; Kaplan, L.; Correa, D. Tissue engineering and cell-based therapies for fractures and bone defects. *Front. Bioeng. Biotechnol.* **2018**, *6*, 105. [CrossRef] [PubMed]

43. Cortese, B.; Palama, I.E.; D'Amone, S.; Gigli, G. Influence of electrotaxis on cell behaviour. *Integr. Biol.* **2014**, *6*, 817–830. [CrossRef] [PubMed]

44. Balint, R.; Cassidy, N.J.; Cartmell, S.H. Conductive polymers: Towards a smart biomaterial for tissue engineering. *Acta Biomater.* **2014**, *10*, 2341–2353. [CrossRef] [PubMed]

45. Ateh, D.D.; Navsaria, H.A.; Vadgama, P. Polypyrrole-based conducting polymers and interactions with biological tissues. *J. R. Soc. Interface* **2006**, *3*, 741–752. [CrossRef] [PubMed]

46. Hoop, M.; Chen, X.Z.; Ferrari, A.; Mushtaq, F.; Ghazaryan, G.; Tervoort, T.; Poulikakos, D.; Nelson, B.; Panel, S. Ultrasound-mediated piezoelectric differentiation of neuron-like PC12 cells on PVDF membranes. *Sci. Rep.* **2017**, *7*, 4028. [CrossRef]

47. Zayzafoon, M. Calcium/calmodulin signaling controls osteoblast growth and differentiation. *J. Cell Biochem.* **2006**, *97*, 56–70. [CrossRef] [PubMed]

48. Martins, P.M.; Ribeiro, S.; Ribeiro, C.; Sencadas, V.; Gomes, A.C.; Gama, F.M.; Lanceros-Méndez, S. Effect of poling state and morphology of piezoelectric poly(vinylidene fluoride) membranes for skeletal muscle tissue engineering. *RSC Adv.* **2013**, *3*, 17938–17944. [CrossRef]

49. Mondal, D.; Gayen, A.L.; Paul, B.K.; Bandyopadhyay, P.; Bera, D.; Bhar, D.S.; Das, K.; Das, N.S. Enhancement of β-phase crystallization and electrical properties of PVDF by impregnating ultra high diluted novel metal derived nanoparticles: Prospect of use as a charge storage device. *J. Mater. Sci. Mater. Electron.* **2018**, *29*, 14535. [CrossRef]

50. Mishra, S.; Unnikrishnan, L.; Nayak, S.K.; Mohanty, S. Advances in piezoelectric polymer composites for energy harvesting applications: A systematic review. *Macromol. Mater. Eng.* **2019**, *304*, 1800463. [CrossRef]

51. Liu, C.; Shen, J.; Liao, C.Z.; Yeung, K.W.K.; Tjong, S.C. Novel electrospun polyvinylidene fluoride-graphene oxide-silver nanocomposite membranes with protein and bacterial antifouling characteristics. *Express Polym. Lett.* **2018**, *12*, 365–382. [CrossRef]

52. Laroche, G.; Marois, Y.; Guidoin, R.; King, M.W.; Martin, L.; How, T.; Douville, Y. Polyvinylidene fluoride (PVDF) as a biomaterial: From polymeric raw material to monofilament vascular suture. *J. Biomed. Mater. Res.* **1995**, *29*, 1525–1536. [CrossRef] [PubMed]

53. Cai, X.; Lei, T.; Sun, D.; Lin, L. A critical analysis of the α, β and γ phases in poly(vinylidene fluoride) using FTIR. *RSC Adv.* **2017**, *7*, 15382–15389. [CrossRef]

54. Wan, C.; Bowen, C.R. Multiscale-structuring of polyvinylidene fluoride for energy harvesting: The impact of molecular-, micro- and macro-structure. *J. Mater. Chem. A* **2017**, *5*, 3091–3128. [CrossRef]

55. Yu, Y.J.; McGaughey, A.J.H. Energy barriers for dipole moment flipping in PVDF-related ferroelectric polymers. *J. Chem. Phys.* **2016**, *144*, 014901. [CrossRef] [PubMed]

56. Prest, W.M., Jr.; Luca, D.J. The morphology and thermal response of high-temperature-crystallized poly(vinylidene fluoride). *J. Appl. Phys.* **1975**, *46*, 4136–4143. [CrossRef]

57. Bao, S.P.; Liang, G.D.; Tjong, S.C. Effect of mechanical stretching on electrical conductivity and positive temperature coefficient characteristics of poly(vinylidene fluoride)/carbon nanofiber composites prepared by non-solvent precipitation. *Carbon* **2011**, *49*, 1758–1768. [CrossRef]

58. He, L.X.; Tjong, S.C. Graphene oxide/polyvinylidene fluoride mixture as a precursor for fabricating thermally reduced graphene oxide/polyvinylidene fluoride composites. *RSC Adv.* **2013**, 22981–22987. [CrossRef]

59. Li, Y.C.; Ge, X.; Wang, L.; Liu, W.; Li, H.; Li, R.K.Y.; Tjong, S.C. Dielectric relaxation behavior of PVDF composites with nanofillers of different conductive nature. *Curr. Nanosci.* **2013**, *9*, 679–685. [CrossRef]

60. Li, Y.C.; Tjong, S.C.; Li, R.K.Y. Dielectric properties of binary polyvinylidene fluoride/barium titanate nanocomposites and their nanographite doped hybrids. *Express Polym. Lett.* **2011**, *5*, 526–534. [CrossRef]

61. Tjong, S.C.; Chen, H. Nanocrystalline materials and coatings. *Mater. Sci. Eng. R Rep.* **2004**, *45*, 1–88. [CrossRef]

62. Tjong, S.C. Polymer nanocomposite bipolar plates reinforced with carbon nanotubes and graphite nanosheets. *Energy Environ. Sci.* **2011**, *44*, 605–626. [CrossRef]

63. He, L.X.; Tjong, S.C. Facile synthesis of silver-decorated reduced graphene oxide as a hybrid filler material for electrically conductive polymer composites. *RSC Adv.* **2015**, *5*, 15070–15076. [CrossRef]

64. He, L.X.; Tjong, S.C. Nanostructured transparent conductive films: Fabrication, characterization and applications. *Mater. Sci. Eng. R Rep.* **2016**, *109*, 1–101. [CrossRef]

65. He, L.X.; Tjong, S.C. Aqueous graphene oxide-dispersed carbon nanotubes as inks for the scalable production of all-carbon transparent conductive films. *J. Mater. Chem. C* **2016**, *4*, 7043–7051. [CrossRef]

66. Ramos, A.P.; Cruz, M.A.; Tovani, C.B.; Ciancaglini, P. Biomedical applications of nanotechnology. *Biophys. Rev.* **2017**, *9*, 79–89. [CrossRef] [PubMed]

67. Falsini, S.; Bardi, U.; Abou-Hassan, A.; Ristori, S. Sustainable strategies for large-scale nanotechnology manufacturing in the biomedical field. *Green Chem.* **2018**, *20*, 3897–3907. [CrossRef]

68. Jiang, J.; Pi, J.; Cai, J. The advancing of zinc oxide nanoparticles for biomedical applications. *Bioinorg. Chem. Appl.* **2018**, *2018*, 1062562. [CrossRef] [PubMed]

69. Liu, C.; Shen, J.; Yeung, K.W.K.; Tjong, S.C. Development and antibacterial performance of novel polylactic acid-graphene oxide-silver nanoparticle nanocomposite mats prepared by electrospinning. *ACS Biomater. Sci. Eng.* **2017**, *3*, 471–486. [CrossRef]

70. Chan, K.W.; Liao, C.Z.; Wong, H.M.; Yeung, K.W.K.; Tjong, S.C. Preparation of polyetheretherketone composites with nanohydroxyapatite rods and carbon nanofibers having high strength, good biocompatibility and excellent thermal stability. *RSC Adv.* **2016**, *6*, 19417–19429. [CrossRef]

71. Liu, C.; Wong, H.; Yeung, K.; Tjong, S.C. Novel electrospun polylactic acid nanocomposite fiber mats with hybrid graphene oxide and nanohydroxyapatite reinforcements having enhanced biocompatibility. *Polymers* **2016**, *8*, 287. [CrossRef]

72. Liao, C.Z.; Wong, H.M.; Yeung, K.W.; Tjong, S.C. The development, fabrication and material characterization of polypropylene composites reinforced with carbon nanofiber and hydroxyapatite nanorod hybrid fillers. *Int. J. Nanomed.* **2014**, *9*, 1299–1310.

73. Chen, H.L.; Ju, S.P.; Lin, C.Y.; Pan, C.T. Investigation of microstructure and mechanical properties of polyvinylidene fluoride/carbon nanotube composites after electric field polarization: A molecular dynamics study. *Comput. Mater. Sci.* **2018**, *149*, 217–229. [CrossRef]

74. Kabir, E.; Khatun, M.; Nasrin, L.; Raihan, M.J.; Rahman, M. Pure β-phase formation in polyvinylidene fluoride (PVDF)-carbon nanotube composites. *J. Phys. D Appl. Phys.* **2017**, *50*, 163002. [CrossRef]

75. Ahn, Y.; Lim, J.Y.; Hong, S.M.; Lee, J.; Ha, J.; Choi, H.J.; Seo, Y. Enhanced piezoelectric properties of electrospun poly(vinylidene fluoride)/multiwalled carbon nanotube composites due to high β-phase formation in poly(vinylidene fluoride). *J. Phys. Chem. C* **2013**, *117*, 11791–11799. [CrossRef]

76. Hosseini, S.M.; Yousefi, A.A. Piezoelectric sensor based on electrospun PVDF-MWCNT-Cloisite 30B hybrid nanocomposites. *Org. Electron.* **2017**, *50*, 121–129. [CrossRef]

77. El Achaby, M.; Arrakhiz, F.Z.; Vaudreuil, S.; Essassi, E.M.; Qaiss, A. Piezoelectric β-polymorph formation and properties enhancement in graphene oxide—PVDF nanocomposite films. *Appl. Surf. Sci.* **2012**, *258*, 7668–7677. [CrossRef]

78. Jiang, Z.Y.; Zheng, G.P.; Zhan, K.; Han, Z.; Yan, J.H. Formation of piezoelectric β-phase crystallites in poly(vinylidene fluoride)-graphene oxide nanocomposites under uniaxial tensions. *J. Phys. D Appl. Phys.* **2015**, *48*, 245303. [CrossRef]

79. Abolhasani, M.M.; Shirvanimoghaddam, K.; Naebe, M. PVDF/graphene composite nanofibers with enhanced piezoelectric performance for development of robust nanogenerators. *Compos. Sci. Technol.* **2017**, *138*, 49–56. [CrossRef]

80. Lee, J.E.; Eom, Y.; Shin, Y.E.; Hwang, S.H.; Ko, H.H.; Chae, H.G. Effect of interfacial interaction on the conformational variation of poly(vinylidene fluoride) (PVDF) chains in PVDF/graphene oxide (GO) nanocomposite fibers and corresponding mechanical properties. *ACS Appl. Mater. Interfaces* **2019**, *11*, 13665–13675. [CrossRef]

81. Abzan, N.; Kharazhiha, N.; Labbaf, S. Development of three-dimensional piezoelectric polyvinylidene fluoride-graphene oxide scaffold by non-solvent induced phase separation method for nerve tissue engineering. *Mater. Des.* **2019**, *167*, 107636. [CrossRef]

82. Haddadi, S.A.; Ghaderi, S.; Amini, M.; Ahmad-Ramazani, S.A. Mechanical and piezoelectric characterizations of electrospun PVDF-nanosilica fibrous scaffolds for biomedical applications. *Mater. Today Proc.* **2018**, *5*, 15710–15716. [CrossRef]

83. Jun, I.; Han, H.S.; Edwards, J.R.; Jeon, H. Electrospun fibrous scaffolds for tissue engineering: Viewpoints on architecture and fabrication. *Int. J. Mol. Sci.* **2018**, *19*, 745. [CrossRef] [PubMed]

84. Sundaray, B.; Bossard, F.; Lati, P.; Orgéas, L.; Sanchez, J.Y.; Lepretre, J.C. Unusual process-induced curl and shrinkage of electrospun PVDF membranes. *Polymer* **2013**, *54*, 4588–4593. [CrossRef]

85. Zheng, J.; He, A.; Li, J.; Han, C.C. Polymorphism control of poly(vinylidene fluoride) through electrospinning. *Macromol. Rapid Commun.* **2007**, *28*, 2159–2162. [CrossRef]

86. Li, M.; Wondergem, H.J.; Spijkman, M.J.; Asadi, K.; Katsouras, I.; Blom, P.W.; de Leeuw, D.M. Revisiting the δ-phase of poly(vinylidene fluoride) for solution-processed ferroelectric thin films. *Nat. Mater.* **2013**, *12*, 433–438. [CrossRef] [PubMed]

87. Weinhold, S.; Litt, M.H.; Lando, J.B. The crystal structure of the γ phase of poly(vinylidene fluoride). *Macromolecules* **1980**, *13*, 5095–5099. [CrossRef]

88. Koga, K.; Ohigashi, H. Piezoelectricity and related properties of vinylidene fluoride and trifluoroethylene copolymers. *J. Appl. Phys.* **1986**, *59*, 2142–2150. [CrossRef]

89. Koga, K.; Nakano, N.; Hattori, T.; Ohigashi, H. Crystallization, field-induced phase transformation, thermally induced phase transition, and piezoelectric activity in P(vinylidene fluoride-TrFE) copolymers with high molar content of vinylidene fluoride. *J. Appl. Phys.* **1990**, *67*, 965–974. [CrossRef]

90. Xia, W.; Xu, Z.; Zhang, Q.; Zhang, Z.; Chen, Y. Dependence of dielectric, ferroelectric, and piezoelectric properties on crystalline properties of p(VDFco-TrFE) copolymers. *J. Polym. Sci. B Polym. Phys.* **2012**, *50*, 1271–1276. [CrossRef]

91. Sun, F.C.; Dongare, A.M.; Asandei, A.D.; Alpay, S.P.; Nakhmanson, S. Temperature dependent structural, elastic, and polar properties of ferroelectric polyvinylidene fluoride (PVDF) and trifluoroethylene (TrFE) copolymers. *J. Mater. Chem. C* **2015**, *3*, 8389–8396. [CrossRef]

92. Issa, A.A.; Al-Maadeed, M.A.; Luyt, A.S.; Ponnamma, D.; Hassan, M.K. Physico-mechanical, dielectric, and piezoelectric properties of PVDF electrospun mats containing silver nanoparticles. *Carbon* **2017**, *3*, 30. [CrossRef]

93. Li, Y.; Sun, L.; Webster, T.J. The investigation of ZnO/poly(vinylidene fluoride) nanocomposites with improved mechanical, piezoelectric, and antimicrobial properties for orthopedic applications. *J. Biomed. Nanotechnol.* **2018**, *14*, 536545. [CrossRef] [PubMed]

94. Alam, M.M.; Sultana, A.; Mandal, D. Biomechanical and acoustic energy harvesting from TiO_2 nanoparticle modulated PVDF nanofiber made high performance nanogenerator. *ACS Appl. Energy Mater.* **2018**, *1*, 3103–3112. [CrossRef]

95. Singh, H.H.; Singh, S.; Khare, N. Enhanced β-phase in PVDF polymer nanocomposite and its application for nanogenerator. *Adv. Polym. Technol.* **2018**, *29*, 143–150. [CrossRef]

96. Ke, K.; Potschke, P.; Jehnichen, D.; Fischer, D.; Voit, B. Achieving β-phase poly(vinylidene fluoride) from melt cooling: Effect of surface functionalized carbon nanotubes. *Polymer* **2014**, *55*, 611–619. [CrossRef]

97. Bormashenko, Y.; Pogreb, R.; Stanevsky, O.; Bormashenko, E. Vibrational spectrum of PVDF and its interpretation. *Polym. Test.* **2004**, *23*, 791–796. [CrossRef]

98. Gregorio, R., Jr.; Cestari, M. Effect of crystallization temperature on the crystalline phase content and morphology of poly(vinylidene fluoride). *J. Polym. Sci. B Polym. Phys.* **1994**, *32*, 859–870. [CrossRef]

99. Li, Y.; Zhang, G.; Song, S.; Xu, H.; Pan, M.; Zhong, G.J. How chain intermixing dictates the polymorphism of PVDF in poly(vinylidene fluoride)/polymethylmethacrylate binary system during recrystallization: A comparative study on core–shell particles and latex blend. *Polymers* **2017**, *9*, 448. [CrossRef]

100. Ribeiro, S.; Ribeiro, T.; Ribeiro, C.; Correia, D.M.; Farinha, J.P.; Gomes, A.C.; Baleizão, C.; Lanceros-Méndez, S. Multifunctional platform based on electroactive polymers and silica nanoparticles for tissue engineering applications. *Nanomaterials* **2018**, *8*, 933. [CrossRef]

101. Nooeid, P.; Salih, V.; Beier, J.; Bocaccini, A.R. Osteochondral tissue engineering: Scaffolds, stem cells and applications. *J. Cell. Mol. Med.* **2012**, *16*, 2247–2270. [CrossRef]

102. Jafari, M.; Paknejad, Z.; Rad, M.R.; Motamedian, S.R. Polymeric scaffolds in tissue engineering: A literature review. *J. Biomed. Mater. Res. B* **2017**, *105*, 431–459. [CrossRef] [PubMed]

103. De Witte, T.M.; Fratila-Apachitei, L.E.; Zadpoor, A.A.; Peppas, N.A. Bone tissue engineering via growth factor delivery: From scaffolds to complex matrices. *Regen. Biomater.* **2018**, *5*, 197–211. [CrossRef] [PubMed]

104. Ribeiro, C.; Costa, C.M.; Correia, D.M.; Nunes-Pereira, J.; Oliveira, J.; Martins, P.; Gonçalves, R.; Cardoso, V.F.; Lanceros-Méndez, S. Electroactive poly(vinylidene fluoride)-based structures for advanced applications. *Nat. Protoc.* **2018**, *13*, 681–704. [CrossRef] [PubMed]

105. Correia, D.M.; Ribeiro, C.; Sencadas, V.; Vikingsson, L.; Gasch, M.O.; Ribelles, J.L.; Botelho, G.; Lanceros-Méndez, S. Strategies for the development of three dimensional scaffolds from piezoelectric poly(vinylidene fluoride). *Mater. Des.* **2016**, *92*, 674–681. [CrossRef]

106. Abzan, N.; Kharaziha, M.; Labbaf, S.; Saeidi, N. Modulation of the mechanical, physical and chemical properties of polyvinylidene fluoride scaffold via non-solvent induced phase separation process for nerve tissue engineering applications. *Eur. Polym. J.* **2018**, *104*, 115–127. [CrossRef]

107. Hu, N.; Xiao, T.; Cai, X.; Ding, L.; Fu, Y.; Yang, X. Preparation and characterization hydrophilically modified PVDF membranes by a novel nonsolvent thermally induced phase separation method. *Membranes* **2016**, *6*, 47. [CrossRef] [PubMed]

108. Jung, J.T.; Kim, J.F.; Wang, H.H.; di Nicolo, E.; Drioli, E.; Lee, M.Y. Understanding the non-solvent induced phase separation (NIPS) effect during the fabrication of microporous PVDF membranes via thermally induced phase separation (TIPS). *J. Membr. Sci.* **2016**, *514*, 250–263. [CrossRef]

109. Turnbull, G.; Clarke, J.; Picard, F.; Riches, P.; Jia, L.; Han, F.; Li, B.; Shu, W. 3D bioactive composite scaffolds for bone tissue engineering. *Bioact Mater.* **2018**, *3*, 278–314. [CrossRef]

110. Bencherif, S.A.; Braschier, T.M.; Renaud, P. Advances in the design of macroporous polymer scaffolds for potential applications in dentistry. *J. Periodontal Implant Sci.* **2013**, *43*, 251–261. [CrossRef]

111. Idris, A.; Man, Z.; Maulud, A.S.; Khan, M.S. Effects of phase separation behavior on morphology and performance of polycarbonate membranes. *Membranes* **2017**, *7*, 21. [CrossRef]

112. Cardoso, V.F.; Botelho, G.; Lanceros-Méndez, S. Nonsolvent induced phase separation preparation of poly(vinylidene fluoride-co-chlorotrifluoroethylene) membranes with tailored morphology, piezoelectric phase content and mechanical properties. *Mater. Des.* **2015**, *88*, 390–397. [CrossRef]

113. Guillen, G.R.; Pan, Y.; Li, M.; Hoek, E.M. Preparation and characterization of membranes formed by nonsolvent induced phase separation: A Review. *Ind. Eng. Chem. Res.* **2011**, *50*, 3798–3817. [CrossRef]

114. Dreyer, D.R.; Park, S.; Bielawski, C.W.; Ruoff, R.S. The chemistry of graphene oxide. *Chem. Soc. Rev.* **2010**, *39*, 228–240. [CrossRef] [PubMed]

115. Eigler, S.; Hirch, A. Chemistry with graphene and graphene oxide—Challenges for synthetic chemists. *Angew. Chem. Int. Ed.* **2014**, *53*, 7720–7738. [CrossRef] [PubMed]

116. Gu, P.; Manu, M.J.; Unnikrishnan, B.S.; Shiji, R.; Sreelekha, T.T. Biomedical applications of natural polymer based nanofibrous scaffolds. *Int. J. Med. Nano Res.* **2015**, *2*, 2. [CrossRef]

117. Kishan, A.P.; Cosgriff-Hernandez, E.M. Recent advancements in electrospinning design for tissue engineering applications: A review. *J. Biomed. Mater. Res. A* **2017**, *105*, 2892–2905. [CrossRef] [PubMed]

118. Kennedy, K.M.; Bhaw-Luximon, A.; Jhurry, D. Cell-matrix mechanical interaction in electrospun polymeric scaffolds for tissue engineering: Implications for scaffold design and performance. *Acta Mater.* **2017**, *50*, 41–55. [CrossRef]

119. Schwinté, P.; Keller, L.; Lemoine, S.; Gottenberg, J.E.; Benkirane-Jessel, N.; Vanleen, M. Nano-engineered scaffold for osteoarticular regenerative medicine. *J. Nanomed. Nanotechnol.* **2015**, *6*, 258. [CrossRef]

120. Liu, W.; Thomopoulos, S.; Xia, Y. Electrospun nanofibers for regenerative medicine. *Adv. Healthcare Mater.* **2012**, *1*, 10–25. [CrossRef]

121. Xie, J.; MacEwan, M.R.; Ray, W.Z.; Liu, W.; Siewe, D.Y.; Xia, Y. Radially aligned, electrospun nanofibers as dural substitutes for wound closure and tissue regeneration applications. *ACS Nano* **2010**, *4*, 5027–5036. [CrossRef]

122. Cozza, E.; Monticelli, O.; Marsano, E.; Cebe, P. On the electrospinning of PVDF: Influence of the experimental conditions on the nanofiber properties. *Polym. Int.* **2013**, *62*, 41–48. [CrossRef]

123. Ribeiro, C.; Sencadas, V.; Ribelles, J.L.; Lanceros-Méndez, S. Influence of processing conditions on polymorphism and nanofiber morphology of electroactive poly(vinylidene fluoride) electrospun membranes. *Soft. Mater.* **2010**, *8*, 274–287. [CrossRef]

124. Ghafari, E.; Jiang, X.; Lu, N. Surface morphology and beta-phase formation of single polyvinylidene fluoride (PVDF) composite nanofibers. *Adv. Compos. Hybrid Mater.* **2018**, *1*, 332–340. [CrossRef]

125. Motamedi, A.S.; Mirzadeh, H.; Hajiesmaeilbaigi, F.; Bagheri-Khoulenjan, S.; Shokrgozar, M. Effect of electrospinning parameters on morphological properties of PVDF nanofibrous scaffolds. *Prog. Biomater.* **2017**, *6*, 113–123. [CrossRef] [PubMed]

126. Yee, W.A.; Kotaki, M.; Liu, Y.; Lu, X. Morphology, polymorphism behavior and molecular orientation of electrospun poly(vinylidene fluoride) fibers. *Polymer* **2007**, *48*, 512–521. [CrossRef]

127. Matabola, K.P.; Moutioali, R. The influence of electrospinning parameters on the morphology and diameter of poly(vinyledene fluoride) nanofibers—Effect of sodium chloride. *J. Mater. Sci.* **2013**, *48*, 5475–5482. [CrossRef]

128. Pillay, V.; Dotte, C.; Choonara, Y.E.; Tyagi, C.; Tomar, l.; Kumar, P.; du Toit, L.C.; Ndesendo, V.M. A Review of the effect of processing variables on the fabrication of electrospun nanofibers for drug delivery applications. *J. Nanometer.* **2013**, *2013*, 789289. [CrossRef]

129. Shao, H.; Fang, J.; Wang, H.; Lin, T. Effect of electrospinning parameters and polymer concentrations on mechanical-to-electrical energy conversion of randomly-oriented electrospun poly(vinylidene fluoride) nanofiber mats. *RSC Adv.* **2015**, *5*, 14345–14350. [CrossRef]

130. Sengupta, D.; Kottapalli, A.G.; Chen, S.H.; Miao, J.M.; Kwok, C.Y.; Triantafyllou, M.S.; Warkian, M.E.; Asadnia, M. Characterization of single polyvinylidene fluoride (PVDF) nanofiber for flow sensing applications. *AIP Adv.* **2017**, *7*, 105205. [CrossRef]

131. Yang, D.; Lu, B.; Zhao, Y.; Jiang, X. Fabrication of aligned fibrous arrays by magnetic electrospinning. *Adv. Mater.* **2007**, *19*, 3702–3706. [CrossRef]

132. Yu, K.; Wang, S.; Li, Y.; Chen, D.; Liang, C.; Lei, T.; Sun, D.; Zhao, Y. Piezoelectric performance of aligned PVDF nanofibers fabricated by electrospinning and mechanical spinning. In Proceedings of the 13th IEEE Conference Nanotechnology (IEEE-Nano 2013), Beijing, China, 5–8 August 2013. [CrossRef]

133. Wang, S.H.; Wan, Y.; Sun, B.; Liu, L.Z.; Xi, W. Mechanical and electrical properties of electrospun PVDF/MWCNT ultrafine fibers using rotating collector. *Nanoscale Res. Lett.* **2014**, *9*, 522. [CrossRef] [PubMed]

134. Wu, C.M.; Chou, M.H.; Zeng, W.Y. Piezoelectric response of aligned electrospun polyvinylidene fluoride/carbon nanotube nanofibrous membranes. *Nanomaterials* **2018**, *8*, 420. [CrossRef] [PubMed]

135. Li, D.; Wang, Y.; Xia, Y. Electrospinning of polymeric and ceramic nanofibers as uniaxially aligned arrays. *Nano Lett.* **2003**, *3*, 1167–1171. [CrossRef]

136. Shehata, N.; Elnabawy, E.; Abdelkader, M.; Hassanin, A.H.; Salah, M.; Nair, R.; Bhat, S.A. Static-aligned piezoelectric poly (vinylidene fluoride) electrospun nanofibers/MWCNT composite membrane: Facile method. *Polymers* **2018**, *10*, 965. [CrossRef] [PubMed]

137. Bhattarai, R.S.; Bachu, R.D.; Boddu, S.H.; Bhaduri, S. Biomedical applications of electrospun nanofibers: Drug and nanoparticle delivery. *Pharmaceutics* **2019**, *11*, 5. [CrossRef] [PubMed]

138. Dong, H.S.; Qi, S.J. Realising the potential of graphene-based materials for biosurfaces—A future perspective. *Biosurf. Biotribol.* **2015**, *1*, 229–248. [CrossRef]

139. Cardenas, L.; MacLeod, J.; Lipton-Duffin, J.; Seifu, D.G.; Popescu, F.; Siaj, M.; Mantovani, D.; Rosei, F. Reduced graphene oxide growth on 316L stainless steel for medical applications. *Nanoscale* **2014**, *6*, 8664–8670. [CrossRef]

140. Park, S.; An, J.; Potts, J.R.; Velamakanni, A.; Murali, S.; Ruoff, R.S. Hydrazine-reduction of graphite- and graphene oxide. *Carbon* **2011**, *49*, 3019–3023. [CrossRef]

141. McAllister, M.J.; Li, J.L.; Adamson, D.H.; Schniepp, H.C.; Abdala, A.A.; Liu, J.; Herrera-Alonso, M.; Milius, D.L.; Car, R.; Prud'homme, R.K.; et al. Single sheet functionalized graphene by oxidation and thermal expansion of graphite. *Chem. Mater.* **2007**, *19*, 4396–4404. [CrossRef]

142. Issa, A.A.; Al-Maadeed, M.A.; Mrlik, M.; Luyt, A.S. Electrospun PVDF graphene oxide composite fibre mats with tunable physical properties. *J. Polym. Res.* **2016**, *23*, 232. [CrossRef]

143. Moraldi, R.; Karimi-Sabet, J.; Shariaty-Niassar, M.; Koochaki, M.A. Preparation and characterization of polyvinylidene fluoride/graphene superhydrophobic fibrous films. *Polymers* **2015**, *7*, 1444–1463. [CrossRef]

144. Damjanovic, D. Ferroelectric, dielectric and piezoelectric properties of ferroelectric thin films and ceramics. *Rep. Prog. Phys.* **1998**, *61*, 1267–1324. [CrossRef]

145. Fulay, P.; Lee, J.K. *Electronic, Magnetic and Optical Materials*; CRC Press: Boca Raton, FL, USA, 2010; pp. 299–362. ISBN 9781498701693.

146. Katsouras, I.; Asadi, K.; Li, M.; van Driel, T.B.; Kjaer, K.S.; Zhao, D.; Lenz, T.; Gu, Y.; Blom, P.W.; Damjanovic, D.; et al. The negative piezoelectric effect of the ferroelectric polymer poly(vinylidene fluoride). *Nat. Mater.* **2016**, *15*, 78–84. [CrossRef] [PubMed]

147. Sharma, M.; Srinivas, V.; Madras, G.; Bose, S. Outstanding dielectric constant and piezoelectric coefficient in electrospun nanofiber mats of PVDF containing silver decorated multiwall carbon nanotubes: Assessing through piezoresponse force microscopy. *RSC Adv.* **2016**, *6*, 6251–6258. [CrossRef]

148. Gebrekrstos, A.; Madras, G.; Bose, S. Piezoelectric response in electrospun poly(vinylidene fluoride) fibers containing fluoro-doped graphene derivatives. *ACS Omega* **2018**, *3*, 5317–5326. [CrossRef]

149. Mahdi, R.I.; Gan, W.C.; Majid, W.H. Hot plate annealing at a low temperature of a thin ferroelectric P(VDF-TrFE) Film with an improved crystalline structure for sensors and actuators. *Sensors* **2014**, *14*, 19115–19127. [CrossRef] [PubMed]

150. Hintzer, K.; Zipplies, T.; Carlso, D.P.; Schmiegel, W. Fluoropolymers, Organic. In *Ullmann's Polymers and Plastics: Products and Processes*; Wiley-VCH: Weinheim, Germany, 2016; Volume 2, pp. 604–656. ISBN 978-3-527-33823-8.

151. Pesano, L.; Dagdevire, C.; Su, Y.; Zhang, Y.; Girardo, S.; Pisignano, D.; Huang, Y.; Rogers, J.A. High performance piezoelectric devices based on aligned arrays of nanofibers of poly(vinylidenefluoride-co-trifluoroethylene). *Nat. Commun.* **2013**, *4*, 1633. [CrossRef] [PubMed]

152. Ico, G.; Showater, A.; Bosze, W.; Gott, S.C.; Kim, B.S.; Rao, M.P.; Myung, N.V.; Nam, J. Size-dependent piezoelectric and mechanical properties of electrospun P(VDF-TrFE) nanofibers for enhanced energy harvesting. *J. Mater. Chem. A* **2016**, *4*, 2293–2304. [CrossRef]

153. Jiang, Y.; Gong, L.; Hu, X.; Zhao, Y.; Chen, H.; Feng, L.; Zhang, D. Aligned P(VDF-TrFE) nanofibers for enhanced piezoelectric directional strain sensing. *Polymers* **2018**, *10*, 364. [CrossRef]

154. Denning, D.; Guyonnet, J.; Rodriguez, B.J. Applications of piezoresponse force microscopy in materials research: From inorganic ferroelectrics to biopiezoelectrics and beyond. *Int. Mater. Rev.* **2016**, *61*, 46–70. [CrossRef]

155. Zhang, B.; Yan, X.; He, H.W.; Yu, M.; Ning, X.; Long, Y.Z. Solvent-free electrospinning: Opportunities and challenges. *Polym. Chem.* **2017**, *8*, 333–352. [CrossRef]

156. Bubakir, M.; Li, H.; Barhoum, A.; Yang, Y. Advances in Melt Electrospinning Technique. In *Handbook of Nanofibers*; Barhoum, A., Bechelany, M., Makhlouf, A.H., Eds.; Springer: Cham, Switzerland, 2018; pp. 1–30.

157. Mota, C.; Puppi, D.; Gazzarri, M.; Bartolo, P.; Chiellini, F. Melt electrospinning writing of three-dimensional star poly(ε-caprolactone) scaffolds. *Polym. Int.* **2013**, *62*, 893–900. [CrossRef]

158. Zaiss, S.; Brown, T.D.; Reichert, J.C.; Berner, A. Poly(ε-caprolactone) scaffolds Fabricated by melt electrospinning for bone tissue engineering. *Materials* **2016**, *9*, 232. [CrossRef] [PubMed]

159. Lian, H.; Meng, Z. Melt electrospinning vs. solution electrospinning: A comparative study of drug-loaded poly (ε-caprolactone) fibres. *Mater. Sci. Eng. C* **2017**, *74*, 117–123. [CrossRef] [PubMed]

160. Asai, H.; Kikuchi, M.; Shimada, N.; Nakane, K. Effect of melt and solution electrospinning on the formation and structure of poly(vinylidene fluoride) fibres. *RSC Adv.* **2017**, *7*, 17593. [CrossRef]

161. Zhou, Z.; Li, W.; He, T.; Qian, L.; Tan, G.; Ning, C. Polarization of an electroactive functional film on titanium for inducing osteogenic differentiation. *Sci. Rep.* **2016**, *6*, 35512. [CrossRef] [PubMed]

162. Rutkovskiy, A.; Stenslokken, A.; Vaage, A.J. Osteoblast differentiation at a glance. *Med. Sci. Monit. Basic Res.* **2016**, *22*, 95–106. [CrossRef] [PubMed]

163. Fukada, E.; Yasuda, I. On the piezoelectric effect of bone. *J. Phys. Soc. Jpn.* **1957**, *12*, 1158–1162. [CrossRef]

164. Eischen-Loges, M.; Oliveira, K.M.; Bhavsar, M.B.; Barker, J.H.; Leppik, L. Pretreating mesenchymal stem cells with electrical stimulation causes sustained long-lasting pro-osteogenic effects. *PeerJ* **2018**, *6*, e4959. [CrossRef]

165. Mobini, S.; Leppik, L.; Parameswaran, V.T.; Barker, J.H. In vitro effect of direct current electrical stimulation on rat mesenchymal stem cells. *PeerJ* **2017**, *5*, e2821. [CrossRef]

166. Leppik, L.; Han, Z.; Mobini, S.; Parameswaran, V.T.; Eischen-Loges, M.; Slavici, A.; Helbing, J.; Pindur, L.; Oliveira, K.M.; Bhavsar, M.B.; et al. Combining electrical stimulation and tissue engineering to treat large bone defects in a rat model. *Sci. Rep.* **2018**, *8*, 6307. [CrossRef] [PubMed]

167. Kulterer, B.; Friedl, G.; Jandrositz, A.; Sanchez-Cabo, F.; Prokesch, A.; Paar, C.; Scheideler, M.; Windhager, R.; Windhager, R.; Preisegger, K.; et al. Gene expression profiling of human mesenchymal stem cells derived from bone marrow during expansion and osteoblast differentiation. *BMC Genom.* **2007**, *8*, 70. [CrossRef] [PubMed]

168. Ribeiro, C.; Panadero, J.A.; Sencadas, V.; Lanceros-Mendez, S.; Tamano, M.N.; Moratal, D.; Salmeron-Sanchez, M.; Ribelles, J.L. Fibronectin adsorption and cell response on electroactive poly(vinylidene fluoride) films. *Biomed. Mater.* **2012**, *7*, 035004. [CrossRef] [PubMed]

169. Nunes-Pereira, J.; Ribeiro, S.; Ribeiro, C.; Gomblek, C.J.; Gama, F.M.; Gomes, A.C.; Patterso, D.A.; Lanceros-Mendez, S. Poly(vinylidene fluoride) and copolymers as porous membranes for tissue engineering applications. *Polym. Test.* **2014**, *44*, 234–241. [CrossRef]

170. Zhao, F.; Wang, J.; Guo, H.; Liu, S.; He, W. The effects of surface properties of nanostructured bone repair materials on their performances. *J. Nanomater.* **2015**, *2015*, 893545. [CrossRef]

171. Pärssinen, H.; Hammarén, H.; Rahikainen, R.; Sencadas, V.; Ribeiro, C.; Vanhatupa, S.; Miettinen, S.; Lanceros-Méndez, S.; Hytönen, V.P. Enhancement of adhesion and promotion of osteogenic differentiation of human adipose stem cells by poled electroactive poly(vinylidene fluoride). *J. Biomed. Mater. Res. A* **2015**, *103*, 919–928. [CrossRef] [PubMed]

172. Rajabi, A.H.; Jaffe, M.; Arinzeh, T.L. Piezoelectric materials for tissue regeneration: A review. *Acta Biomater.* **2015**, *24*, 12–23. [CrossRef]

173. Arinzeh, T.L.; Collins, G.; Lee, Y.S. Method of Tissue Repair Using a Piezoelectric Scaffold. U.S. Patent 9,476,026 B2, 25 October 2016.

174. Kitsara, M.; Blanquer, A.; Murillo, G.; Humblot, V.; Vieira, S.; Nojues, C.; Ibanez, E.; Esteve, J.; Barrios, L. Permanently hydrophilic, piezoelectric PVDF nanofibrous scaffolds promoting unaided electromechanical stimulation on osteoblasts. *Nanoscale* **2019**, *11*, 8906–8917. [CrossRef]

175. Wang, A.; Hu, M.; Zhou, L.; Qiang, X. Self-powered well-aligned P(VDF-TrFE) piezoelectric nanofiber nanogenerator for modulating an exact electrical stimulation and enhancing the proliferation of preosteoblasts. *Nanomaterials* **2019**, *9*, 349. [CrossRef]

176. Zhang, X.; Zhang, C.; Lin, Y.; Hu, P.; Shen, Y.; Wang, K.; Meng, S.; Chai, Y.; Dai, X.; Liu, X.; et al. Nanocomposite membranes enhance bone regeneration through restoring physiological electric microenvironment. *ACS Nano* **2016**, *10*, 7279–7286. [CrossRef]

177. Augustine, R.; Dan, P.; Sosnik, A.; Kalarikkal, N.; Tran, N.; Vincent, B.; Thomas, S.; Menu, P.; Rouxel, D. Electrospun poly(vinylidene fluoride-trifluoroethylene)/zinc oxide nanocomposite tissue engineering scaffolds with enhanced cell adhesion and blood vessel formation. *Nano Res.* **2017**, *10*, 3358–3376. [CrossRef]

178. Saburi, E.; Islami, M.; Hosseinzadeh, S.; Moghadam, A.S.; Mansour, R.N.; Azadian, E.; Joneidi, Z.; Nikpoor, A.R.; Ghadiani, M.H.; Khodaii, Z.; et al. In vitro osteogenic differentiation potential of the human induced pluripotent stem cells augment when grown on graphene oxide-modified nanofibers. *Gene* **2019**, *696*, 72–79. [CrossRef] [PubMed]

179. Ribeiro, C.; Correia, D.M.; Rodrigues, I.; Guardao, L.; Guimareas, S.; Soares, R.; Lanceros-Méndez, S. In-vivo demonstration of the suitability of piezoelectric stimuli for bone reparation. *Mater. Lett.* **2017**, *209*, 118–121. [CrossRef]

180. Wang, A.; Liu, Z.; Hu, M.; Wang, C.; Zhang, X.; Shi, B.; Fang, Y.; Cui, Y.; Li, Z.; Ren, K. Piezoelectric nanofibrous scaffolds as in vivo energy harvesters for modifying fibroblast alignment and proliferation in wound healing. *Nano Energy* **2018**, *43*, 63–71. [CrossRef]

181. Bruzauskaite, I.; Bironaite, D.; Bagdonas, E.; Bernotiene, E. Scaffolds and cells for tissue regeneration: Different scaffold pore sizes—Different cell effects. *Cytotechnology* **2016**, *68*, 355–369. [CrossRef] [PubMed]

182. Genchi, G.G.; Ceseracciu, L.; Marino, A.; Labardi, M.; Marras, S.; Pignatelli, F.; Bruschini, L.; Mattoli, V.; Ciofani, G. P(VDF-TrFE)/BaTiO$_3$ nanoparticle composite films mediate piezoelectric stimulation and promote differentiation of SH-SY5Y neuroblastoma cells. *Adv. Healthc. Mater.* **2016**, *5*, 1808–1820. [CrossRef]

183. Swindle-Reilly, K.E.; Paranjape, C.S.; Miller, C.A. Electrospun poly(caprolactone)-elastin scaffolds for peripheral nerve regeneration. *Prog. Biomater.* **2014**, *3*, 20. [CrossRef]

184. Xu, J.; Wu, T.; Xia, Y. Perspective: Aligned arrays of electrospun nanofibers for directing cell migration. *APL Mater.* **2018**, *6*, 120902. [CrossRef]

185. Corey, J.M.; Lin, D.Y.; Mycek, K.B.; Chen, Q.; Samuel, S.; Feldman, E.L.; Martin, D.C. Aligned electrospun nanofibers specify the direction of dorsal root ganglia neurite growth. *J. Biomed. Mater. Res. A* **2007**, *83*, 636–645. [CrossRef]

186. Koppes, A.N.; Zaccor, N.W.; Rivet, C.J.; Williams, L.A.; Piselli, J.M.; Gilbert, R.J.; Thompsom, D.M. Neurite outgrowth on electrospun PLLA fibers is enhanced by exogenous electrical stimulation. *J. Neural Eng.* **2014**, *11*, 046002. [CrossRef]

187. Yoshida, M.; Onogi, T.; Onishi, K.; Inagaki, T.; Tajitsu, Y. High piezoelectric performance of poly(lactic acid) film manufactured by solid-state extrusion. *Jpn. J. Appl. Phys.* **2014**, *53*. [CrossRef]

188. Lee, Y.S.; Arinzeh, T.L. The Influence of piezoelectric scaffolds on neural differentiation of human neural stem/progenitor cells. *Tissue Eng. A* **2012**, *18*, 2063–2072. [CrossRef] [PubMed]

189. Meng, Y.Z.; Hay, A.S.; Jian, X.G.; Tjong, S.C. Synthesis and properties of poly (aryl ether sulfone)s containing the phthalazinone moiety. *J. Appl. Polym. Sci.* **1998**, *68*, 137–143. [CrossRef]

190. Tjong, S.C.; Meng, Y.Z. Morphology and mechanical characteristics of compatibilized polyamide 6-liquid crystalline polymer composites. *Polymer* **1997**, *38*, 4609–4615. [CrossRef]

191. Meng, Y.Z.; Tjong, S.C. Rheology and morphology of compatibilized polyamide 6 blends containing liquid crystalline copolyesters. *Polymer* **1998**, *39*, 99–107. [CrossRef]

192. Cao, X.; Zhao, L.; Song, Z.B.; Zhang, X.Z.; Qin, J.Q. The influence of the alignment of electrospun fibrous scaffolds on the biological behavior of RSC96 cells. *J. Biomater. Tissue Eng.* **2014**, *4*, 488–491. [CrossRef]

193. Kim, S.; Maynard, J.C.; Strickland, A.; Burlingame, A.L.; Milbrandt, J. Schwann cell O-GlcNAcylation promotes peripheral nerve remyelination via attenuation of the AP-1 transcription factor JUN. *Proc. Natl. Acad. Sci. USA* **2018**, *115*, 8019–8024. [CrossRef]

194. Wu, S.; Lee, Y.; Bunge, M.B.; Arinzeh, T.L. Investigating Schwann cell growth and neurite extension on PVDF-TrFE scaffolds in vitro. In *Frontiers in Bioengineering and Biotechnology Conference Abstract, Proceedings of the 10th World Biomaterials Congress, Montréal, QC, Canada, 17–22 May 2016*; Frontiers: Lausanne, Switzerland, 2016. [CrossRef]

195. Wu, S.; Chen, M.S.; Maurel, P.; Lee, Y.; Bunge, M.B.; Arinzeh, T.L. Aligned fibrous PVDF-TrFE scaffolds 1554 with Schwann cells support neurite extension and myelination in vitro. *J. Neural Eng.* **2018**, *15*, 056010. [CrossRef]

196. Vigani, B.; Rossi, S.; Sandri, G.; Bonferoni, M.C.; Ferrari, F. Design and criteria of electrospun fibrous scaffolds for the treatment of spinal cord injury. *Neural Regen. Res.* **2017**, *12*, 1786–1790. [CrossRef]

197. Frost, H.K.; Andersson, T.; Johansson, S.; Englund-Johansson, U.; Ekstrom, P.; Dahlin, L.B.; Johansson, F. Electrospun nerve guide conduits have the potential to bridge peripheral nerve injuries in vivo. *Sci. Rep.* **2018**, *8*, 16716. [CrossRef]

198. Aebischer, P.; Valentini, R.F.; Dario, P.; Domenici, C.; Galletti, P.M. Piezoelectric guidance channels enhance regeneration in the mouse sciatic nerve after axotomy. *Brain Res.* **1987**, *436*, 165–168. [CrossRef]

199. Lee, Y.; Wu, S.; Arinzeh, T.L.; Bunge, M.B. Enhanced noradrenergic axon regeneration into schwann cell-filled PVDF-TrFE conduits after complete spinal cord transection. *Biotechnol. Bioeng.* **2017**, *114*, 444–456. [CrossRef] [PubMed]

200. Wu, J.; Hong, Y. Enhancing cell infiltration of electrospun fibrous scaffolds in tissue regeneration. *Bioact. Mater.* **2016**, *1*, 56–64. [CrossRef] [PubMed]

201. Hutmacher, D.W. Scaffold design and fabrication technologies for engineering tissues—State of the art and future perspectives. *J. Biomater. Sci. Polym. Ed.* **2001**, *12*, 107–124. [CrossRef] [PubMed]
202. McCann, J.T.; Marquez, M.; Xia, Y. Highly porous fibers by electrospinning into a cryogenic liquid. *J. Am. Chem. Soc.* **2006**, *128*, 1436–1437. [CrossRef] [PubMed]
203. Rnjak-Kovacina, J.; Weiss, A.S. Increasing the pore size of electrospun scaffolds. *Tissue Eng. Part B* **2011**, *17*, 365–372. [CrossRef]
204. Nam, J.; Huang, Y.; Agarwal, S.; Lannutti, J. Improved cellular infiltration in electrospun fiber via engineered porosity. *Tissue Eng.* **2007**, *13*, 2249–2257. [CrossRef]
205. Sun, D.; Chang, C.; Li, S.; Lin, L. Near-field electrospinning. *Nano Lett.* **2006**, *6*, 839–842. [CrossRef]
206. Middleton, R.; Li, X.; Shepherd, J.; Li, Z.; Wang, W.; Best, S.M.; Cameron, R.E.; Huang, Y.Y. Near-field electrospinning patterning polycaprolactone and polycaprolactone/collagen interconnected fiber membrane. *Macromol. Mater. Eng.* **2018**, *303*, 1700463. [CrossRef]
207. Chen, H.; Malheiro, A.B.; van Blitterswijk, C.; Mota, C.; Wieringa, P.A.; Moroni, L. Direct writing electrospinning of scaffolds with multidimensional fiber architecture for hierarchical tissue engineering. *ACS Appl. Mater. Interfaces* **2017**, *9*, 38187–38200. [CrossRef]
208. Liu, Z.H.; Pan, C.T.; Lin, L.W.; Lai, H.W. Piezoelectric properties of PVDF/MWCNT nanofiber using near-field electrospinning. *Sens. Actuators A Phys.* **2013**, *193*, 13–24. [CrossRef]
209. Lee, T.H.; Chen, C.Y.; Tsai, C.Y.; Fuh, Y.K. Near-field electrospun piezoelectric fibers as sound-sensing elements. *Polymers* **2018**, *10*, 692. [CrossRef] [PubMed]
210. Kennedy, G.L. Toxicology of dimethyl and monomethyl derivatives of acetamide and formamide: A second update. *Crit. Rev. Toxicol.* **2012**, *42*, 793–826. [CrossRef] [PubMed]
211. Hochleitner, G.; Jungst, T.; Brown, T.D.; Hahn, K.; Moseke, C.; Jakob, F.; Dalton, P.D.; Groll, J. Additive manufacturing of scaffolds with sub-micron filaments via melt electrospinning writing. *Biofabrication* **2015**, *7*, 035002. [CrossRef] [PubMed]
212. Brown, T.D.; Edin, F.; Detta, N.; Skelton, A.D.; Hutmacher, D.W.; Dalton, P.D. Melt electrospinning of poly(ε-caprolactone) scaffolds: Phenomenological observations associated with collection and direct writing. *Mater. Sci. Eng. C* **2014**, *45*, 698–708. [CrossRef] [PubMed]
213. Johnson, C.T.; Garcia, A.J. Scaffold-based anti-infection strategies in bone repair. *Ann. Biomed. Eng.* **2015**, *43*, 515–528. [CrossRef] [PubMed]
214. Gao, J.; Huang, G.; Liu, G.; Liu, Y.; Chen, Q.; Ren, L.; Chen, C. A biodegradable antibiotic-eluting PLGA nanofiber-loaded deproteinized bone for treatment of infected rabbit bone defects. *J. Biomater. Appl.* **2016**, *31*, 241–249. [CrossRef] [PubMed]
215. Ausbaugh, A.G.; Jiang, X.; Zheng, G.; Tsai, A.S.; Kim, W.S.; Thompsom, J.M.; Miller, R.J.; Shahbazian, J.H.; Wang, Y.; Dillen, C.A.; et al. Polymeric nanofiber coating with tunable combinatorial antibiotic delivery prevents biofilm-associated infection in vivo. *Proc. Natl. Acad. Sci. USA* **2016**, *113*, 6919–6928. [CrossRef] [PubMed]
216. Barros, C.H.; Fulaz, S.; Stanisic, D.; Tasic, L. Biogenic nanosilver against multidrug-resistant bacteria (MDRB). *Antibiotics* **2018**, *7*, 69. [CrossRef] [PubMed]
217. Inkielewicz-Stepniak, I.; Santos-Martinez, M.J.; Medina, C.; Radomski, M.W. Pharmacological and toxicological effects of co-exposure of human gingival fibroblasts to silver nanoparticles and sodium fluoride. *Int. J. Nanomedic.* **2014**, *9*, 1677–1687. [CrossRef]
218. Zielinska, E.; Tukaj, C.; Radomski, M.W.; Inkielewicz-Stepniak, I. Molecular mechanism of silver nanoparticles-induced human osteoblast cell death: Protective effect of inducible nitric oxide synthase inhibitor. *PLoS ONE* **2016**, *11*, e0164137. [CrossRef] [PubMed]
219. McGovern, J.A.; Grifin, M.; Hutmacher, D.W. Animal models for bone tissue engineering and modelling disease. *Dis. Models Mech.* **2018**, *11*, dmm033084. [CrossRef] [PubMed]

MDPI

St. Alban-Anlage 66

4052 Basel

Switzerland

Tel. +41 61 683 77 34

Fax +41 61 302 89 18

www.mdpi.com

Nanomaterials Editorial Office

E-mail: nanomaterials@mdpi.com

www.mdpi.com/journal/nanomaterials

9 783039 432264